The Story of Science

Einstein Adds a
New Dimension

Published by Smithsonian Books

Produced by American Historical Publications

Byron Hollinshead	President
Sabine Russ	Managing Editor, Picture Editor, Production Manager
Lorraine Jean Hopping	Editor
Monique Claire Vescia	Copy Editor
Marleen Adlerblum	Designer and Illustrator
Nancy Rose	Production Assistant

Library of Congress Cataloging-in-Publication Data

Hakim, Joy.
  The story of science: Einstein adds a new dimension / by Joy Hakim.
        p.cm.
        Includes bibliographical references and index.
        ISBN 978-1-58834-162-4 (alk. paper)
        1. Science—History. I. Title. II. Title: Einstein adds a new dimension.

Q125.H263 2007
509—dc22

2007014096

British Library Cataloguing-in-Publication Data available

Manufactured in the United States of America
12 11 10 09 08   2 3 4 5

The paper used in this publication meets the minimum requirements of the American
National Standard for Information Sciences—Permanence of Paper for Printed Library
Materials ANSI Z39.48-1984.

Dedication Page

Albert Einstein composed a handwritten message to the young daughter of a fellow professor. (It was 1921, and he was in Bologna, Italy. The professor was Federigo Enriques; his daughter was Adriana Enriques.) This is what Einstein wrote:

Study and, in general, the pursuit of truth and beauty is a sphere of activity in which we are permitted to remain children all our lives.

Permitted to remain children? Yes, he saw it as a privilege. Einstein also understood that the intense curiosity that is part of being a child is a key to creativity. Those who pursue truth and do it well—in any field—almost always hold on to the inquisitiveness and imagination that is part of being young.

Einstein believed this all his life. In 1947, when he was what many would call an old man, he wrote to a friend, Otto Juliusburger, who was celebrating his eightieth birthday,

People like you and me, though mortal, of course, like everyone else, do not grow old no matter how long we live. What I mean is that we never cease to stand like curious children before the great Mystery into which we are born. This interposes a distance between us and all that is unsatisfactory in the human sphere—and that is no small matter.

So this book is written for young thinkers—of all ages. I hope it will answer some questions and lead to new ones. It is written for you, dear reader, for my children and grandchildren, for Sabine Russ's son, and Byron Hollinshead's grandsons, and for some bright rising stars, who all happen to be Taylors. They are:

Meredith Christine Taylor   Cynthia Grace Taylor
Bradley James Taylor   Abigail Claire Frank Taylor
Samantha Marie Taylor   Samuel Bennett Frank Taylor
Katherine Rose Taylor   Mao Mao Andrew den Heeten
Victoria Lynne Taylor

Contents

About Quarks, Red Giants, and Why This Book Got Written

You are a quark warehouse. Me, too. The desk I'm leaning on is, too. What's a quark? Ha! You'll have to read this book to find out. Actually, that's why I wrote the book: to find out for myself. When I want to learn something, I write a book. (Not a bad way to learn.) I'd been hearing scientific terms, like relativity and quantum theory, and I didn't know what to make of them. Then I read about neutrinos, the Big Bang, and red giants—and I was really lost. So I wrote this book for me, and for people like me, who are curious about the world around us.

That world is strange—and it seems to be getting stranger. Or maybe it is just that modern science is telling us about things—like dark energy—that no one yet understands. As for today's science of cosmology? Ah, if only Galileo were here and could know what we now know. Cosmology has brought us solid data, beyond conjecture, that tells us the universe has an unfolding story. We now know, for instance, that ours is an expanding universe and that the expansion has begun to speed up.

There was a time when everyone seemed to believe that science was sober and serious, and that if you wanted to be imaginative you had to turn to fantasy. When it came to imagination, I didn't think science could touch *Star Wars* or Hollywood's special effects.

But, next to modern science, the big screen is no big deal. Today's science is more astonishing than anything science-fiction writers have ever invented. (My editor doesn't agree. She says good sci-fi takes the best of cutting-edge science and runs with it for a relationship that is mutually mind-bending. Maybe, but I'm awed by the real stuff.)

We now know (thanks to Albert Einstein) that time on your watch may tick differently from time on mine. It depends on how fast we're traveling relative to each other. And quarks? There isn't much chance you'll see a quark—even with our most sophisticated microscopes. Quarks are so small they make atoms seem mountainous.

But let's talk about big. Our universe is so vast that it would take some 13.7 billion years for a beam of light to travel from its farthest regions to a telescope on your roof. (How did we figure out 13.7 billion years? How old is the Earth? How fast does light travel? You'll find out in this book.)

Lewis Carroll, the nineteenth-century mathematician who wrote *Alice in Wonderland* and *Through the Looking-Glass*, had the White Queen say to Alice, "It's a poor sort of memory that only works backwards." Carroll may have had hints of what we know today: Time doesn't have to go in only one direction.

As for gravity, well, it isn't what Newton thought it was when that legendary apple hit his head. Albert Einstein came along and told us that gravitation isn't a force at all.

When Einstein was a young man, he didn't know what we now know: Our galaxy, with its billions of stars, is just one of perhaps 100 billion galaxies. Yet Einstein's

On the grand universal scale, here's the largest shock wave we've ever seen—that green-colored swath in the center of a galaxy cluster. The wave, made of hydrogen, is larger than our Milky Way. What created it? That bright pink dot immediately to the right of it. It's a galaxy racing at super-high speed toward a neighbor. For the complete blow-by-blow, research the Stephan's Quartet galaxy cluster.

theories predicted the black holes that seem to lie at the center of those galaxies. Black holes? Fall into one and you might see the past—and maybe the future—whiz by on the way to having the nuclei in your atoms recycled.

But black holes are old news. You're a citizen of the twenty-first century, and cosmic wormholes are beckoning. Wormholes? They could be the way future astronauts will travel from this universe to another. Other universes? Just keep reading; you may be surprised at what you'll learn.

How did we get where we are today? In a Berlin interview, just after his most productive years, Albert Einstein famously said, "Imagination is more important than knowledge." But, as you'll learn in this book, Einstein had solid Germanic schooling to feed his astonishing imagination. Start with information, ask the right questions, apply some imaginative leaps, and you are on the path to creativity. Einstein had the background knowledge to tread that path. He also had a character trait—call it audacity, nerve, or maybe impudence—that made him a freethinker. He often thumbed his nose at authority—and in the long run got away with it—which made him enormously aware of the importance of freedom in all fields of endeavor but especially in science. Einstein was not a narrow thinker. He loved music, was a sophisticated reader of philosophy, and spoke out on issues of peace and politics.

Einstein, of course, was ahead of his times. The twentieth century was an age of specialization. But in today's information-based world, broad thinking has become essential. To not know science means being out of touch with the basic ideas that underlie our fast-paced existence. It also means missing out on some of the most exciting creativity in human history.

So in this, the greatest scientific era ever, scientific illiteracy is no longer acceptable. These books were written, in part, to address that issue. My hope is that they will be read by anyone who wants to join the adventure that is modern science. In addition, they are intended as a new kind of classroom tool (we need many more) that will stimulate thoughtful reading and lead to Socratic teaching. *The Story of Science* concerns itself, primarily, with the human quest to understand the way the universe works. These books are not meant as a replacement for the focused teaching of subjects such as energy and matter—but by telling the stories that brought us those disciplines, the aim is to give them connections and greater meaning.

Now, about the making of this book: Once again Byron Hollinshead and his superb staff have done all the work that goes into producing a book. That means finding pictures, checking facts, editing, proofreading, and then putting the whole thing together. It's a huge job, and they do it very well. You won't see many books as beautiful as this one.

Sabine Russ is the amazing picture researcher and put-it-all-in-place person. Lorraine Hopping Egan, the experienced, perceptive editor (she's written many very good science books herself), is a key player on the team. Marleen Adlerblum, the book designer, is responsible for the look of these handsome pages. Monique Vescia, the copy editor, brought magic to the task (she filled that same role for the Harry Potter books in the U.S.).

Robert Fleck, a professor of physics at Embry-Riddle Aeronautical University, who is also writing a science history, read the whole manuscript and offered wonderfully helpful comments. MIT physicists Alan Guth, Josh Winn, and Seth Lloyd generously read or responded to questions on their specialties. Edmund Bertschinger answered an important query. Bob Stair, co-author of a fine science book titled *Force, Motion, and Energy*, caught an error. Jeff Hakim, the head of the math department at American University, patiently answered his mom's queries.

I owe much to some author/scientists who have, in recent years, written very good science books for a general audience. I read voraciously. A few of the names are: Stephen Hawking, Timothy Ferris, Brian Greene, Hans Christian von Baeyer, Pedro Ferreira, Richard Wolfson, Lee Smolin, Michio Kaku, Paul Davies, John Gribbin, Paul Hewitt, Dennis Overbye, Alan Lightman, Marcia Bartusiak—and that's just for starters. Browse the science shelves at your library or bookstore; you'll find treasures.

Science educators John Hubisz and Juliana Texley cheered me on. Both read and commented on an early version of the text. Texley offered critical insights and additions to the coordinated teaching materials developed in classrooms for *Aristotle Leads the Way* and *Newton at the Center* by Doug McIver, Maria Garriott, and Cora Teter at Johns Hopkins University. Educator/writer Dennis Denenberg, who knows how to make learning fun, taught me much. Barbara Dorff, the 2002 Texas Teacher of the Year, is just one of the inspiring classroom teachers and administrators I know who encouraged me to do the best job I could for their students. (Our great teachers, and I know many, are national treasures and should be celebrated as such.) No-nonsense Stephanie Harvey, a reading guru, shared her insights on nonfiction reading, the essential reading in the information age. Richard Halls put books in the hands of students at La Academia, a small inner-city school in Denver that educates well. Sue Lubeck, owner of The Bookies, a bookstore in Denver that focuses on children and teachers, awed me with her energy and intelligence. Carolyn Gleason and Severin White at the Smithsonian supported the effort. T. J. Kelleher at Smithsonian Books read the final page proof and made thoughtful suggestions. Robert Noyes at Smithsonian Astrophysical Observatory helped with chapter 48 and Charles Whitney, retired from the same organization, read final proofs and gave us the benefit of another perspective from a distinguished scientist. Gerry Wheeler and David Beacom, at the NSTA, encouraged. And Byron Hollinshead, in addition to producing the books, shared a deep commitment to children, schools, and the search for new ways of teaching and learning.

But if this book has unusual strengths as a science book for the general reader and for school students (and I believe that it does), one person is most responsible. Edwin F. Taylor heard that I was attempting to write about contemporary science for young readers (of all ages) and E-mailed me some suggestions. Now, Edwin Taylor is an MIT physicist and the author of a number of distinguished books, especially two he wrote with Princeton physicist John Archibald Wheeler. Those books, *Spacetime Physics* and *Exploring Black Holes*, intended for college physics students (it helps to have high school calculus when you read them), are the best textbooks I've encountered (and I claim some authority as an expert on textbooks). Imagine a textbook that begins with a parable, and you'll get the idea. (A new edition of *Exploring Black Holes* adds Edmund Bertschinger as a co-author.)

So when Edwin Taylor offered physics suggestions, I listened. (If you write for young readers, the best people want to help you.) Before long, Edwin was involved, volunteering his time to read, reread, and comment on every chapter and put me on track when I went astray.

He did more than that. I got a private tutorial with one of the greatest physics teachers this country has produced. It was an intellectual adventure unlike anything I'd ever encountered. (Sometimes my head hurt with all the stretching.) In this book, I've tried to share some of that journey with you. Errors and fuzziness in the book are my responsibility, but where there are insights and acuity, you can be sure that much of it is due to Edwin F. Taylor's guidance.

—Joy Hakim

1 A Boy with Something on His Mind

One thing I have learned in a long life: that all our science, measured against reality, is primitive and childlike—and yet it is the most precious thing we have.

—Albert Einstein (1879–1955), who also said:

The most beautiful thing we can experience is the mysterious. It is the source of all true art and science. He who knows it not and can no longer wonder, no longer feel amazement, is as good as dead.

And, when he was asked for a message to put in a time capsule, this is what he wrote:

Dear Posterity, If you have not become more just, more peaceful, and generally more rational than we are—why then, the Devil take you.

ifteen-year-old Albert Einstein is miserable. He is trying to finish high school in Germany, but he hates the school; it's a strict, rigid place. To make things worse, his parents have moved to Italy. They think he should stay behind until his schooling is completed. It isn't long, though, before he is on his way over the Alps, heading south to join them. Why does he leave Germany? Today, no one is quite sure, but a letter from the school offers a powerful clue: "Your presence in the class is disruptive and affects the other students."

What are the Einsteins to do with their son? He is a high school dropout who has arrived without warning.

There, to start this book, is a world-famous scientist having fun on a bike in southern California, in 1933. Having fun helped make Albert Einstein successful. He never lost his boyish enthusiasm for new ideas. As a great scientist, he found the quest to understand the universe to be an adventure, a privilege, and a grand game.

"Dear **POSTERITY**"? In the third quotation (above), Einstein is addressing people who will live after him. *Post-* is Latin for "after." To *postpone* means to "place after" or put off. (Think homework or chores.) A *postscript* (or P.S.) comes after the signature on a letter. Einstein's note is aimed at future generations—which means you. So what do you think of his message?

Six-year-old Albert is with his sister, Maja. He was not quite three when his parents told him he would have a baby to play with. He thought they meant a new toy. He took one look at her and, puzzled, said, "Where are the wheels?" Later, Maja remembered that he had a temper as a boy and sometimes threw things. She said, "The sister of a thinker needs a sound skull."

Albert was born in the small city of Ulm, but a year later the family moved to fast-growing Munich, the intellectual capital of Bavaria in southern Germany. This hand-colored photo shows Munich's New City Hall in the 1890s. Hermann Einstein (Albert's father) and his brother, Jakob, were business partners, and their families shared a comfortable house with a tree-filled garden. The Einstein brothers had big plans. They wanted to produce a dynamo that Jakob had invented.

What makes Albert Einstein focus on this dilemma? No one knows for sure, but 15 is a good age for questioning. And Einstein, at that age, is already well grounded in mathematics and the new sciences. He is fortunate; he was born to parents who are interested in books and ideas and conversation. Einstein says his father is "very wise." (But he isn't much of a businessman; his factories keep failing.)

When Albert was a five-year-old in Germany and sick in bed, his father gave him a compass. The needle always pointed in the same direction. His father said that was because of magnetism. It made the little boy so excited, he said he "trembled and got cold." How could an invisible force work through the empty space between the magnetic north pole and his home? No one could explain that to him. It made him start thinking about nature's forces.

Einstein's uncle Jakob introduced him to mathematics. He turned it into a game. "Algebra is a merry science," said Uncle Jakob. "When we are hunting an animal that we can't find, we call him x until we catch him." Einstein's mother read aloud

An animal called x? Uncle Jakob is talking about variables, the unknown numbers in algebraic equations. Here's an easy one to find: $x + 3 = 8$. Happy hunting.

Here's Albert in a school photograph taken in Munich. He's the small boy near the right end of the front line, next to the boy in the light jacket. Einstein showed a talent for math and for Latin. His teachers found him hopeless at most other subjects.

from the best books she could find, and she introduced him to the violin. His fiddle became more than a friend; it fueled a lifetime passion for music. His younger sister, Maja, described the violin as his "dear child."

And then there was a regular dinner guest. His name was Max Talmud (later called Talmey); he was studying to be a doctor. It was a tradition for Jewish families to invite poor students to dinner. Max came every Thursday, bringing the latest ideas in science and mathematics to the dinner table. Together, Max and Albert pored over a science series called Popular Books on Natural Science. After dinner, Albert's father, Hermann, sometimes read Shakespeare or Goethe to the assembled family.

When Albert was 12, Uncle Jakob gave him a volume on Euclid's math, and Einstein could hardly contain his excitement when he delved into it. He called it his "holy geometry book." Max wrote that his eager young friend was soon far beyond him in mathematical knowledge.

At about the same time, Einstein was discovering a traditional world of holiness. His parents were nonpracticing Jews, but the German state required that all children have religious training. A distant relative was enlisted to provide lessons for Albert. No one expected the boy to find religious ecstasy, but he did. Captivated by the

Johann Wolfgang von Goethe (1749–1832) was a German poet, playwright, and scientist. Along with William Shakespeare, he is one of the great figures in world literature.

Euclid taught math in Alexandria, Egypt, in the fourth century B.C.E. In his famous 13-volume text, *The Elements*, he boiled down geometry to five postulates—unproven statements that are accepted as true. You can read more about Euclid in book one of *The Story of Science*.

I remember that an Uncle [Jakob] told me [about] the Pythagorean theorem before the holy geometry book had come into my hands. After much effort I succeeded in "proving" this theorem on the basis of similarity of triangles; in doing so it seemed to me "evident" that the relations of the sides of the right-angled triangles would have to be completely determined by one of the acute angles. Only something which did not in similar fashion seem to be "evident" appeared to me to be in need of any proof at all.

—Albert Einstein, *Autobiographical Notes*

Here's the Pythagorean Theorem pictured: In a right triangle, the square on the hypotenuse equals the sum of the squares of the other two sides. Or, as here, $3^2 + 4^2 = 5^2$.

wisdom and ethics of his faith, he was soon composing and singing songs in praise of God. When he tried to get his parents to take their religion more seriously, they were indulgent— until he asked them to give up eating pork.

Meanwhile Albert began studying higher math on his own, and he and Max kept sharing books. When Albert was 13, Max lent him a book by the German philosopher Immanuel Kant. Kant is tough reading at any age, but the boy was excited by the challenge. Kant tried to connect all of the great ideas of philosophy into one embracing system. Later, Einstein would attempt to do the same thing in science.

Geometry Without a Doubt

Looking back on his childhood, Einstein described his wonder at a compass and then wrote:

At the age of 12 I experienced a second wonder of a totally different nature: in a little book dealing with Euclidean plane geometry, which came into my hands at the beginning of a schoolyear. Here were assertions, as for example the intersection of the three altitudes of a triangle in one point which—though by no means evident—could nevertheless be proved with such certainty that any doubt appeared to be out of the question. This lucidity and certainty made an indescribable impression upon me. That the axiom had to be accepted unproved did not disturb me....I could peg proofs upon propositions the validity of which did not seem to me to be dubious.

Try it: No matter what shape the triangle, the three altitudes always meet at a point (called the orthocenter). An altitude line runs through a triangle tip and perpendicular (at right angles) to a side.

The challenging reading he was doing gradually brought Einstein to a philosophical view of religion. According to a biographer, Denis Brian, "Attaching himself to no sect, repulsed by the rigid rules and compulsory behavior dictated by most organized religions, he was still considered by those who knew him to have been deeply religious."

But his deep reading didn't help at the stern, rigid German school—called a gymnasium—where no one dreamed that what the questioning young Einstein was doing would lead to a new model of the universe. There, he was treated as a problem and a misfit, in good part because he had no interest in practicing sports or memorizing lessons or serving in the German army (mandatory at 16 for German boys). A doctor, recognizing symptoms of depression, gave Albert a note saying that he might suffer a breakdown if he didn't spend some time with his family. That helped get him released. The gymnasium authorities seemed happy to send young Einstein on his way.

> The infinitude of the creation is great enough to make a world, or a Milky Way of worlds, look in comparison with it what a flower or an insect does in comparison with the Earth.
>
> —Immanuel Kant (1724–1804), German philosopher, *Universal Natural History and Theory of the Heavens*

Your (Probable) Future Is in the Data

In school, Einstein was fascinated with statistics, a branch of mathematics about analyzing data. His close attention to detail allowed him (and will allow you) to see not only the specifics but, surprisingly, the big picture. How? Here's a simple example: Scientists have painstakingly counted the number of sunspots on the Sun, year after year, for several centuries. If you plot the average annual number on a long graph, you can easily see a pattern: Sunspots peak in number about every 11 years. You can then use the pattern to predict the most likely year of the next peak. "Most likely" means probably—and probability is the term for predicting outcomes based on statistics. Probably, not certainly. A sunspot peak could happen from 9 to almost 14 years after the previous one.

Einstein pondered the probability and statistics of something much tinier than sunspots. He predicted how atoms and molecules are likely to behave, on average, based on temperature and pressure and velocity. There's more on his brand of statistical mechanics to come.

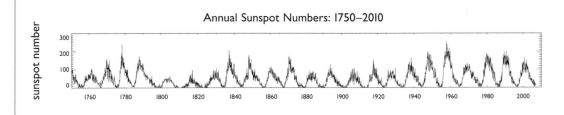
Annual Sunspot Numbers: 1750–2010

The Swiss Federal Institute of Technology, nicknamed Zurich Polytechnic, has an international reputation for excellence. Einstein graduated in 1900 with a teaching degree in mathematics and physics. The school's superb laboratory equipment (like that above) was supplied by Siemens, the giant conglomerate that helped put Hermann Einstein out of business.

After 15-year-old Einstein arrives in Italy, his parents suggest he come down to earth. The family factory isn't doing well. Albert has to find a career. He says he wants to be a high school teacher, so he is sent off to Switzerland to finish high school and prepare for a university. There, he boards with a friendly family, and the Swiss school—in a town named Aarau—turns out to be just right for him. It has outstanding teachers, high standards, and an informal atmosphere. Students are expected to ask questions and search for answers. And there, he and a friend play Mozart sonatas in the school refectory. Fifty years later, he still will remember it as a place where everyone joined in "responsible and happy work."

From Aarau, Einstein goes to Zurich, Switzerland, to the Swiss Federal Institute of Technology (one of Europe's leading technical universities), where he studies physics and mathematics. Zurich, in the heart of Europe, is a lively city with cafés and conversations that are attracting an energetic mix of artists, writers, and political thinkers. Vladimir Lenin (who will one day rule Russia) and James Joyce (an Irish writer who will change the modern novel) are two of them. Einstein has a favorite café, the Metropole, where he discusses ideas and books with friends. His favorite drink is iced coffee. (Einstein won't drink alcohol; he thinks it slows his brain.) Whenever he gets a chance, he goes sailing on Lake Zurich, usually in his landlady's sailboat. (Sailing will become a lifetime hobby.) With big brown eyes, curly hair, a quick wit, and an intense intellect, he attracts attention. A female friend describes him as "irresistible." A male friend says he thinks Einstein will become a great man.

There is only one woman in his class, a Serbian, Mileva Maric. She is a pioneer, one of the first women anywhere in the world to study advanced physics. Einstein is impressed. Later, they will fall in love and marry. (It will be a marriage beset with problems. It will fail.)

Meanwhile, he manages to annoy most of his professors. It is clear that Albert Einstein is bright, but he has an attitude problem. He has little patience with schoolwork and often doesn't appear in class. He seems to learn best by talking about ideas and problems with friends. So when he graduates and needs a job recommendation, he is the only one in his class who doesn't get one. One of his teachers calls him a "lazy dog" because he doesn't always do his assignments.

Albert and Mileva were deeply in love, as his passionate letters attest. But life's complications and his work got in the way. They had two sons, but the marriage didn't last. This is their wedding picture.

The professor is wrong. Einstein isn't lazy. His mind is working hard. It is concentrating on that light beam. For more than 10 years, the question of what happens at the speed of light never seems to leave his head. "In all my life I never labored so hard," Einstein writes to a friend about one occasion of deep thinking.

When, finally, in 1905 he is able to answer his own questions about light, he has developed one of the most important scientific theories in all of history—the Special Theory of Relativity. It is just one of the things he will write about in that *annus mirabilis* (AN-uhs mi-RAB-uh-lis), which is Latin for "miracle year." Einstein is 26, and he is about to set the direction of the twentieth century.

Traveling Ahead

What is relativity? And what is it like to ride on a beam of light? Are there really atoms—bits of matter too small to be seen by any ordinary microscope? When the twentieth century began, no one was sure of the answers to those questions. If you finish reading this book, you will know more than anyone did then. You will know things that even Einstein didn't know.

To understand atoms and relativity, we'll need to go backward into the past as well as forward into the future. Just climb into a time machine, and we can get started. We now know, thanks to Albert Einstein and others, that time and space are woven together—and

It takes you back, doesn't it?

so are past and future. Is time travel really possible? Maybe. Some scientists are taking the idea seriously. Others are skeptical (which means they're not willing to bet on it)—but don't let that stop you. Anything is possible in your mind.

Time on Replay

William Gilbert was the dividing point between medieval thinking
and modern science....He was often wrong in his conclusions, but
he began a new age....He also introduced experimental
observations into science in a systematic way.
—David P. Stern, American physicist, as quoted in the *New York Times*,
  (June 13, 2000)

In the discovery of secret things and in the investigation of hidden causes, stronger reasons are
obtained from sure experiments and demonstrated arguments than from probable conjectures and
the opinions of philosophical speculators of the common sort.
—William Gilbert (1544–1603), English physician, *De Magnete*

The important thing is not to stop questioning. Curiosity has its own reason for existing.
—Albert Einstein, "What I Believe" lecture, 1930

Put on your running shoes. In this chapter and the next,
we're going to sprint through 400 years of science—
which is a bit of a trick. So take a deep breath. You're
going to see the way knowledge builds on itself. Then
you'll watch Albert Einstein take ideas from the past,
look at them with an open, imaginative mind, and come up
with a spectacular new picture of the universe.

The year 1600 is a good place to begin, although it wasn't
a good year for Giordano Bruno. A contemporary of Galileo,
Bruno was burned alive in Rome for, among other things,
saying that the Earth orbits the Sun. That idea was being
whispered about in his time. A Polish church official,
Nicolaus Copernicus, had published a book in 1543
promoting that revolutionary concept. But most church
leaders (from Protestant Martin Luther to Rome's Catholic

Nicolaus Copernicus, shown
in a 1575 portrait, turned
the solar system inside out
and the astronomy world
upside down.

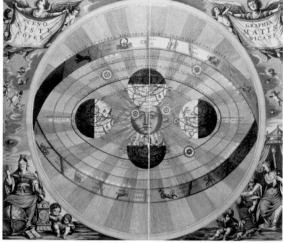

Andreas Cellarius (ca. 1596–1665), a superb Dutch-German celestial mapmaker, was born half a century after Copernicus died. Yet the Sun-centered, Copernican model of the solar system (right) hadn't entirely replaced the old, Earth-centered one. Cellarius drew it both ways.

pope) thought it was the other way around: They believed that the Sun orbits the Earth. Since those leaders were involved in religious wars and serious intellectual turmoil, they didn't need an aggravating distraction. Besides, Copernicus's theory appeared to contradict both the revered Greek philosopher Aristotle and the Bible. So when Bruno trumpeted the same concept, along with ideas about God in every blade of grass and in all of nature (ideas then seen as heretical), he wasn't well received in most of Christian Europe. Only in England, an up-and-coming nation on Europe's geographical fringe, was Bruno able to get his scientific writings published. (In France, he did have two works on mnemonics—memory tricks—published, along with a play he wrote.)

The same year that Bruno was charred, Queen Elizabeth's personal physician, William Gilbert, published a scholarly book on magnetism, *De Magnete* (Latin for *"About Magnets"*). Gilbert had spent 18 years studying the subject. Magnetism was important in a world where most travel was by water and a compass could point the way. The big question was: Why does a compass needle always point in the same direction? (It was the same question that puzzled five-year-old Albert Einstein almost 300 years later.)

Gilbert said that the round Earth is a giant magnet with north and south poles of attraction. He was close; actually

Someone who is **REVERED** is held in awe, respect, esteem, and perhaps devotion. *Worship, adore, venerate,* and *idolize* can sometimes be synonyms for *revere.*

HERETICAL means going against accepted beliefs, especially religious ones. In Christian history, heretics have often been church members who dissented from established Christian doctrine. Bruno's ideas may not seem heretical today, but they were in his time.

Magnetic Attraction

DE MAGNETE, LIB. V.   185
Instrumentum declinationis.

Rocks that are natural magnets—lodestones—turned up in China as early as the fourth century B.C.E. The Chinese lined graves with them so that they could tell which direction to lay the corpses. The proper direction, they believed, helped the dead enter the next world. Later, lodestones were found to be useful for navigation.

In the early fifth century C.E., Saint Augustine wrote of someone he knew who "produced a magnet and held it under a silver plate on which he placed a bit of iron. The intervening silver was not affected at all, but, precisely as the magnet was moved backward and forward below it, no matter how quickly, so was the iron attracted above." What was to be made of that? No one knew; William Gilbert intended to find out.

Why do compass needles point north-south? In 1600, William Gilbert revealed the answer in De Magnete: *Earth itself is a giant magnet!*

Get the flu, sneeze, and you've spread a **CONTAGION**. Like germs, ideas can spread rapidly.

The invention of the Leyden jar sparked an eighteenth-century craze for electrical party tricks that sometimes turned deadly.

the Earth is a giant electromagnet. He got it right in believing that the Earth revolves around the Sun.

Living in England, Gilbert could read Copernicus. But that concept of a revolving Earth was so distasteful to most Europeans that when the Italian Galileo Galilei was given a copy of Copernicus's book, it was from someone who said he wanted "to free his library of its contagion." (In Italy, the book was forbidden property; Galileo quietly read it anyway.)

Gilbert was also interested in a strange phenomenon that causes sparks and sizzles; he named it "electricity." It was fascinating stuff, sometimes shocking, yet it seemed impossible to capture.

In the next century (the eighteenth), scientists (then called natural philosophers) were able to hold on to some of it. Pieter van Musschenbroek and his colleagues in Holland invented what was called a Leyden jar. That jar allowed you to build up electric potential (which is an imbalance of positive and negative charges) and then release it all at once, producing a spark of static electricity at will. It gave experimenters something to play with. Several, like van Musschenbroek, were jolted while using the device. "I thought I was done for," he wrote in a letter to a friend. He was lucky to survive; some others didn't.

Static, Moving, Shocking

Electrons (tiny subatomic particles) account for most (but not all) of electric current, which can be described as a flow of electrons through a wire or other conductor. Why? Because electrons move more easily than, say, protons (other subatomic particles).

Each electron carries one unit of negative charge, and each proton has the same amount of positive charge. So an oversupply of loose electrons means there's a negative charge, and a deficit means there's a positive charge (proton charges predominate).

Electrons tend to flow from a region where there are lots of them to a region where there are fewer—or from negative to positive, a case of "opposites attracting." When the imbalance between negative and positive is strong enough, electrons jump the gap between objects, causing a spark of static electricity.

Comb your hair or rub your feet on a carpet, and you pile up the electrons, causing an imbalance of charge. Then touch a metal doorknob, and you might rebalance the electrons, creating an "ouch" that is static electricity.

Sparks flew, literally, when eighteenth-century electricity fanatics played a parlor game called "Electric Kiss." In this French engraving, a woman is electrically charged by a generator just before her lips touch his—with a jolt.

Think of lightning as a big electric shock across the sky. Rain separates the charge in the clouds (often negative, with too many electrons) from the charge on the wet surface of Earth (often positive, with too few electrons). When the electric field between them gets too large, the air breaks down, becomes a conductor, and—boom! The electric shock you see, hear, and feel results from those opposite charges getting in balance.

When it was understood that electricity could move in the form of electric current, we were on our way to electrical wonders. There's more on electrons in chapters 5 and 6 and on protons in chapter 15.

The Key to Science

Bacon, Galileo, and Newton helped set the foundation for the modern scientific method. Here's how Richard Feynman, a twentieth-century physicist, described that method in *The Character of Physical Law*:

In general, we look for a new law by the following process. First we guess it. Then we compute the consequences of the guess to see what would be implied if this law that we guessed is right. Then we compare the result of the computation to nature, with experiment or experience, compare it directly with observation, to see if it works. If it disagrees with experiment it is wrong. In that simple statement is the key to science.

Ben Franklin discovered that lightning results from static electric charge on clouds, just as sparks can jump across the terminals of a Leyden jar. That realization tied action in the heavens to that on Earth (a startling idea and an important scientific concept). Franklin was a teenager when he first went to London from Philadelphia. In England he found himself surrounded by the new science. Everyone was talking about Galileo and Johannes Kepler and, especially, about the aged English idol—Isaac Newton.

Isaac Newton made British men and women proud. He was their very own genius. Newton, who had few social skills,

This chapter, and others in this book, are full of names. Should you remember all of them? How can you keep them straight? This timeline will be helpful in making those decisions. You'll find that some people are more important historically than others. But the big names—such as Copernicus, Galileo, Newton, Maxwell, and Einstein—should be in every well-informed brain.

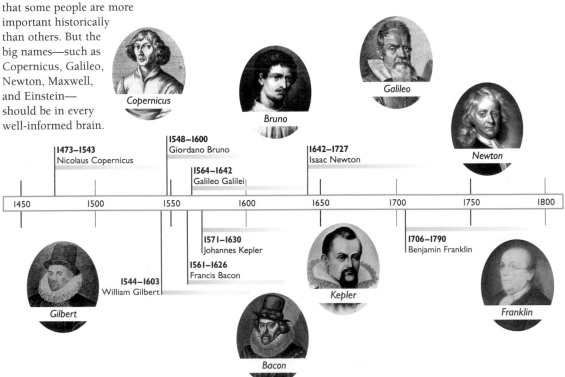

Copernicus

Bruno

Galileo

Newton

1473–1543
Nicolaus Copernicus

1548–1600
Giordano Bruno

1564–1642
Galileo Galilei

1642–1727
Isaac Newton

| 1450 | 1500 | 1550 | 1600 | 1650 | 1700 | 1750 | 1800 |

1571–1630
Johannes Kepler

1561–1626
Francis Bacon

1544–1603
William Gilbert

1706–1790
Benjamin Franklin

Gilbert

Kepler

Bacon

Franklin

An apple made Newton "think of gravity extending to the orb of the Moon," but the concept of universal gravitation took a couple decades to ripen.

Newton's Three Laws of Motion

1. Objects tend to stay at rest or move in a straight line at a constant speed unless acted upon by an outside force.

2. The force acting on a body is directly proportional to, and in the same direction as, its acceleration. Mathematically, that's $F = ma$, which means "force equals mass times acceleration." (Actually, that modern formula was devised by the Swiss mathematician Leonhard Euler using Newton's concept.)

3. For every action there is an equal and opposite reaction.

had isolated himself at Cambridge during his productive years. But by the time the eighteenth century arrived, he was a scientific superstar and a public figure. His ideas were influencing art, music, and even political theory.

Nature is guided by laws that can be understood by ordinary people, said Newton. Natural laws can guide human societies and governments, said Newton's friend, the political thinker John Locke. Newton's scientific method (which depends on human intelligence) was empowering to anyone bright enough to learn. Thinking people, not just kings and queens, began taking responsibility for the changing times.

Isaac Newton, who died in 1727, was said to be the greatest scientist of all time—surpassing even the amazing Aristotle. (Between Aristotle and Albert Einstein, Newton had no serious competition except perhaps from Galileo.)

According to the often-told tale, Newton was sitting in his mother's backyard when an apple hit him on the head, which got him thinking about what causes apples to drop, and that led to his theory of universal gravitation and to a new view of the cosmos.

The apple didn't actually fall on his head—he just imagined it—but Newton did make a connection between the apples on the ground under his mother's tree and the

In Cambridge in 1833, William Whewell suggested to the British Association for the Advancement of Science that their members be called scientists, a term analogous to the word artist for those who practice the arts....The word gradually caught on...and began to displace natural philosopher.
—Alexander Hellemans and Bryan Bunch, science writers, *The Timetables of Science*

Moon way beyond its top branches. He realized that the force that makes an apple fall and the force that keeps the Moon and planets in their orbits is the same. Aristotle had said that the heavens follow different rules from those that work on Earth. Newton (with gravitation) and Franklin (with lightning) showed that isn't so.

Newton was 23 in 1666, the year of the apple incident. That same year he began to work out calculus—mathematics that measures motion and change. He later wrote three laws that govern all motion (page 15) and he refracted (bent) light by shining it through a prism, which told him that it is composed of rays of different colors. The rest of the world, in considering that awesome year (actually it was 18 months), called it an "*annus mirabilis.*"

Meanwhile, in two productive centuries (Newton's seventeenth and Franklin's eighteenth), chemists were evolving from less-than-scientific alchemists. Some "chymists" analyzed air and discovered it is composed of several gases. Others analyzed water and realized that it is a compound, made from two gases, hydrogen and oxygen. It then became clear that earth, air, fire, and water are not elements— as everyone had once believed.

A Change of Terms

In the nineteenth century, chemists classified elements by atomic weight. After we found that a single chemical element could have different atomic weights, we began using atomic numbers to identify elements.

Then we ventured into space, and scientists began using mass instead of weight for clarity. Weight is a measure of the force exerted on an object by gravity—such as Earth's gravity. If you want to lose weight, go to the Moon. It has less gravity than Earth, so you will weigh less there. Your mass, however, won't change.

Earth's gravity keeps the space shuttle (left), an astronaut, and the International Space Station (right) in orbit.

At the very beginning of the nineteenth century, John Dalton, a quiet, hardworking fellow, rediscovered the idea of ultimate particles of matter. He called them *atoms*, which means "uncuttables." He got the word and the concept from Democritus, an ancient Greek. The atoms Dalton considered would not turn out to be uncuttable; still smaller particles would be found. But his atoms were smaller and more basic than anything previously known. And Dalton figured out that in each element, all atoms are identical, but that they are different from the atoms in every other element. That was a huge step.

Later in the century, a Russian professor, Dmitry Mendeleyev (men-duh-LAY-uhf), made a chart of the newly discovered elements, which showed that their atoms have characteristics (like the ability to bond, or combine, with other atoms) that fall naturally into an underlying order or arrangement. The periodic table of the elements was an awesome concept.

But the scientific journey isn't all straight ahead. When Albert Einstein was a boy (about one-half century after Dalton), scientists were having second thoughts about atoms. No one could be sure that they actually existed. Atoms are much too small to be seen by any magnifying microscope. So how did the scientists figure out that they are real?

The evidence slowly grew: For example, when elements combine—such as hydrogen and oxygen to make water—they always do it in fixed proportions. This suggests that elements come in basic units. And they do!

But that evidence for atoms wasn't good enough for many scientists. They said if you can't see something, you can't say that it exists. They just wouldn't believe in atoms. Would you, if you thought there was no hope of ever seeing one?

Lightweight hydrogen leads Dalton's early nineteenth-century list of "elements" (top). (Some in fact are compounds.) In 1869, Mendeleyev's first periodic table left hydrogen as "number 1" but featured dozens of elements grouped in columns by chemical properties. Those columns are now rows (page 123).

Electrifying Thoughts and Magnetic Reasoning

In the heavens we discover by their light, and by their light alone, stars so distant from each other that no material thing can ever have passed from one to another; and yet this light, which is to us the sole evidence of the existence of these distant worlds, tells us also that each of them is built up of molecules of the same kinds as those which we find on earth...Each molecule, therefore, throughout the universe, bears impressed on it the stamp of a metric system as distinctly as does the metre of the Archives at Paris.
—James Clerk Maxwell (1831–1879), Scottish physicist, "Molecules" lecture, 1873

No one would have thought up magnetism and electricity if their effects had never been detected; they are not a consequence of Newton's laws. So it's worth keeping an open mind regarding the existence of as-yet-undiscovered forces.
—Brian L. Silver (d. 1997), physical chemist and science historian, *The Ascent of Science*

Science has always been a search for connections and patterns. But for a long time no one made a connection between magnetism and electricity. How could anyone have figured out that these two phenomena are fraternal twins, linked in the birthing process?

In 1800 in Italy, Alessandro Volta did an experiment—he was trying to find out things about electricity. He took two metal disks (they were zinc and copper), separated them with cardboard soaked in brine (salt water), and made an electrolyte cell. When he connected two or more cells in a

A voltaic pile is a battery in its essence. Zinc and copper disks interact chemically, causing loose electrons—electric current—to flow around a circuit the instant the two wires touch.

magnetic needle

stack, he had what we now call a voltaic pile, or early battery. And experimenters finally had a chemical source of something Volta called "electric current."

While demonstrating electric current (a steady flow of charge) to a class, Hans Christian Oersted, a Dane, saw a nearby magnetic needle turn. He didn't expect that to happen. But Oersted knew right away that somehow the current had reached the needle and that electricity and magnetism are related. He understood the importance of what he had seen, and he spread the news. Oersted invited a group "of most learned men" to witness a repetition of the experiment. It was 1820, and he *electrified* a whole generation of scientists, set off a wave of experiments, and helped generate a new field of science—electromagnetism.

The very next year, England's Michael Faraday built an electric motor, and 10 years after that, Faraday designed the first electric generator. Inside a generator—another name for a dynamo—a coil rotates in a magnetic field. That creates electric current.

Faraday was fascinated with electricity and magnetism and their linkage. He watched iron filings (tiny bits of iron) and saw that when placed near a magnet, they formed a curved pattern that he called a "field." Electric current produced the same effect. It, too, created fields (or lines of force). Most people thought those fields were mysterious and left it

Oersted observed a magnetic needle pivot near electric current.

An **ELECTROLYTE** is a chemical compound that, when dissolved in water, can conduct electricity. All acids, bases, and salts are electrolytes, and all electrolytes contain ions (atoms or molecules that have lost or gained one or more electrons).

Electromagnetism gave physics a great unifying concept: Magnetism, electricity, and light were tied in one package.

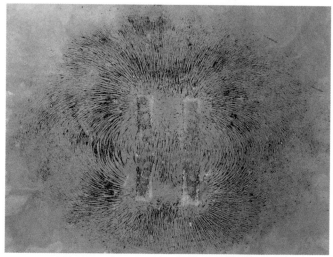

The magnetic field around a magnet is invisible, but Michael Faraday illustrated what he called a "sphere of attraction" by allowing a magnet to shape iron filings into this pattern of curves.

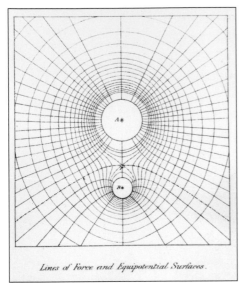

Lines of Force and Equipotential Surfaces.

In *Treatise on Electricity and Magnetism*, Maxwell envisioned "lines of force" weaving a web across the sky.

at that. To Faraday's inquiring mind, nothing in this world was beyond analysis. He said the force field must be part of space itself. Hardly anyone paid attention to that idea except for a Scotsman named James Clerk Maxwell.

In 1857, Maxwell wrote a letter to Michael Faraday. He said, "Your lines of force can 'weave a web across the sky' and lead the stars in their courses without any necessarily immediate connection with the objects of their attraction." Read that a few times and think about it. (Einstein thought hard about it.)

Isaac Newton had described the gravitational force as "action at a distance," but he wasn't able to explain it. Newton was frustrated because he couldn't figure out what made gravity work. Could Faraday's lines of force—his webs—provide a clue? And is there such

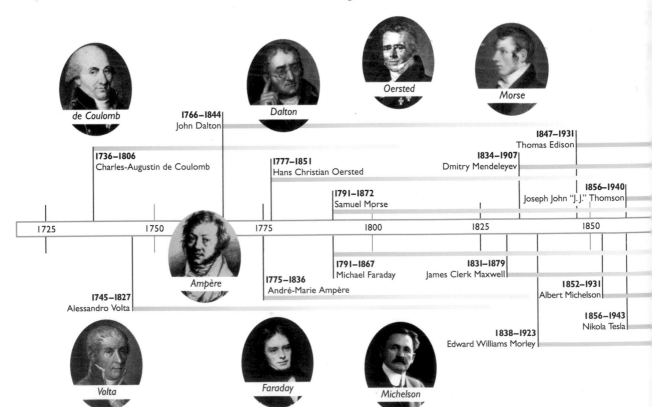

de Coulomb

Dalton

Oersted

Morse

Ampère

Volta

Faraday

Michelson

1766–1844
John Dalton

1736–1806
Charles-Augustin de Coulomb

1777–1851
Hans Christian Oersted

1791–1872
Samuel Morse

1847–1931
Thomas Edison

1834–1907
Dmitry Mendeleyev

1856–1940
Joseph John "J. J." Thomson

1791–1867
Michael Faraday

1831–1879
James Clerk Maxwell

1745–1827
Alessandro Volta

1775–1836
André-Marie Ampère

1852–1931
Albert Michelson

1856–1943
Nikola Tesla

1838–1923
Edward Williams Morley

1725   1750   1775   1800   1825   1850

a thing as "action at a distance"?

While he was thinking about those lines of force, Maxwell learned of some experiments by two Germans, Wilhelm Weber and Rudolph Kohlrausch, showing that the speed of electric current moving in a wire is close to the measured speed of light. That got Maxwell wondering. Might there be a light-electricity-magnetism connection?

According to Newton's laws of motion, there is no *normal* limit to how fast light or anything can go—if enough force is applied to it. But there exists a boundary, or limit, that goes beyond normal. That's infinity. Mathematically, when you reach infinity, you have instantaneous action, which means light would be here and there at the same time.

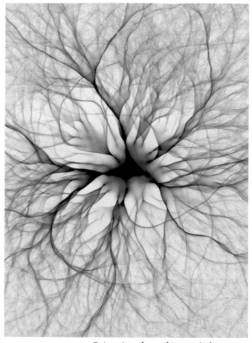

Scientists later learned that electric current (flowing electrons) can cause magnetism. What about everyday magnets, like those on your refrigerator? Is anything flowing there? You bet. Electric charge is circulating inside the atoms in such a way as to create the magnet. Physicist Eric Heller at Harvard University has discovered that electron flow is a beautiful thing. He creates digital models for scientific research and sells the resulting images as fine art. The unexpected flower pattern in *Transport II* (above) happens when some 200,000 electrons in an area the size of a bacterium fan out from the center in a two-dimensional gas (it's sandwiched between two solids). The negatively charged particles are attracted to and diverted by positive charges they encounter. The dark branches are the well-traveled paths.

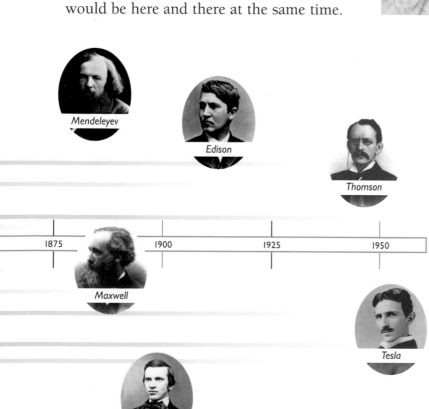

Mendeleyev

Edison

Thomson

1875 1900 1925 1950

Maxwell

Tesla

Morley

Isaac Newton's work described a world that was mechanical. James Clerk Maxwell showed there is more to it than that. The universe has energy at its core; electromagnetic energy is one of its forms.

An **INFERENCE** is a conclusion based on indirect facts or observations. If you see human footprints in the sand, you can infer that someone walked on the beach, even though you never saw it happen. **TRANSVERSE** means a crosswise.

But Maxwell questioned that instantaneous idea. He wondered if light could have a set, unchanging velocity.

Maxwell observed a magnet's effect on iron filings and thought about that, too: Can magnetism be made to go faster, or slower, as Newton's theories implied? Or does magnetism have a set velocity?

"The idea of the *time* of magnetic action…seems to have struck Maxwell like a bolt out of the blue," writes his biographer, Martin Goldman.

Maxwell had a mathematical mind. To prove things, he turned to numbers. "All the mathematical sciences are founded on relations between physical laws and laws of numbers," he wrote. He started with experimental results, especially Faraday's, and then began to pin down the electricity-magnetism-light relationship. He came up with four mathematical formulas that define electromagnetic theory.

Maxwell could hardly believe what his own mathematical equations showed. "We can scarcely avoid the inference that *light consists in the transverse undulations of the same medium which is the cause of electric and magnetic phenomena.*" (Those are his italics.) In other words: *Light is a form of electromagnetism* (my italics), and it undulates (travels in waves) in the form of *fields that it creates.* Imagine that you are holding a small rug. Shake it, and you will see waves undulate.

Maxwell published a paper that said light is electromagnetic. He also said that light has a set speed (in a vacuum), and that all electromagnetic waves travel at the speed of light (in a vacuum). Whew—if that was true, it was astonishing. It meant that light is just one form of electromagnetism, akin to gamma rays and X rays and radio waves. It also meant that the great Isaac Newton's theories don't tell the whole story. Light does *not* travel instantaneously. It cannot be made to go faster or slower (in a vacuum), so it doesn't obey Newton's laws of motion.

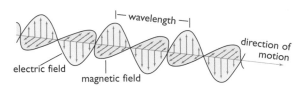

EM (electromagnetic) waves are an electric field that oscillates (wavers back and forth) with a changing magnetic field at right angles.

Math Illuminates Light

Maxwell took the experiments of Michael Faraday and expressed them in four (now-famous) equations. When he found that the computed speed of electromagnetic (EM) waves is the same as the speed of light, he correctly deduced that light is electromagnetic.

He understood that although all light moves at the same speed (in a vacuum), there must be many kinds of EM waves (he was right) and that one differed from another in being either tightly packed or drawn out (right again). Waves that are tight (called short waves) pass a stationary point in short, rapid vibrations (or high frequencies). Those that are drawn out pass a point as long waves (low frequencies). To remember this, just think:
- long waves mean low frequencies;
- short waves mean high frequencies.

Most EM waves are invisible—such as radio waves, ultraviolet light, and X rays. Light is the visible part of the electromagnetic spectrum.

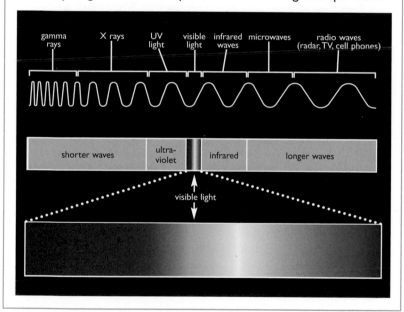

At first almost no one believed that. It contradicted everything they thought they knew. But the behavior of light (which always shows the same speed) was contrary to the behavior of matter and didn't square with the motion of ripples in a rug. An explanation was needed.

Just how do EM (electromagnetic) waves travel through space? What carries them? That was the big question scientists wanted answered. They thought that waves need

When we give the velocity of light, we usually add "(in a vacuum)." Why? When light travels through a medium—such as air or water—it hits the molecules in that medium and slows a bit.

An Electrifying Lawmaker

Charles-Augustin de Coulomb (1736–1806), a Frenchman who was a contemporary of Ben Franklin, wanted to measure the attraction and repulsion between two charged spheres. This electric force is so tiny that no measuring instrument existing at the time could track it. So Coulomb invented his own device. He found that electric and magnetic attraction follow an inverse square law similar to the one Isaac Newton discovered for gravitation. It was an example of mathematical harmony in the universe. His discovery is called Coulomb's Law.

Today, a unit of electric charge is termed a *coulomb*, and its symbol is *C*. It is the amount of charge that flows past a given point in a wire in 1 second if the current is 1 *ampere*. The ampere, or amp, measures the flow of electricity. A typical household current might be 5 amps. Consider a point on a lamp filament: in 1 second, 5 coulombs of charge will pass it by.

Inverse square law? It's a law of physics stating that the value of the electric field of a charge or the gravitational force of mass is in inverse (opposite) proportion to the square of the distance from its source. If it were in direct proportion to the square of the distance from the source, the electric field would get stronger the farther from its source. But the opposite happens.

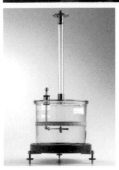

Charles-Augustin de Coulomb, born to wealthy parents, served as an engineer in the French army until 1791 (the peak of the French Revolution). His practical experience in mechanics helped him devise the torsion balance (left) to measure extremely tiny amounts of electric attraction and repulsion.

Maxwell's equations seem to give a complete description of the way light travels, but that wave picture is only half the light story. Albert Einstein will come up with another description, and it too will be valid. As you'll soon see, light (and electromagnetic waves) are two-faced.

something to wave—like ocean water. What could be out there in space to carry waves? What's the celestial rug? Maxwell said it is the "ether."

Ether? What's that?

It was thought to be an invisible, jelly-like substance that fills space. The ancient Greeks had come up with the idea of ether (sometimes spelled *aether*). It made more sense than having a void—or nothingness or a vacuum—in the heavens. There had to be something out there. Without a medium to disturb, how can waves be waves? To get on with things—scientifically—someone needed to find and measure the ether. Finding the ether should also help answer Newton's questions about gravity. Newton never could figure out how gravitation actually works.

When Einstein imagines himself on a light beam, he realizes it must be stationary beneath him. But the EM equations say that that can't happen. There is a paradox here to be solved.

What's it like to ride on a light beam? With an ingenious leap of imagination, Einstein's answer led science in a new direction.

Inside Albert's Head

Einstein was born the year that Maxwell died (1879). He studied and restudied Maxwell's equations. He was aware of a problem that was troubling the foremost scientists: Why don't Maxwell's electromagnetic equations agree with Newton's laws of motion?

Newton's laws had been around for almost 200 years, and they had passed every test put to them.

Maxwell must have made a mistake, skeptical scientists said. But no one could find the mistake. Einstein was the kind of student who liked to reason things out for himself. He read and read and read. Then he concentrated hard and worked out his thoughts by talking with friends. He was a new kind of scientist—or maybe he was a throwback to the ancient Greeks—because he did most of his science in his head. Others would prove his thoughts experimentally. Einstein was sure that if he could only ride that light beam, which had now taken residence in his head, he would solve the conflict between Maxwell's equations and Newton's laws of mechanics.

The first step is to understand how light gets through space. Young Einstein sets up an experiment to try to detect the ether. He has use of the superb equipment at the Federal Institute of Technology. But his experiment backfires; he is lucky not to be permanently injured. This is a challenge. It is where those with scientific curiosity and lab skills are putting their energy. Albert Einstein isn't the only one trying to solve the paradoxes that have come with knowledge of the speed of light.

When something ceases to be mysterious, it ceases to be of absorbing concern to scientists. Almost all the things scientists think and dream about are mysterious.

—Freeman Dyson, English-born American physicist, *Science Week*, March 2000

Three
Charged
Americans

The first message ever sent in Morse code (top) spells out "What hath God wrought" in a series of dots and dashes that stand for letters—w is dot-dash-dash, for example. An operator (below) skillfully taps out words on a telegraph key, pressing three times longer for a dash than a dot and pausing slightly between letters.

The public was fascinated with the new science of electricity and the technology that it brought. When Samuel Morse (1791–1872) sent an electric current from Washington, D.C., to Baltimore in 1844, it turned a magnet on and off. When the magnet was on, an iron lever pressed a rolling paper tape against an inked wheel. A long burst of current produced a dash; a short one made a dot. Morse used that technology to send a coded message (made up of dots and dashes). That telegraph message was the birth certificate of the world of fast long-distance communication.

Morse was an artist as well as an inventor, but as an inventor he couldn't begin to match Thomas Edison (1847–1931), who was known as "the inventor of inventing." Edison had a regular invention factory in Menlo Park, New Jersey. The phonograph and the stock ticker and an invention

"Genius...is 99 percent perspiration," Thomas Edison said. Below, he reflects after toiling 72 hours straight on a phonograph. With his kinescope, an early motion-picture recorder, he immortalized a man's sneeze, frame by frame (right).

Nikola Tesla (bottom) switched the world from DC (direct current) to AC (alternating current). The hard rock band AC/DC took the electricity craze to an interesting level with its first album, *High Voltage* (1975).

that led to motion pictures earned three of his thousands of patents. But it was his creation of the incandescent electric lightbulb that changed the way we humans live. It turned night into day.

Nikola Tesla (1856–1943) was an Austrian-born Serbian who moved to the United States and got a job with Thomas Edison. A big problem at the time was how to transport electrical power on wires. Tesla worked out a way to move high-voltage power and then transform it to low-voltage power at its destination. His transformers worked with alternating current (AC for short), in which the flow of current switches direction periodically. Edison was using direct current (it flows in a single direction only), and he had a hard time admitting that another method could be more efficient. (Today's household current is AC.)

Tesla got angry with Edison and went out on his own. (Among other things, Edison didn't pay Tesla promised money for an invention.) Tesla was a recluse who stayed away from most people but lavished affection on his pigeons. He was also unforgiving. These two great men never settled their feud. This is an interesting story, worth researching on your own.

The M. and M.'s of Science

4

[Albert] Michelson talked Alexander Graham Bell, newly enriched inventor of the telephone, into providing the funds to build an ingenious and sensitive instrument of Michelson's own devising called an interferometer, which could measure the velocity of light with great precision. Then, assisted by the genial but shadowy [Edward] Morley, Michelson embarked on years of fastidious measurements... [and] by 1887 they had their results. They were not at all what the two scientists had expected to find.
—Bill Bryson, American author, *A Short History of Nearly Everything*

When air is removed from a glass jar containing a bell and a flashlight, the sound of the bell disappears, yet light continues to shine out.
—Hans Christian von Baeyer, German-born American physicist, *The Fermi Solution*

A lbert Einstein is eight years old in 1887 when two American scientists, Albert Abraham Michelson and Edward Williams Morley, attempt to answer a big scientific question: What is the ether?

Almost everyone believes it is an invisible substance that fills space. James Clerk Maxwell had said the ether is "the largest, and probably the most uniform body of which we have any knowledge."

Of which we have knowledge? Actually, no one has proof of the ether. There is just a general certainty that it exists. Faraday visualized electric

A French map of the universe includes a *"Region Aetherée,"* an outer space filled with a mysterious ether. Belief in the ether persisted well into the nineteenth century, but no one could prove its existence.

and magnetic fields, and those fields must travel through something. That something has to be the ether, say all the experts. But it has never been detected directly. Maxwell's equations show light as a traveling electromagnetic wave. So light waves should hold the clue to solving the ether problem.

In a well-known experiment, a clock is covered with a glass dome. The clock can be heard ticking through the glass. Then the air is pumped out of the dome, creating a vacuum. The evacuated dome is silent. *Observers can see the hands of the clock moving*, but no ticks can be heard.

Sound waves need to travel through a medium (like air or the skin of a drum); in a vacuum, there is no medium and therefore no sound. Everyone assumes that light travels the same way as sound. So why can we *see* the clock in the evacuated dome when we can't hear it? Conventional logic says something in that vacuum must carry light waves from the clock to our eyes.

The same logic tells thinkers that when light journeys from the Sun and other stars to Earth, it has to be vibrating something. What is out there in the vacuum of space beyond Earth's atmosphere? That's what everyone wants to know.

Those two Americans, M. and M., decide to try to find out.

Michelson was born in Prussia in 1852 and, as a boy, came to the United States, where he attended the U.S. Naval

Light is going to turn out to be more complicated than just being wave action. So think of light as waves, but don't be limited by that picture. A bigger one is coming.

Hold on to Your Hat

You may think you're sitting still reading a book, but this Earth of ours is doing some zooming, and you're going with it. Earth travels around the Sun at the astonishing average speed of about 107,000 kilometers per hour, or almost 67,000 miles per hour. That's about 1,000 times faster than cars on a highway. (It's important to say "average" speed, because Earth speeds up and slows down as it orbits.)

In addition, Earth is rotating on its axis. The speed that the surface is moving varies according to your latitude. At the equator, it's 1,670 kilometers per hour (a little more than 1,000 miles per hour)—which is the length of the equator (roughly 40,000 kilometers or 25,000 miles) divided by 24 hours (the time it takes Earth to make one spin). The rotation speed of the surface is slower at other latitudes—and zero at the poles.

Meanwhile, our solar system is orbiting the center of the Milky Way galaxy, which is also moving through space. Why don't we feel any of this? Chapter 28 will tell you.

Albert Michelson graduated from the U.S. Naval Academy in 1873 and spent two years at sea before returning as a professor of chemistry and physics.

Big Is Big but So Is Small

As you read this book, keep in mind: There are two stories here. One deals with the science of **cosmology**, which is about the cosmos (another word for universe), the biggest thing there is. That's where Michelson and Morley's search for the ether belongs.

But scientists have begun to realize that to understand the very big, they need to look at the world of the very small. And so they come up with a new science—**particle physics**—which is about the tiniest things there are. Albert Einstein will help point the way in both fields. He even tries to bring them together.

Academy at Annapolis, Maryland. Now, in the 1880s, he is a physics professor at the Case School of Applied Science in Cleveland, Ohio.

Morley, who was born in Newark, New Jersey, in 1838, has studied at Williams College and is a chemistry professor at Adelbert College, part of Western Reserve University in Cleveland. Both men are world-class experts when it comes to precise measurement. No one is any better.

Like James Clerk Maxwell, they are quite sure ether fills space and carries the wave vibrations of light. They believe that ether must transmit the Sun's rays in the same way that water carries waves.

These two scientists intend to confirm the existence of the ether. They have designed a measuring device meant to do just that. It will measure light as it travels through space.

Michelson and Morley know that a swimmer goes much faster downstream—with the water's current—than upstream or across a stream. They reason that when light travels in the direction of Earth's movement, it will be swimming upstream against the ether, which will resist it like a wind or water current—slowing it a bit. Moving with the current, it will go faster. So if they can measure the speed of light in the direction of the Earth's rotation, and also the speed of light moving across that rotation, the difference—no matter how slight—will help them confirm the ether. That difference will also give them an exact measurement of Earth's speed in space.

They set up an experiment with a series of measurements that treats light as if it were a swimmer going with, against, and across the current.

To do that, they build an apparatus called an interferometer that splits a beam of light into two parts and then reunites them after they have traveled different paths. M. and M. build their interferometer with metal arms placed at right angles to each other (see diagram, opposite page). Then they put the crossed arms on a big stone block floating in mercury to

Physicist Dayton Miller (1866–1941) built the most accurate ether-drift interferometer of the day. His early 1903–1905 model (left), which he used with Edward Morley, has a mirror on each of the four arms. The split light beams bounced back and forth between opposing mirrors, extending their distance traveled before rejoining. In 1926, Miller later reported a slight positive result for the presence of ether, but his data and methods were questioned, and his results weren't duplicated elsewhere.

reduce vibrations that might skew the results. One part of the split beam goes down an arm pointed in the direction that Earth moves; the other goes at right angles to Earth's path. Each hits a mirror and turns around. The professors expect to see one beam take longer for the round-trip to the mirrors and back, falling behind by a fraction of a wavelength compared to the other beam. That means the wave crests and troughs of the two beams should end up slightly out of step with each other, which will show up as an interference pattern in the recombined beam. The interferometer is exquisitely accurate. It can measure time differences as small as one-thousandth of one-trillionth of a second.

But the light beams always stay perfectly matched. There is no interference pattern. Michelson and Morley are sure they have goofed. Perhaps their experiment is flawed. They must be doing something wrong. They try again and again. They try for *20 years*. But the results never change.

Galileo argued that the Earth is moving through space; by showing no time differences, Michelson and Morley's experiments seem to show Earth standing still. They know that isn't so. M. and M. decide they have failed. But they don't know why.

Even failed experiments can be important. And this one turns out to be momentous. Put the M. and M. experiment in your head; we'll get back to it. You'll find out what they proved without even knowing it.

Here's an overhead view of an interferometer with two mirrors. The light source (left) emits a beam (1) that strikes a splitter in the center. The beam divides into two beams, each of which hits a mirror and bounces back (2 and 3). The two beams rejoin into one (4 and 5), and an observer checks their wave pattern (bottom) for interference—mismatched crests and valleys.

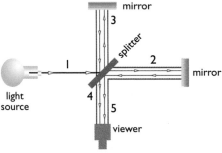

Something **MOMENTOUS** is outstanding, significant, crucial, historic, earth-shattering—in other words, very important. *Momentum*, a science term from the same root, is the "oomph" of a moving object—its mass multiplied by its velocity (speed and direction).

If You Want Something, Go for It!

In the nineteenth century, when few Americans went to college, Albert Michelson was determined to do so. And the college he wanted to go to was the U.S. Naval Academy in Annapolis, Maryland. He picked that school because it was known for its excellence in science and engineering—his special interests. Besides, those chosen to go to the U.S. military academies don't have to pay for their schooling, and his family didn't have extra money.

Michelson's father, Samuel, had emigrated to the United States from Strzelno, a city claimed and fought over by both Prussia and Poland, which made living there difficult. In addition to that, the Michelsons were Jews, and that didn't make their life in Europe easy. So, in 1855, when Albert was three years old, his family headed for the land of promise. When they got to New York, his parents heard tales of the gold rush and of a brother-in-law who had made a success in California; they decided that was where they wanted to go.

Michelson's birthplace, the medieval port town of Strzelno, was once the German Strelno but is now in Poland, near Gdansk.

Gold-mining towns weren't built for comfort. A daguerreotype (an early photographic process) captures Murphy's Camp in the summer of 1853, two years before the Michelsons lived there and at the peak of the California gold rush. The boomtown was "terrific rich," as the locals liked to say. Miners stripped the surrounding Sierra Nevada mountains of gold, iron, quartz, granite, limestone, lava, and gravel.

There were three ways to get to California—across the country by wagon and stagecoach (long and dangerous); around South America by clipper ship (the most expensive choice); or across Panama—by canoe, muleback, train, and ship. They chose the Panama route and made it to Murphy's Camp in Calaveras County in the Sierra Nevada. Two American writers, Bret Harte and Samuel Clemens (who was later known as Mark Twain), were attracted to that same rugged gold-rush camp.

Samuel Michelson set up a small store and stocked it with shovels, boots, blankets, and tents. The town prospered, and so did he. When the Civil War came along (1861–1865), Samuel, like most of the men in town, paraded and drilled in support of the Union. When news of Lincoln's assassination reached the town, he gave his son Albert a middle name. It was *Abraham*.

The boy had been sent to live with cousins in San Francisco when he was 12. That was so he could go to the Lincoln Grammar School and on to San Francisco Boys' High School. Meanwhile, the gold strikes were petering out in Murphy's Camp, and silver had been found at the Comstock Lode in Virginia City, Nevada. So the Michelsons piled their belongings into a mule wagon and moved to Nevada.

Most of the family loved the excitement and drama of life in a mining town, but Albert wasn't much interested. He was an outstanding high school student with a special talent for science.

He wanted to keep studying. It was his father who saw a notice from Congress announcing that two Nevada boys would be chosen to go to the Naval Academy in Annapolis. Applicants were to take an exam. Albert not only passed the exam but, with two other boys, tied for first place. The other two were picked to be naval midshipmen. (Midshipman is the rank given to students at the academy; they are eligible to become lieutenants upon graduation.)

But Michelson had made up his mind: He was going to Annapolis. He knew that each year, in addition to the midshipmen picked from each state and territory, there were 10 openings kept for the President of the United States to fill. So he bought a ticket for Washington, D.C., on the transcontinental railroad. Just a month earlier, on May 10, 1869, the Union Pacific Railroad had raced the Central Pacific Railroad to a rendezvous

Sierra Nevada means "snowy mountain range" in Spanish, a formidable barrier to explorers, gold seekers, and a certain future physicist traveling across the Nevada-California border. Naturalist John Muir wrote, "[T]he mighty Sierra, miles in height, and so gloriously colored and so radiant, it seemed not clothed with light but wholly composed of it, like the wall of some celestial city."

spot at Promontory, Utah, where a golden spike was used to nail the two lines together. Michelson was one of the first coast-to-coast passengers. Armed guards sat in each car to deal with possible Indian or bandit attacks. (There were none on Albert's journey.)

When Michelson got to Washington, he learned that President Grant took a walk every afternoon. He decided to join him on that stroll. (It was possible for citizens to do that then.) Albert pleaded his case to the commander in chief, but Grant had already used up his 10 appointments. He said there was nothing he could do. The president's naval aide told the boy to go to Annapolis and talk to the commandant of midshipmen. Albert did and waited three days for an interview. He was turned away.

Discouraged, with little money left in his pockets, Albert Abraham Michelson boarded a train for San Francisco. He was in his seat when he heard a messenger from the White House calling his name. The president had received several letters urging him to make an exception in Michelson's case. He did so. President Grant appointed Michelson a midshipman-at-large. And once he did that, he also did it for two other young men.

Years later, Michelson would chuckle that he had begun his career with "Grant's illegal act." It turned out to be a wise one. In 1907, Michelson became the first American to win a Nobel Prize. Albert Einstein would write of him, "I always think of Michelson as the artist in Science. His greatest joy seems to come from the beauty of the experiment itself, and the elegance of the method employed."

U.S. Navy Cadet Albert Michelson excelled in optics (the science of light and vision) and drawing at the academy. "It seems to me that scientific research should be regarded as a painter regards his art, a poet his poems, and a composer his music," he later said. Michelson also played the violin and a skillful game of billiards.

If a poet could at the same time be a physicist, he might convey to others the pleasure, the satisfaction, almost the reverence, which the subject inspires. The aesthetic side of the subject is, I confess, by no means the least attractive to me. Especially is its fascination felt in the branch which deals with light.

—Albert A. Michelson, *Light Waves and Their Uses*

Invisible Bits of Electricity

> It is the charm of physics that there are no hard and fast boundaries, that each discovery is not a terminus [end] but an avenue leading to country as yet unexplored, and that however long the science may exist, there will still be an abundance of unsolved problems.
>
> —Joseph John "J. J." Thomson (1856–1940), English physicist, Royal Institution lecture, 1897

An experiment is a question which science poses to Nature, and a measurement is the recording of Nature's answer.

—Max Planck (1858–1947), German physicist, *Scientific Autobiography and Other Papers*

Joseph John Thomson, the son of a Manchester bookseller, is a rumpled, absentminded Cambridge University professor who is so clumsy in the laboratory that his assistants try to keep him from touching any of the equipment. Here is what one student says of him:

> *Along would shuffle this remarkable being who, after cogitating . . . over his funny old desk in the corner and jotting down a few figures and formulae in his tidy handwriting on the back of somebody's Fellowship Thesis or an old envelope, or even the laboratory checkbook, would produce a luminous suggestion like a rabbit out of a hat.*

If you think deeply about something, you are **COGITATING**. *Pondering* is a synonym. If your thoughts are brilliant and enlightening, they can be described as **LUMINOUS**. Scientifically, something that is luminous emits light.

Thomson, who is called J. J. by everyone including his son, does not think big. He thinks small—very small. But the rabbit that comes out of his hat in 1897 turns out to be mountainous in importance.

J. J. Thomson was a scientific superstar. "Students from all over the world looked to work with him," wrote physicist Abraham Pais. Pais added that Thomson's ability to understand intricate apparatus without handling it was "something verging on the miraculous, the hallmark of a great genius."

That's when J. J. Thomson announces that electric charge (what most people call electricity) **is carried by tiny particles**. He calls them "corpuscles of electricity." (They will soon be known as **electrons**.) Many years later, someone will label that pronouncement the biggest revolution in physics since Newton. That may be overdoing things—but it is a breakthrough that will change the way we look at the world, help set a foundation for a new era of science and technology, and send us searching for the other innards of the atom.

When a physicist uses the word **CORPUSCLE**, he or she means a discrete particle. To a doctor, a corpuscle is a blood cell. An **ATOM** is the smallest unit of an element having all the characteristics of that element. There are 92 natural elements, from hydrogen to uranium. Each atom consists of a dense nucleus of positively charged protons and electrically neutral neutrons surrounded by negatively charged electrons. All the atoms of a given element have the same number of protons—for instance, every uranium atom has 92 protons.

Charge It!

Charge is a measure of the strength with which some elementary particles interact. It comes in two varieties: positive and negative. Positives repel positives, and negatives repel negatives. Benjamin Franklin understood that in the eighteenth century, saying, "Opposites attract, and likes repel."

Franklin's contemporary Charles-Augustin Coulomb put electrically charged spheres different distances from one another and measured their attraction or repulsion on a torsion balance. Coulomb found that electric force follows an inverse square rule similar to that of gravitational force. The standard unit of charge, the coulomb (*C*), is named for him (see page 24).

In the twentieth century, thanks to J. J. Thomson, electricity came to be associated with subatomic particles. We now know it takes more than 6 billion billion electrons to make 1 coulomb (6.24×10^{18}). As a standard unit of charge, the coulomb is ridiculously large.

At the time, though, the idea seems absurd. Electricity flows like a river's current; how can it be made up of particles? J. J. wonders. But he has a genius for discovery and is quite sure of himself. He doesn't know how or why those minuscule particles exist, but he says they are almost 2,000 times lighter than an atom of the lightest element, hydrogen—and very fast, too.

Minuscule is more than just small; it's very, very small.

Who is this man who is about to change physics from a science that no one thinks has much of a future to one that will dominate the new century?

J. J. Thomson grew up in the 1860s, the golden age of railroads. He dreamed of designing locomotives, like the one in this 1870 British watercolor. "The railway locomotive was powerful, noisy, dramatic, frightening, exciting, a focus of exhilaration and of awe; it redefined time and reordered space," writes Ralph Harrington, a railroad historian at the University of York. "There was something deeply stirring about its size and power, its speed, its use of the elemental forces of fire and water."

"I have had good parents, good teachers, good colleagues, good pupils, good friends, great opportunities, good luck and good health," says J. J. Thomson after he is famous, but there's more to his story than that.

As a boy, J. J. intends to become a mechanical engineer and build locomotives. Everyone believes trains are the future. In 1870, when J. J. is 14 and ready to tackle a career, train tracks are crisscrossing England and every other progressive nation. (The Wright brothers won't fly a heavier-than-air machine until 1903. Even then, no one imagines that mass transportation will one day take to the air.) J. J.'s father has talked to an important railroad man in Manchester about hiring his son, but the locomotive firm has filled all its apprenticeships (and there is a long waiting list for the positions as well as a significant fee). So young Thomson will have to wait to become an apprentice. Meanwhile, he takes courses at Owens College in Manchester, where there are some gifted professors, especially in physics, history, law, and math.

Then, unexpectedly, his 39-year-old father dies. J. J. is 16. Becoming an engineer is now out of the question. There is no family money for apprenticeship fees.

One of his professors helps him get a scholarship so he can continue at Owens College. After that he encourages J. J. to compete for an entrance scholarship to Trinity College at the University of Cambridge (where Newton studied and taught).

Lord Rayleigh said of J. J. Thomson (circled), "Nothing delighted him more than the opportunity of going to a good football match or watching the performance of Trinity men on the river." (Competitive boat races were and are important in Cambridge. And soccer is called "football" in Britain.)

J. J. doesn't make it the first time he takes the exam, but at 19 he is admitted to Cambridge. The great James Clerk Maxwell is head of Cambridge's Cavendish Lab, England's most distinguished physics laboratory. Maxwell dies in 1879 (before J. J. can study with him), and Lord Rayleigh follows Maxwell in the prestigious position. (J. J. does study with Rayleigh.) Then, in 1884, when J. J. Thomson is 27, he succeeds Lord Rayleigh as head of the Cavendish Lab.

"Matters have come to a pretty pass when they elect mere boys [as] professors," says a disgruntled older physicist, but both J. J. and the Cavendish Lab flourish. One of J. J.'s pupils, F. W. Aston, writes, "[H]is boundless, indeed childlike, enthusiasm was contagious...." Science students flock to Cambridge from all over the world.

"I have never seen a lab in which there seemed to be so much independence and so little restraint on the man with ideas," says another of his students. Yet another student writes, "J. J. spent a good part of most days sitting in the armchair that had belonged to Maxwell, doing mathematics."

J. J. has no doubt that mathematics along with deep thinking can lead the human mind to an understanding of the physical world and its phenomena.

Thomson doesn't limit his interests to math and physics,

The motto of Cambridge University's Cavendish Lab, carved over the entrance, is *Cavendo tutus*. The Latin phrase means "safe by being cautious" or, more loosely, "always on the lookout."

Hermann von Helmholtz (below, upper left) was an outstanding German physicist, doctor, mathematician, student of languages, and the inventor of the ophthalmoscope (for peering into the eye). But he was best known for his Law of Conservation of Energy, also known as the first law of thermodynamics (the change in all forms of energy of any system, plus the change in all forms of energy in its surroundings, always adds up to zero). Von Helmholtz suggested that "atoms of electricity" (in other words, electrons) would be found. J. J. Thomson was one of the few who took this notion seriously.

A **SYSTEM** could be a box of gas or your body or objects orbiting a star in a solar system. A car is a system—its parts interact and work together to create a whole. It's part of a larger transportation system that includes roads, gas stations, oil wells, traffic laws, etc. Scientifically, the idea behind a system is that there's input, output, interaction between parts, and processes (like energy conversion) in action.

however. He never misses a Gilbert and Sullivan operetta, he plays golf, he reads poetry, he gardens, he is fascinated by American politics (J. J. lectures at Yale and Princeton). And of course he cheers his students' achievements. (Seven of those students, including his son, will win Nobel Prizes.)

How does J. J. Thomson manage to discover the electron? Well, it has to do with some glass tubes. For several decades, physicists have been trying to figure out the nature of the rays produced when an electric current is sent through a vacuum. It all begins when English scientist William Crookes (1832–1919) devises a special sealed glass tube. At one end he places a heated, negatively charged electrical terminal called a cathode. At the other end, another plate is given a positive electrical charge to form an anode. (Crookes's cathode-ray tube is not very different

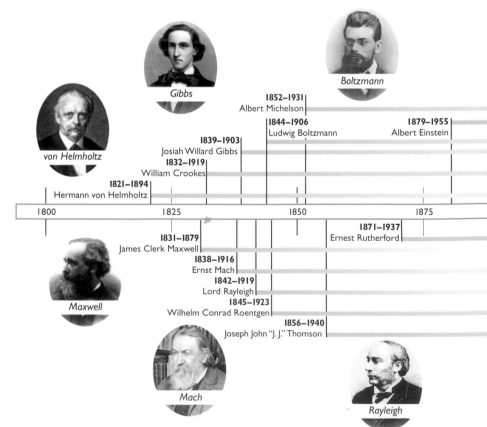

Gibbs

Boltzmann

von Helmholtz

| 1852–1931 |
| Albert Michelson |

| 1844–1906 |
| Ludwig Boltzmann |

| 1879–1955 |
| Albert Einstein |

| 1839–1903 |
| Josiah Willard Gibbs |

| 1832–1919 |
| William Crookes |

| 1821–1894 |
| Hermann von Helmholtz |

1800   1825   1850   1875

| 1871–1937 |
| Ernest Rutherford |

| 1831–1879 |
| James Clerk Maxwell |

| 1838–1916 |
| Ernst Mach |

| 1842–1919 |
| Lord Rayleigh |

| 1845–1923 |
| Wilhelm Conrad Roentgen |

| 1856–1940 |
| Joseph John "J. J." Thomson |

Maxwell

Mach

Rayleigh

from the picture tube used in early television sets.)

J. J. Thomson and others begin experimenting with the tube. They pump air out of the device and send electric current through it. The whole tube glows in colors that change as the tube is emptied and the pressure is lowered. At sufficiently low pressure, the glow stops—except at the far end of the tube near the anode, where, especially if the glass is coated with zinc sulfide (or some other fluorescent material), it glows as soon as the current is turned on. What is causing the glow? Something must be going from the cathode to the anode. But how can electricity (the current) travel in an empty tube? Does the phenomenon change if a gas fills the tube? What about different gases? All this is baffling—and fascinating, too. The invisible, unexplained current is called "cathode rays."

What are cathode rays? That's the puzzle J. J. Thomson

Sir William Crookes was the eldest of 15 children of a London tailor. He became a famous chemist and editor of *Chemical News*. He discovered that cathode rays cast a shadow and that they heat obstacles put in their way. He concluded that they might be negatively charged particles, but no one paid attention to that idea until J. J. Thomson came along.

We now know that a **CATHODE RAY** is a stream of electrons (which carry a negative charge) projected from the surface of a cathode (the negative pole) in a vacuum tube. That stream is what paints a picture, line by line, on a TV tube with an electron gun. Inside an electron-gun TV (one without a plasma or liquid-crystal display), electrons travel at about 64,000 kilometers (about 40,000 miles) per second.

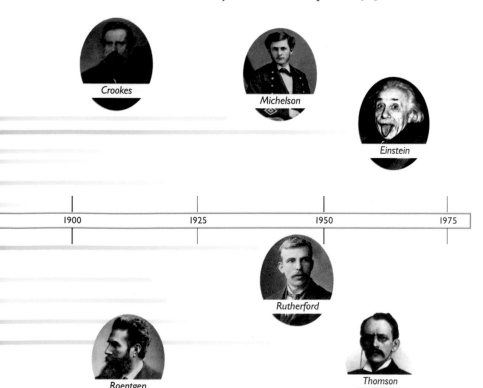

Crookes

Michelson

Einstein

1900   1925   1950   1975

Rutherford

Roentgen

Thomson

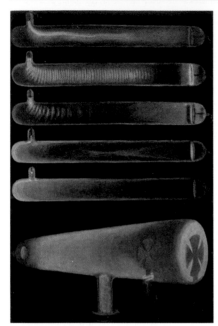

Thomson's experiments with cathode rays inspired this colorful French lithograph, *The Phenomenon of Electrical Luminosity II* (ca. 1900), from a series of books by Hans Kraemer called *l'Univers et l'Humanité* ("The Universe and Humanity").

Something that's **ETHEREAL** is so light and airy that it's not really there. By material, Thomson means the opposite—solid or physical in form. To chemists, ethereal describes any substance dissolved in ether, a colorless liquid once used to knock out patients before surgery.

sets out to solve. One of his students, Ernest Rutherford, writes to his fiancée that his professor "of course, is trying to find out the real cause and nature of the waves, and the great object is to find the theory of matter before anyone else, for nearly every professor in Europe is now on the warpath."

"Every professor in Europe"? That means Thomson isn't the only one experimenting with cathode rays. Crookes's tube is very popular with scientists. Some just play around with the device, as if it were a toy, but a few use it for serious experiments.

"According to the almost unanimous opinion of German physicists they [the rays] are due to some process in the aether," says J. J. (In other words, this radiation is the result of something, perhaps wave action, in the mysterious ether—as we now spell it— say the distinguished German physicists.)

"Another view of these rays," says J. J., "is that, so far from being wholly ethereal, they are in fact wholly material." That's where he stands. J. J. Thomson believes he is dealing with "rays" of matter. He comes up with two hypotheses:

Hypothesis One: The size of the carriers (the particles of electric charge) must be small compared with the dimensions of ordinary atoms and molecules.

Hypothesis Two: These carriers are the same whatever the gas used in the discharge tube.

If he can prove these hypotheses, it will show that what he calls the "carriers" of electricity (those corpuscles/electrons) are elementary bits of *matter*. "The assumption of a state of matter more finely subdivided than the atom of an element is a somewhat startling one." Read that sentence again. This is an understatement!

Thomson begins experimenting with Crookes's glass tube. He knows that the rays can be bent by a magnet; Jean Baptiste Perrin in Paris discovered that. Thomson repeats and improves on Perrin's experiment.

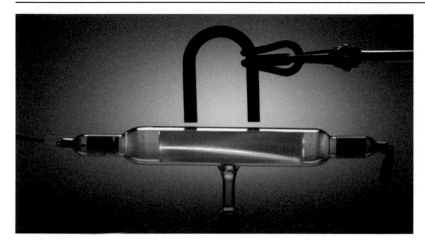

A beam of electrons, traveling from left to right, glows green in the partial vacuum of a cathode ray tube. In this popular physics demonstration, the horseshoe magnet is causing the negatively charged electron beam to deflect downward, perpendicular to the magnetic field.

He sends a thin stream of current through the tube to a fluorescent screen. When the beam hits the screen, it creates a pinpoint, a dot of light, which stays in place. Then J. J. puts a glass tube between the pole pieces of a large magnet, and the dot jumps downward. Waves don't respond to magnetism, but particles do. That clue helps convince J. J. Thomson that he is dealing with particles.

J. J. does not stop there. He goes on to compare the trajectory of the dot (which is responding to the magnetic force) to the path of a cannonball (which is matter reacting to the force of gravity). Using this comparison, he is looking for a ratio—the ratio of the corpuscle's electric charge to its mass (e/m). Scientists have figured out the charge-to-mass (e/m) ratio for the tiny hydrogen ion, and because of that, they have been able to come up with a number for the mass of the hydrogen atom. It is .0000000000000000000000017 grams. That makes the hydrogen atom the smallest thing known (at the time) and so small that many scientists can't accept the atomic concept.

J. J. Thomson believes in atoms. If he can now get that e/m ratio for the unknown corpuscles—well, then he can work on mathematical formulas, have a number to compare with the hydrogen atom, and figure out the size of the particles of electricity (those corpuscles).

He begins his next experiment by deflecting (bending) cathode rays in a magnetic field and in an electric field.

Thomson used a cloud chamber to estimate the charge and mass of an electron. The chamber produces a visible trail for the passing of elementary particles in somewhat the same way that a vapor trail forms behind a high-flying jet. The photo (below) shows one of the first such trails recorded on film, at Cambridge's Cavendish Lab in 1911. (Those are alpha particles emitted by radium.) Thomson's value for e (the elementary unit of charge) was way off—about three times too high—but it was a start. The actual value was found by American Robert Millikan in 1913.

Carefully adjusting the currents, he makes the deflections equal. Then he switches off the source of current for the magnetic field. He now has only the deflection caused by the electric field. He can measure its angle easily. He does the same with the magnetic field. With that information, he can figure out the ratio he needs.

Being a careful scientist, Thomson keeps experimenting. Sometimes he puts gases in the cathode tubes. It doesn't matter; the rays don't change. Then (through some inspired calculations) he figures out the velocity of the electron beam. It moves at about 32,000 kilometers (20,000 miles) per second (faster than any known object). But—and this is very important—*it doesn't move anywhere close to the speed of light*. That means these "corpuscles" are not EM waves, as X rays turn out to be.

Thomson has discovered something previously unknown—a beam of energetic particles. With his ratio, Thomson can determine the mass (*m*) of a particle. He compares the mass for the cathode-ray particles and for hydrogen ions. The mass of a cathode-ray particle is 1,000 times lighter than the mass of a hydrogen ion. (He soon corrects that to 2,000 times lighter.)

What he comes up with is smaller than the smallest atom—much smaller! He calls it "some primordial substance X" and "a negative ion." We know that it is an electron.

On Friday, April 29, 1897, in a speech at the Royal Institution, J. J. Thomson says, "I have lately made some experiments which are interesting." Interesting? Hmm. *Astonishing* would be a better word. *Revolutionary* would be still better. J. J. has done experiments that explain electricity and show that atoms are not hard solid

Ions Sing—to the Tune of "Clementine"

Physicists speak of beautiful experiments the way artists talk about beautiful paintings or composers speak about beautiful music. J. J. Thomson's experiments were elegant. When he put magnets around a cathode tube, he sometimes made the ray spin. Because the vacuum in the tube wasn't perfect, ions (charged atoms) would often hit a stray atom, causing it to glow. Thomson and his colleagues wrote a song about all this, called "Ions Mine." Here's part of it:

Ions Mine

In the dusty lab'ratory,
'Mid the coils and wax and twine,
There the atoms in their glory,
Ionize and recombine.

   (Chorus—sung after each verse)
Oh my darlings! Oh my darlings!
Oh my darling ions mine!
You are lost and gone forever
When just once you recombine.

In a tube quite electrodeless,
They discharge around a line,
And the glow they leave behind them
Is quite corking for a time.

…In the weird magnetic circuit
See how lovingly they twine,
As each ion describes a spiral
Round its own magnetic line.

Everyone Learns of X Rays

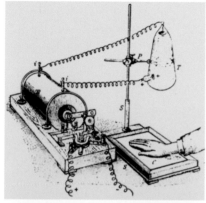

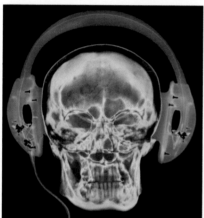

A German physicist, Wilhelm Roentgen (1845–1923), studied the luminescence (the glow) on the walls of Crookes's tube. When he covered the tube with black paper, the luminescence went right through the dark paper and onto a screen in his laboratory. Then he found that when the cathode rays hit tungsten or some other heavy metals, they gave off a ray that passed through rubber, wood, and even his fingers! The year was 1895, and Roentgen had accidentally discovered X rays.

Just three months later, schoolboy Eddie McCarthy of Dartmouth, New Hampshire, had a broken arm set with the help of an X-ray image. You can imagine the excitement. You didn't have to be a doctor to understand the potential of these rays. Lots of people wanted to see their bones. Not since Galileo's first telescope in the seventeenth century had the general public been so fascinated by a scientific announcement. (We now know that X rays can damage human cells, and, today, medical exposure is carefully limited.)

X RAYS are by-products given off when cathode rays (electrons) strike a metal target or when inner electrons in heavy elements transition. "They are a form of electromagnetic (EM) radiation.

X-ray technology has come a long way from Roentgen's early machine (diagram at top), which took minutes per picture, to a colorized X ray of a person wearing headphones.

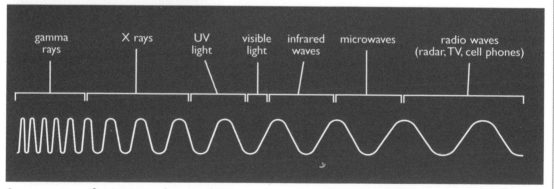

Seventeen years after Roentgen discovered X rays, Max von Laue, a German physicist, found that they are electromagnetic waves, like visible light, but with shorter wavelengths. Since all EM waves travel at the same speed in a vacuum—roughly 300,0000 kilometers (about 186,000 miles) per second—short waves must go up and down (undulate) much more often to reach the same speed as long waves. Radio wavelengths can be many miles long. And a gamma ray (on the other end of the spectrum) can be as small as an atom's nucleus. The wavy line gives the idea but fails to show this radical difference in wavelength.

In reading this book, keep the very small, nano world separate from the big, macro world, and be ready to have electromagnetic forces link them. *Macro* is short for "macroscopic," or big enough to be seen by the unaided eye. *Micro* applies to things seen through an optical microscope. At right is a colorized, scanning electron microscope (SEM) image of plant pollens. On the even smaller atomic and subatomic scale, we speak of the *nano* world, meaning objects that are about 1 nanometer (one-billionth of a meter) in size.

J. J. Thomson, writing in England's *Philosophical Magazine* in 1897, conjectured as to what else might be in the atom: "In a neutral atom the negative effect is balanced by something which causes the space through which the corpuscles are spread to act as if it had a charge of positive electricity equal in amount to the sum of the negative charges of the corpuscles." That means there must be some positive something in an atom to balance the negative electrons. What can it be? J. J. thinks there might be an equal number of tiny positive particles spread out through the atom. Is he right? His challenge is to find out.

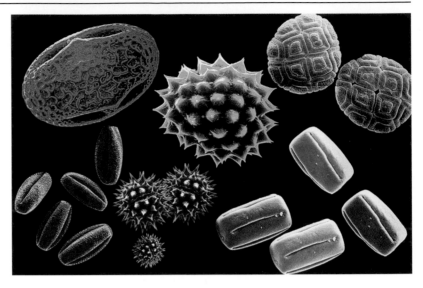

spheres. But his audience isn't ready for his announcement.

He starts by telling them that the stream of current in a cathode tube is made up of tiny negatively charged particles or corpuscles of electricity (electrons). He adds that these corpuscles are each of the same mass and that one of those particles is some 2,000 times lighter than an atom of hydrogen.

Thomson's electron idea doesn't have a chance with the traditionalists. For most scientists, the notion of subatomic, or smaller-than-an-atom, particles is too fantastic to believe. When they do believe it, most don't think it's particularly important information. And the general public yawns. Unlike X rays, electrons have no dramatic appeal. What can you do with them? At Zurich Polytechnic, where Albert Einstein is an 18-year-old student in 1897, the physics-history professor Heinrich Weber *doesn't even include Maxwell* in his lectures. Herr Weber focuses on Newton. So, of course, he doesn't tell his students about Thomson's findings. Einstein, who has antennae that search out new things, finds Maxwell and Thomson on his own. Albert will pay attention to J. J. Thomson's corpuscles/electrons well before most others take them seriously.

As for J. J. Thomson's colleagues, even when they get it, they joke that the electron will never be of use to anyone.

Smaller than an Atom!

With electrons, Thomson has opened the door to a subatomic world—which will turn out to be full of complexities and surprises. No one before him knew that world existed. He has found within the atom a particle much less massive than any atom. (Note: In actuality, the electron cloud spreads across the atom, so it cannot be "smaller" than the atom in dimension. But it is less massive. Thomson didn't yet know the shape that electrons take.)

Electrons are not only real, they are versatile critters. They can move on their own through empty space, or they can move through electrical conductors (like wire) as current, or they can exist inside an atom. Electric current is just a river of electrons. Its strength is in proportion to the number of electrons that pass a specific place, like a wall socket, every second.

Electrons carry a negative charge. It is the behavior of electrons, which is two-faced, that will be a real shock (pun intended). Everyone thinks that particles are particles and waves are waves and you have to be one or the other. But electrons have it both ways—they exist in a dual particle-wave world.

It will take a while for all that to be discovered. Our friend Albert Einstein will do much of the headwork to make it clear that matter and energy are related through tiny particles—electrons and others. It will be one of the big breakthrough insights of the new century.

A sharp blow to a special block of plastic causes electrons inside it to "branch out," forming a lovely tree pattern.

Oh my, are they wrong. Very wrong. **To begin with, electric current *is* made up of moving electrons. J. J. helped give us a world powered by electrons and controlled by electronics.** Scientists soon learn that electrons are in every atom in your body. Electrons are in every atom of everything around you. By 1906, when J. J. Thomson gets the Nobel Prize for his discovery of the electron, the skeptics have become convinced.

Charging on—to *e*

Yes, J. J. Thomson discovered that electricity consists of negatively charged particles—electrons. And he knew the ratio of the electric charge of the electron to its mass. But he didn't know the actual charge of an electron (e). Was e important to know?

You bet. When atoms and molecules interact, the dealings are electric, so knowing the charge that each electron carries is basic. Could that charge be found? Not easily. An American scientist, a professor at the University of Chicago, Robert Andrews Millikan (1868–1953), used a perfume sprayer (an atomizer) to track it down.

Millikan began by spraying tiny drops of water and letting them fall in an electric field where he could measure them. That just didn't work. Then Millikan hit on oil. He sprayed tiny oil droplets into a capacitor (a device for storing electric charge). He said, "These droplets...were found in general to have been strongly charged by the frictional process involved in blowing the spray." (Friction can add or

The young, athletic Robert Millikan toyed with the idea of a career in physical education and excelled at Greek and mathematics while at Oberlin College in Ohio. He opted instead to study advanced physics at Columbia University and later in Germany.

Millikan's oil-drop experiment is often cited as one of the greatest experiments of all time. This photo, from Caltech, shows the setup, including the cylindrical chamber in which oil drops were suspended. The goal was to measure the charge of an electron by balancing the upward pull of an electric field on an oil drop against the downward pull of gravity—a known quantity.

remove electrons from an oil drop—converting some of its atoms into ions—thus giving the drop an electric charge.)

Now, two forces were acting on those oil droplets: the electric force up and gravitation down. Millikan could watch the droplets with a microscope and control the speed of their fall by changing the voltage of the capacitor. When he balanced the two forces, an oil droplet would stay suspended in midair, "like a brilliant star on a black background."

Millikan sprayed away and watched one drop after another, changing the voltage and recording what happened. No matter how many droplets he examined, the charge always came in multiples of a basic unit of electricity, which he labeled e.

He realized e was the charge of a single electron. (The value of e is about 1.602×10^{-19} coulombs. All electric charges, positive or negative, that we observe directly are integer multiples of this quantity.)

Millikan wrote, "I have observed… the capture of many thousands of ions [atoms that have a charge]… and in no case have I ever found one, the charge of which… did not have either exactly the value of the smallest charge… or else a very small multiple of that value."

Trained by Albert Michelson, who devoted much of his physics career to developing extremely precise instruments of measurement, Robert Millikan was a highly skilled experimenter.

Smaller than Atoms?
Subatomic? Is This a Joke?

Ironically it turned out that the first sub-atomic particle to be discovered was also one of the most fundamental. Of the hundreds of so-called elementary particles, the electron is one of the very few that has remained impervious to further subdivision. It is truly elementary.
—Hans Christian von Baeyer, German-born American physicist, *Taming the Atom: The Emergence of the Visible Microworld*

An electrical current in a wire is nothing but a flow of electrons. Electrons participate in the nuclear reactions that produce the heat of the sun. Even more important, every normal atom in the universe consists of a dense core (the nucleus), surrounded by a cloud of electrons.
—Steven Weinberg, American physicist, *The Discovery of Subatomic Particles*

J. has two contemporaries, Ludwig Boltzmann (in Germany) and Josiah Willard Gibbs (in America), who are convinced that atoms explain many questions about matter, but not everyone agrees with them. The atomic concept doesn't seem like solid science. As the well-respected physicist Ernst Mach keeps asking with infuriating regularity, "Have you ever seen one?" So when Thomson announces his discovery of electrons, most scientists laugh—especially those who are still insisting that molecules and atoms don't exist.

There is no real proof of the existence of atoms—although some nineteenth-century scientists have come up with statistical evidence that they must be there. Statistics is a new way of looking at science, but many physicists aren't ready to acknowledge

You can see ketchup—it's in the macro world. You can even see the structure of its cell fragments and pectin fibers (below), thanks to scanning electron microscope (SEM) technology. But how do you know ketchup (and everything else) is made of atoms and molecules—which are far too small to see with an SEM?

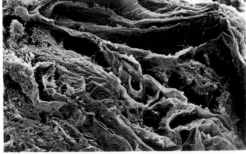

The Proof Is in the Numbers

Statistics is about using numbers to predict a probable outcome, as you may recall from chapter 1, page 7. So how can numbers predict that atoms exist?

Book two in this series, *Newton at the Center*, answers that question in detail. Here's the gist: In the nineteenth century, Ludwig Boltzmann and others calculated the kinetic energy of gas molecules ping-ponging around inside a container. (Kinetic energy is the energy of motion.) They found that, if you accept that atoms and molecules exist and that they have different masses (*m*), this formula works:

$$kinetic\ energy = \tfrac{1}{2}mv^2$$

The *v* stands for velocity (or, more accurately, speed). Using this formula, the scientists calculated the probable average speed of the gas molecules *without actually seeing them*. In other words, the proof of their existence was in the numbers, not in the eye.

statistics as a tool of science. They won't accept atomic theory, either. And electrons? How can any sane person believe in something so small that we have no chance of actually seeing it?

Even Thomson has a hard time accepting the results of his experiments. "It was only after I was convinced that the experiment left no escape from it that I published my belief in the existence of bodies smaller than atoms."

Most physicists treat Thomson's particles of electricity as a wild conjecture. But they are no conjecture.

Later J. J. will recall, "At first there were very few who believed in the existence of these bodies smaller than atoms. I was even told long afterwards by a distinguished physicist who had been present at my lecture at the Royal Institution that he thought I had been 'pulling their legs.' I was not surprised at this, as I had myself come to this explanation of my experiments with great reluctance."

When the scientific community realizes Thompson isn't kidding, it is stunned. Those negatively charged particles—the electrons—are enough to startle anyone. We now know

A **CONJECTURE** is a guess, but solid conjectures are guesses based on information.

One electron is tiny—very, very tiny. A proton is about 2,000 times greater than an electron in mass. Yet the electron is way bigger in dimension because it is whizzing around. (More on this ahead.)

that they can't even be described exactly in terms of size. A hydrogen atom essentially occupies the room filled by its single electron, which spreads out in its tiny bit of space as if it were a cloud. **It would take about 4 *million* hydrogen atoms to cover the period at the end of this sentence.**

As for the behavior of electrons, the more you know about them, the more bizarre (and intriguing) they seem. So if you find the subatomic world hard to imagine, you're not alone. Many of Thomson's colleagues just couldn't do it.

When asked what an atom looks like, J. J. Thomson guesses. He says, "a spongy-ball" or a "plum pudding." He believes electrons are like bits of fruit distributed evenly throughout the pudding. How do they stay in there? No one has any idea. (It's the electromagnetic force that holds them; within atoms it is stronger than gravity. That is yet to be understood.)

Does an atom actually have a structure? Is it like a fruit-filled pudding? No optical microscope can see an atom (not then, not now). It will take decades of experimenting for scientists to answer these questions. When they do, they will understand that atoms are complex entities.

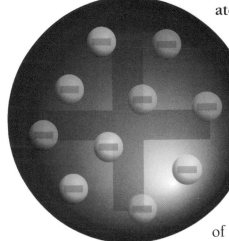

Thomson envisioned an atom as having negatively charged electrons (–) spread evenly throughout, balanced by the uniform positive charge of the rest of the atom (large +).

Atoms are nature's building blocks. They're bits of matter. Pile up enough of them, and you get a lump of stuff that you can see or break or hit someone over the head with. Atoms make up every *thing*, starting with the elements. When you combine elements you get all else—you, me, apples, stones, mountains.

Once scientists began to suspect that electrons might be associated with atoms, a problem arose. Electrons carried a negative electric charge, but atoms were electrically neutral.... [T]here had to be positive electric charges located somewhere in the atom that served to neutralize the charges of the electron.
—Isaac Asimov, American science and science-fiction writer, *Atom: Journey Across the Subatomic Cosmos*

(What is Asimov saying?)

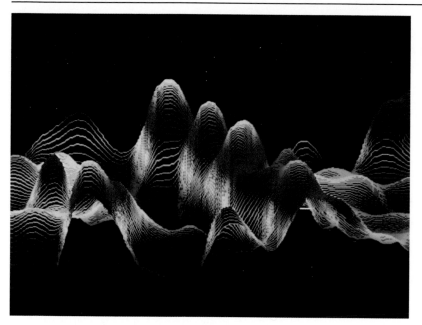

Optical microscopes (the kind most of us know) have lenses that focus the light that bounces off objects, making the objects appear bigger. You look at the enlarged objects directly. What you see is what is actually there. With nonoptical microscopes, what you see is more like a map or data plot or scan. A scanning electron microscope (above) shoots out electrons to create a black-and-white image, line by line. To "see" individual atoms, you need a scanning tunneling microscope (STM) or an atomic force microscope (AFM), which have extremely tiny tips that move across a surface. The STM image (above left) is a section of a DNA molecule magnified 1,500,000 times and colorized. Those orange peaks are the coils of DNA's famous helix structure.

The electron is a different affair. It's one of the atom's building blocks, but you can't gather millions of electrons together and have a lump of anything that can be tasted or smelled or held in your hand. Their mutually repulsive charge would blow any lump to smithereens. You can *do* things with electrons; you can't *make* things. Yet these charged particles do add mass to the atom when they speed around its nucleus. (More information on the atom ahead.)

Since Thomson's colleagues are not yet sure about atoms, it's no wonder that some think he is joking when he describes the electron. And no one, not even J. J., has any idea of the electronics wizardry these minute particles will spawn. Even science-fiction writers can't guess at the computers and electronic games their great-grandchildren will enjoy.

Nobel Marie

Life is not easy for any of us. But what of that? We must have perseverance and above all confidence in ourselves. We must believe that we are gifted for something, and that this thing, at whatever cost, must be attained.
—Marie Curie (1867–1934), Polish-born French physicist and chemist, in a letter to her brother

Her strength, her purity of will, her austerity towards herself, her objectivity, her incorruptible judgment—all these were of a kind seldom found joined in a single individual.
—Albert Einstein, *Ideas and Opinions*

Marya Sklodowska is born in Poland in 1867. It is a year that many see as full of promise.

In Paris, at a splendid World's Fair, nations from around the globe proudly display treasures and inventions; Europeans are awed by what for most is a first look at Japanese art.

In Germany, political philosopher Karl Marx publishes volume one of *Das Kapital*, a book urging an end to private ownership of most means of production.

In England, Michael Faraday dies, mourned by an appreciative nation.

Marya Sklodowska was born in Warsaw, Poland, then under the repressive thumb of Czar Alexander II of Russia.

In the United States, in New York City, an elevated railway is set up on a single track from Battery Place (at the southern tip of Manhattan) to 30th Street. Its cable car outpaces the horse-drawn carriages below. This same year the United States buys Alaska from Russia (for $7.2 million in gold). And in the U.S. Congress, some visionaries, known as Radical Republicans, try to bring fairness to the nation after the devastating Civil War by passing a Reconstruction Act that grants equal rights to former slaves.

But in Poland things aren't going well. A few years earlier, the Russians had conquered the country, renamed it "Vistula Land," and announced that Polish schools had to teach in the Russian language. Their intent was to wipe out the Polish past, replacing it with Russian culture.

The Sklodowska family stuck close together after the deaths of Marya's mother and her sister Sophie. Marya, standing behind her father, is about 18 years old in this portrait (ca. 1885). She was working as a governess, taking care of children, to put Bronya (middle) through medical school. That's her sister Hela on the right. Her only brother, Joseph, is missing from the picture. Marya was the youngest of the five children, the baby of the family.

Their laws have the opposite effect: They make the Polish people fiercely patriotic. Marya's mother runs a school in her house. Quietly, she teaches her students Polish history along with other subjects. Marya's father teaches high school science.

The Sklodowskas have five children—four daughters and a son. When Marya is eight, a sister dies. Two years later, Marya's mother dies. Her father doesn't have much money, but he loves his children and teaches them to encourage and support one another. All become outstanding students.

Marya has curly blond hair, gray eyes, high cheekbones, and a broad forehead. When she finishes high school at age 15 she is first in her class. But she can't go to college. Poland's universities won't accept women.

She joins a "Flying University," a group of patriotic Poles who teach one another. It is against the law for them to have classes—especially in the Polish language—so they "fly" from location to location to avoid the Russian authorities. (This is all before there are airplanes, so the "flying" is just an expression. Or maybe it is wishful

What was Paris like in the fall of 1891, when Marya—make that *Marie*—eagerly began her university studies at the Sorbonne? The City of Light had a new iron Eiffel Tower (built in 1888, captured on canvas by Georges Seurat the following year) that stood as a raw and controversial symbol of industry and technology. Seurat, who coincidentally died in the spring of 1891, created the illusion of lines and shapes with tiny, uniformly sized dots of color that vary in hue, tone, and density.

Pierre and Marie Curie married in July 1895. He was 35; she was 27. Marie graduated the following year and had a daughter, Irène, the year after that. Pierre gave up his research into crystals to join Marie in pursuing radioactivity.

thinking.) It is also against the law to teach reading and writing to peasant children. Marya does it anyway.

Marya and her older sister Bronya make a pact. Marya will work as a governess (taking care of children) and use her earnings to help send Bronya to medical school in Paris (women *can* go to universities there). Then Bronya will repay her sister with her earnings as a doctor and bring her to France. And that's exactly what happens.

In 1891, Marya steps off a train into what seems the world's most exciting metropolis—Paris. The newly built Eiffel Tower soars above the other Parisian buildings. Electric street lamps illuminate the grand boulevards. Automobiles, called horseless carriages, terrify real horses. And a few artists—labeled impressionists—perhaps reacting to the achievements of electromagnetic science, are painting light in new ways.

Marya, who is not quite 24, has waited nine years for the chance to be a university student. She is older than many of her classmates, and her French isn't very good, but she intends to do well in her studies. She begins by giving herself a French name—Marie. Then she rents a tiny, sixth-floor room; she has little heat or food and no elevator. Despite the hardships, it is a place where she can study with few distractions.

Two years later (in 1893), she gets a degree in physics; she is first in her class. A year after that (in 1894) she receives a degree in mathematics; this time she is second in her class. And she meets Pierre Curie.

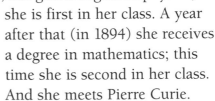

Pierre is slim and auburn-haired with a mustache, a trimmed beard, and a shy, thoughtful manner. A talented professor of science, he has invented a device that measures the hard-to-measure electric charges given off by minerals.

When Pierre proposes marriage, Marie isn't sure. She intends to go back to Poland and help Poles overthrow the Russian czar. Pierre writes love letters. He tells her that she can contribute more to the world as a scientist than as a political activist. He wins the argument.

In 1895, they marry and climb on bikes for their honeymoon trip. Marie is now Madame Curie. When they return, everyone they know is talking about Roentgen's newly discovered X rays. But hardly anyone pays attention when, in 1896, French physicist Antoine-Henri Becquerel (1852–1908) accidentally finds *strange, unexpected* rays.

It happens when Becquerel puts a lump of uranium on top of a photographic plate that is wrapped in dark paper. He intends to place that package in the sunlight to capture electromagnetic rays. But it is rainy, and he sticks the package in a dark drawer. To his astonishment, mysterious rays mark the photographic plate while it is *in the dark*. These are not X rays—which, like visible light, are electromagnetic. These rays must be coming from the uranium.

MADAME (Mme.) is the French equivalent of "Mrs." A mademoiselle is a single woman.

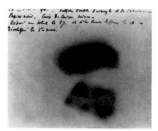

Antoine-Henri Becquerel was surprised to find two dark blobs where he had placed uranium on a photographic plate tucked in a lightless drawer. If not light, what had exposed the plate? Some form of unknown rays coming from the uranium, he concluded.

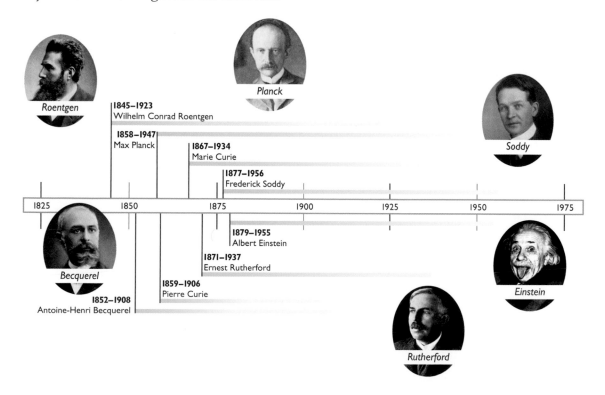

Roentgen

Planck

Soddy

1845–1923
Wilhelm Conrad Roentgen

1858–1947
Max Planck

1867–1934
Marie Curie

1877–1956
Frederick Soddy

1825   1850   1875   1900   1925   1950   1975

1879–1955
Albert Einstein

1871–1937
Ernest Rutherford

Becquerel

1859–1906
Pierre Curie

1852–1908
Antoine-Henri Becquerel

Einstein

Rutherford

Mysterious Rays

[The Curies'] work on radioactivity led both inward to the secrets of the atom, and outward to the depths of space and time. Radioactivity pointed the way toward understanding the energy source of the stars, provided a natural clock that showed the earth was a hundred times older than the best nineteenth-century estimates suggested, and helped cosmologists understand the particles and forces governing the evolution of the universe at large.
—Laurence A. Marschall, American astronomer and science writer, *The Sciences*, March/April 1999

The ancient alchemists were doomed to fail because it is simply not possible to transmute the elements using chemical energy (that is, the energy involved in the making and breaking of bonds between atoms). Everything changed, however, with the discovery of radioactivity at the end of the nineteenth century.
—Philip Ball, English science writer, *The Ingredients: A Guided Tour of the Elements*

J. J. Thomson's student Ernest Rutherford looks at the mysterious rays that Becquerel found and names them alpha rays, after the first letter in the Greek alphabet.

The alpha rays are a puzzle. No one knows what they are. Uranium—a metal—seems to be shooting out rays of energy! This is completely unexpected. There is no scientific explanation for it. These are not rays from the Sun. They are coming from inside an element.

Marie Curie decides to go on with her education and to get a doctorate in science (she will be the first woman in France to attempt it in any field). For her research project, Curie chooses to study Becquerel's rays and to see what she can find out about them.

At the Sorbonne, shown here in an 1886 painting, Marie Curie graduated first in her class in physics and second in math. While pursuing her Ph.D., she urged Pierre to finish his, too.

Curie believes the alpha rays are coming from the atoms of uranium. If that is so, then atomic radiation—the ejection of rays from inside an atom—must change the atom in some way. That means something is going on inside atoms. They are more interesting than anyone has imagined. If she can understand those tiny alpha rays, her research may lead to an understanding of the structure of the atom. (Most scientists who are willing to accept the idea of atoms think atoms are either hard, impenetrable bodies or squishy plum puddings, as you'll learn in chapter 13.)

Is she right? Are the alpha rays escaping from atoms? If so, what are they? No one has a clue. This is a totally new field. It demands laboratory work, not library reading, and lab experimentation is what Marie Curie likes to do.

She soon discovers that uranium is not the only element that emits spontaneous rays. Thorium, another chemical element, is also capable of radiation. She names the phenomenon "radioactivity."

Then she finds a mineral ore—pitchblende—that contains uranium but gives off even more alpha rays than pure uranium itself. How can that be? Is there some *other* radioactive element in pitchblende? This radiation is far stronger than that coming from the element uranium. Curie is quite sure there must be an unknown element in pitchblende, but she doesn't know what she is looking for. How can she find it?

It is the kind of quest that most scientists believe is almost impossible. Hardly anyone wants to spend years searching for some mysterious something that may never be found.

Later, physicist Ernest Merritt will describe Curie's search:

The task undertaken by Mme. Curie in attempting to separate [an unknown element] from pitchblende was somewhat similar to that of a detective who starts out

When particles or gamma rays leave an atom, the atom changes. We call that process *radioactive decay*. An atom that undergoes radioactive decay is described as "unstable." (See "Spitting Allowed?" on page 66.)

RADIOACTIVITY is the property of some elements (such as uranium) to spontaneously emit alpha and beta particles and sometimes gamma rays as the nuclei of atoms break apart. Spontaneous particles coming from an element? An atom's nucleus breaking apart? Who could have imagined all this action inside an atom?

Pitchblende is uranium oxide, but the natural ore also contains thallium, cerium, lead, and—Marie Curie was certain—some unknown radioactive element. In 1898, she and Pierre announced they had isolated not one but two new elements.

The lab in which Marie and Pierre discovered radium was a "cross between a stable and a potato shed," said one chemist. Russian artist Valerian Gribayedoff captured the harsh bleakness in a black-and-white photo.

CONSECRATE usually means "to set something aside as sacred." That's the way Abraham Lincoln used the word in the Gettysburg Address. Marie Curie wrote in French, and her translator used a secondary meaning of *consecrate*: "to dedicate to a goal."

to find a suspected criminal in a crowded street. Pitchblende is one of the most complex of minerals, containing twenty or thirty different elements, combined in a great variety of ways. The chemical properties of the suspected new element were entirely unknown; in fact, except for its one property of radioactivity, nothing whatever was known about it. The problem was one of extreme difficulty; but it had all the fascination of a journey into an unexplored land.

Marie and Pierre go to work in an old, unheated shed. They have little money for supplies, and they have a new baby, named Irène, at home. "Sometimes I had to spend a whole day mixing a boiling mass with a heavy iron rod nearly as large as myself. I would be broken with fatigue at the day's end," Marie will write, telling of those years. "If we had had a fine laboratory, we should have made more discoveries and our health would have suffered less.

"And yet, it was in this miserable old shed that the best and happiest years of our life were spent, entirely consecrated to work."

Of Pierre, she writes to her relatives, "He was as much and much more than all I had dreamed at the time of our union.

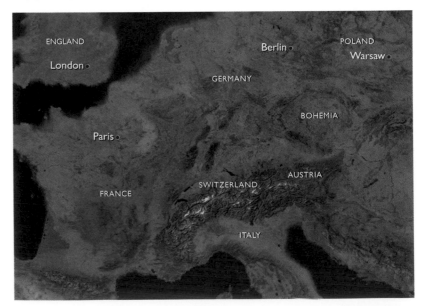

My admiration of his unusual qualities grew continually."

They think the new substance, whatever it is, will make up about 1 percent of the pitchblende. They are very wrong. "Its proportion did not reach even a millionth percent!" says Marie.

In other words, they need a huge amount of pitchblende to get a tiny bit of this radioactive element. Pitchblende is expensive. But it happens that at a mine in Bohemia, where uranium salt is extracted from pitchblende ore, the leftover residue is thrown in a pine forest. The government agrees to give the two young scientists 1 ton of it. (Someone describes them as "lunatics." No one can understand why they want the stuff.) The dull brown ore arrives in Paris in sacks mixed with pine needles. The Curies clean and process it themselves in the old shed.

The 1902 Nobel Prize in physics went jointly to Hendrik Lorentz of Leiden University and Pieter Zeeman of Amsterdam University for their research "into the influence of magnetism upon radiation." Lorentz said that light waves are the product of the to-and-fro movement of charged particles (electrons). A magnetic field will affect those electron oscillations and thus the frequency of the light, he said. Zeeman confirmed this experimentally.

Lorentz was also known for a set of mathematical formulas, called *Lorentz transformations*, which paved the way for Einstein's Special Theory of Relativity.

You can't see alpha particles, but you can make their trails visible in a cloud chamber. In this photo, water and alcohol vapor formed droplets where the particles traveled as they escaped from polonium.

Finally, in 1898, they discover a new element—polonium (they name it for Marie's homeland, Poland). Polonium is 400 times more radioactive than uranium. That isn't all. They realize there is still another radioactive element in the pitchblende and that it is still more powerful—although they haven't yet been able to separate it. They name it "radium" before they even find it.

Discovering a way to extract radium from the pitchblende

RADIUM comes from the Latin word *radius*, meaning "ray."

is far more difficult than finding polonium. It is much more difficult than even Marie and Pierre imagine it will be. It turns out that in *10 million* parts of ore, there is only 1 part of radium.

Many years later, their second daughter, Eve, will write of her parents' struggle:

Marie homeschooled her daughter Irène (left), who followed in her parents' footsteps to study nuclear physics. Eve (right) loved music and became a writer.

Marie continued to treat, kilogram by kilogram, the tons of pitchblende residue. . . . With her terrible patience, she was able to be, every day for four years, a physicist, a chemist, a specialized worker, an engineer and a laboring man all at once. Thanks to her brain and muscle, the old tables in the shed held more and more concentrated products—products more and more rich in radium. . . . But the poverty of her haphazard equipment hindered her work more than ever. It was now that she needed a spotlessly clean workroom and apparatus perfectly protected against cold, heat and dirt. In this shed, open to every wind, iron and coal dust was afloat which, to Marie's despair, mixed itself into the products purified with so much care. Her heart sometimes constricted before these little daily accidents, which absorbed so much of her time and strength.

ELUSIVE means "hard to pin down." A synonym for it is *mysterious*.

A **DECIGRAM** is one-tenth (.1) of a gram, or .00353 ounces. That's about the mass of a house spider.

Pierre wants to give up. Marie refuses to stop. She is sure they will find that elusive element. They talk about it most of the day and dream about it at night. Often they wonder what radium will look like. Pierre hopes it will have a beautiful color.

Here are Eve Curie's words telling the story:

In 1902, forty-five months after the day on which the Curies announced the probable existence of radium, Marie finally carried off the victory in this war of attrition: she succeeded in preparing a decigram of pure radium, and made a first determination of the atomic weight of the new substance, which was 225.

The incredulous chemists—of whom there were still a few—could only bow before the facts, before the superhuman obstinacy of a woman.
Radium officially existed.

INCREDULOUS means "disbelieving or doubting."

A Daughter Remembers

In her book *Marie Curie*, Eve Curie writes of the day her parents discovered radium:

The door squeaked, as it had squeaked thousands of times, and admitted them to their realm, to their dream.

"Don't light the lamps!" Marie said in the darkness. Then she added with a little laugh: "Do you remember the day when you said to me 'I should like radium to have a beautiful color'?"

The reality was more entrancing than the simple wish of long ago. Radium had something better than "a beautiful color"; it was spontaneously luminous. And in the somber shed, where, in the absence of cupboards, the precious particles in their tiny glass receivers were placed on tables or on shelves nailed to the wall, their phosphorescent bluish outlines gleamed, suspended in the night.

"Look... Look!" the young woman murmured.

She went forward cautiously, looked for and found a straw-bottomed chair. She sat down in the darkness and silence. Their two faces turned toward the pale glimmering, the mysterious sources of radiation, toward radium—their radium....

Her companion's hand lightly touched her hair.

She was to remember forever this evening of glowworms, this magic.

This flask held Marie Curie's radium salts. Radiation damaged the once-clear glass, causing it to scatter light differently and take on a violet hue.

The Curies sit in their dark shed and admire its luminous blue glow. Radium proves to have 2 million times more radioactivity than uranium.

They guess, but don't know for sure, that the glowing blue element will soon have significant industrial, medical, and scientific uses. But they refuse to patent their findings. If they do, it might make them wealthy, but they believe that scientific discoveries belong to everyone. Marie says, "If our discovery has a commercial future, that is an accident by which we must not profit. And radium is going to be of use in treating disease.... It seems to me impossible to take advantage of that."

In 1903, Marie becomes the first woman to receive a doctorate from the Sorbonne at the University of Paris. That

Radioactivity used to be measured in curies (named for Pierre and Marie). One curie equals the radioactivity of 1 gram of radium. But that term has become obsolete. Today, the unit of radioactivity is called a *becquerel*. (And you know who that is named for.)

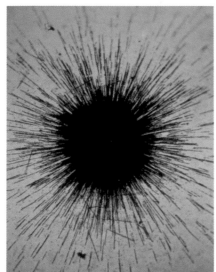

A speck of radium kicks out surprisingly intense streams of alpha particles in all directions. Their presence was recorded here as a burst of dark rays on a specially coated photographic plate.

same year, she, Pierre, and Antoine-Henri Becquerel win Nobel Prizes for physics.

Meanwhile, Becquerel (a slim, balding man with a Vandyke beard) has found another surprise. Other rays in addition to alpha rays are spewing out of uranium's atoms. Ernest Rutherford has a name for them. He calls them **beta rays**. But he doesn't know what they are. He says, "The cause and origin of the radiation continuously emitted by uranium and its salts still remain a mystery." Then he says that the radiation is "similar to Roentgen rays" (X rays). That is a *big goof*, as he will discover, but nobody yet knows that X rays are electromagnetism.

Meanwhile, the elements that the Curies discovered, polonium and radium, can be used in experiments. Polonium and radium emit strong alpha and beta rays (much stronger than those emitted from the uranium Rutherford and Becquerel used for earlier experiments). That opens up opportunities.

They discover that a lump of radium decaying radioactively spews alpha and beta rays in all directions. To do a controlled experiment, it helps to have a focused beam. So they put the lump in a lead box with a single hole in it. A beam emerges. Now they expose the beam to a magnetic field.

The scientific history of radium is beautiful. The properties of the rays have been studied very closely. We know that particles are expelled from radium with a very great velocity near to that of light. We know that the atoms of radium are destroyed by expulsion of these particles, some of which are atoms of helium. And in that way it has been proved that the radioactive elements are constantly disintegrating and that they produce at the end ordinary elements, principally helium and lead.

—Marie Curie, from a speech at Vassar College, 1921

When this experiment is tried, the beta part of the beam bends. Whoops—that is unexpected. Bending in response to magnetism is what charged particles do. This can mean only one thing: **Beta "rays" aren't rays at all. They are particles.** Ernest Rutherford figures out that beta rays are electrons—negatively charged particles—just like those that his mentor, J. J. Thomson, found in cathode tubes. But this time, the electrons are coming from inside atoms. The beta particles (electrons) aren't at all similar to X rays.

What about alpha "rays"? Pierre Curie discovers that the alpha entities coming from radium travel 6.7 centimeters into the air and then seem to disappear. He knows they can't actually disappear. Marie Curie finds that they are attaching themselves to stray electrons in the air and turning into helium atoms. So they too must be particles. All this is unexpected, surprising, and important enough to repeat: Alpha and beta are particles, not rays.

Scientists will learn that alpha particles are helium nuclei (with two neutrons and two protons) **and beta particles are high-energy electrons.**

Meanwhile, Rutherford, who is now at McGill University in Canada, builds an apparatus with a very strong magnetic field. His magnetic field deflects (bends or scatters) the

The ABGs of Particle Physics

Alpha, *beta*, and *gamma* are the first three letters of the Greek alphabet. All three particles with those names are found in the atom. If you put a magnet across a radioactive beam, the three types of particles—alpha, beta, and gamma—will separate.

Alpha particles are positively charged, high-energy particles that are identical to helium's nuclei, each with two protons and two neutrons. Think of an alpha particle as a naked helium atom. It lacks its two negatively charged electrons and has a double positive charge (because of the two protons). Alpha particles come charging out of uranium atoms at more than 16,000 kilometers (about 10,000 miles) per second. Elements that release alpha particles are radioactive. They are not stable. Over time they will change and become different elements.

Beta particles are fast-moving electrons, similar to those in cathode rays but carrying much more energy. They have a negative charge.

Gamma rays are photons—particles of high-energy light, or very high-frequency electromagnetic radiation. They have no charge at all. Gamma rays are similar to X rays but with shorter wavelengths and higher frequencies. They aren't like alpha and beta particles. Understanding that photons can be particles too will be much harder for scientists to accept than the idea of alpha and beta particles. Hold on for an explanation.

An April 25, 1986, explosion at the Chernobyl nuclear power station in Ukraine sent clouds of radioactivity across northern Europe. Months later, this piglet was born with a deformed eye due to lingering contamination.

Remember Your ABGs
• **Alpha particles** are helium ions. They can be stopped by a few sheets of paper.
• **Beta particles** are electrons. They go through paper but not aluminum.
• **Gamma rays** are photons of electromagnetic radiation. They can penetrate lead.

Spitting Allowed? Half-Lives? That's Chemistry!

Ninety-two elements (with atomic numbers 1 to 92) occur naturally in the universe. Eighty-one of them are stable; the others are radioactive, spitting out particles as the energy level of the nucleus changes (called radioactive decay). The original radioactive nucleus, known as the parent, becomes a nucleus of another element, known as the daughter. Sometimes the daughter is also radioactive, in which case decay continues until there is a stable nucleus (one that will not decay further).

Ernest Rutherford was a professor in Montreal when he discovered and named the two kinds of particles emitted by radioactive elements—alpha and beta. He also found that gamma rays—high-energy electromagnetic radiation—are ejected.

In a sample of a radioactive element, the amount of time between spits is random and unpredictable, but the process is governed by an average time in which half the atoms of an element would be expected to decay. That's called a half-life, and it can vary from one-millionth of a second to billions of years. But the decay process is never complete. There is always some leftover radioactivity. After every half-life, only half of the nuclei decay. Then half of the remainder decay, and so on and so on and so on. That's why we measure the half-life rather than the total decay time.

RADIUM 1300 yrs.   EMAN. 4 dys.   RAD. A 3 mins.   RAD. B 21 mins.   RAD. C 28 mins.   RAD. D 40 yrs. RADIO-LEAD   RAD. E 6 dys.   RAD. F 143 dys. RADIO-TELLURIUM. POLONIUM

ACTIVE DEPOSIT RAPID CHANGE   ACTIVE DEPOSIT SLOW CHANGE

In 1906, Rutherford pictured the decay of a radium-226 atom (left) through "Emanation" (radon gas) into a chain of elements ending with lead.

We happen to live in one of those rare parts of space that has lots of matter. Just as a fish might look around its immediate environment and conclude that the universe is made of water, we intuitively sense that our peculiar circumstances are generic. They aren't. Most of space is almost devoid of matter, but the radiation bath is everywhere.

—Lawrence M. Krauss, *Atom: An Odyssey from the Big Bang to Life on Earth…and Beyond*

alpha beam, which means it must be a stream of particles. When Rutherford studies the scattering pattern, he is able to measure and analyze the alpha radiation. Then he writes about his surprising discovery of particles in what were believed to be continuous waves.

This means abandoning previous ideas. It means the scientific detectives have new things to puzzle over. The atom has been thought to be the smallest entity there is, but clearly that isn't so. There are electrons inside atoms. Electrons have a negative electrical charge. Atoms can be neutral (without a net charge).

That means there must be something else in the atom—

What's in a Word?

That word *radiation* has more than one meaning.

1. If you see a sign that says STAY AWAY; DANGEROUS RADIATION, you can guess that ionizing radiation is present. *Ionizing* means that the radiation can knock electrons out of atoms (ionize them), including atoms in your body. Ionizing radiation is emitted in the normal process of radioactive decay, but if you get too close to it, some of your cells might undergo a destructive change.

2. Radiation also describes the action of energy traveling through space. (Light is described as electromagnetic radiation. Heat radiation is the movement—or propagation—of infrared rays through space. Sound is a form of energy that radiates through air and other materials.) To most physicists, radiation is the emission of radiant energy as particles or waves—for example, heat, light, and alpha, beta, or gamma particles.

3. When ecologists use the word *radiation*, they mean the spread of something from a central point into the surrounding area. Germs, for instance, can radiate from one person to a neighborhood.

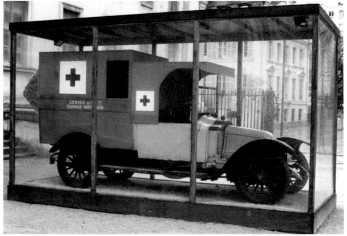

Marie Curie had 20 Red Cross trucks loaded with X-ray equipment for use in battlefield hospitals during World War I. "She climbed in beside the driver, on the seat exposed to the wind, and soon the stout car was rolling at full speed [then 20 miles per hour]," wrote Eve Curie. Before an examination, "she prepared the radioscopic screen." Then "she darkened the room . . . with black curtains she had brought." In that darkroom "were placed the baths of chemicals where the plates would be developed. Half an hour after Marie's arrival, everything was ready." Marie Curie also installed X-ray rooms in 200 French hospitals. More than 1 million wounded men were examined in those mobile and fixed stations.

something with a positive charge—to balance the negative electrons. What is it? The challenge is to find those positive particles. But it isn't the only challenge.

The radioactive particles shooting out of uranium, polonium, and radium show that something unexplained is going on inside some atoms. Radioactive elements seem to be spontaneously creating energy. According to known scientific laws, that can't happen. The Law of Conservation of Energy states that energy cannot be created or destroyed. What is going on here?

Just what does the inside of an atom look like? And how can you explain the energy coming from some atoms? Does it change an atom to give off radiation? Why are most atoms

A radioactive face cream made of thorium and radium? It's unthinkable today, but Tho-Radia debuted in 1933 with eerily glowing beauty ads and claims that the product "boosted circulation, toned and firmed up the skin, got rid of oil, and suppressed wrinkles." This highly hazardous substance didn't vanish from the market until about 1960.

not radioactive? The Curies give the scientific sleuths a new quest: to find the structure of the atom and answer those questions.

What the Curies don't know, or perhaps don't want to face, is the danger of radioactive elements. They put test tubes with radium in their pockets; they touch the element with their fingers. Marie keeps radium next to her bed so she can admire its blue glow. They work with radium in closed rooms. They are often ill. Some of the people who work with them die. Manufacturers, excited by the new element, will soon use radium to paint glow-in-the-dark faces on clocks and doorknobs. Before the danger is recognized, radium is added to some "cure-all" patent medicines.

Then another kind of tragedy strikes. Pierre, coming home from a meeting in 1906, is run over and killed by a horse-drawn carriage in a traffic accident. The great collaboration is over.

Marie continues her research, raises their two daughters, and writes books. In 1911, she receives a second Nobel Prize. She is the first woman professor in the Sorbonne's

Still Glowing After All These Years

Roy Lisker, editor of *Ferment Magazine*, describes a well-worthwhile visit to the Curie Museum in Paris:

Everything on the premises is a replacement owing to the extreme radioactive contamination over the decades in which they housed the offices and research lab of the Curies. Only one piece of paper [right], in a glass case hanging from the wall, is from a notebook of the period. The [tour] guide passed a Geiger Counter detector over the page to show that even three-quarters of a century had done little to diminish its potency.

Scientific Climber

In 1912, Victor Hess (1883–1964) put an apparatus that measured radiation into a balloon and sent it 5,300 meters (17,400 feet) above the Earth. He found that the radiation at that altitude was about four times that on Earth's surface. "The results of my observation are best explained by the assumption that a radiation of very great penetrating power enters our atmosphere from above." That high-speed radiation came to be called "cosmic rays." Hess won a Nobel Prize in 1936 for this work of discovery.

Since then, physicists have found that most cosmic rays are protons or helium nuclei, but a small fraction are electrons, gamma rays, or high-energy neutrinos. They may originate on the Sun or they may come from the far reaches of the universe; therefore, their magnitude varies widely. As they approach Earth and hit the nuclei of atoms in the outer atmosphere, subatomic particles (especially particles called muons) are created in the collisions.

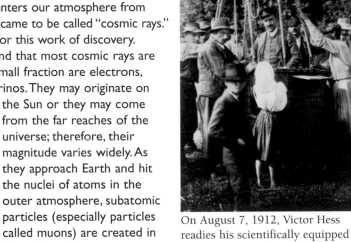

On August 7, 1912, Victor Hess readies his scientifically equipped balloon for one of 10 flights to investigate the source of radiation several miles above Earth.

A cylinder deep inside a Japanese mountain captured a cosmic ray muon—a subatomic particle from space. It zipped out the side of the container in 120 nanoseconds.

These subatomic particles decay before they reach the surface of Earth. In this and other ways, the atmosphere protects us from direct cosmic rays and their products, which can be dangerous to humans. Understanding cosmic rays is important if we hope to become space travelers.

650-year history—and an international celebrity.

When she dies of leukemia in 1934, almost certainly caused by exposure to radioactivity, Marie Curie has done far more than she set out to do. She has not only made a place for women in science, she has discovered a new world inside the atom.

Leukemia is a group of cancers caused by abnormal blood cells—in this case, too few red ones and too many white (tinted blue in this SEM).

Making Waves

By nature I am peacefully inclined and reject all doubtful adventures....However, a theoretical interpretation had to be found at any cost, no matter how high....I was ready to sacrifice every one of my previous convictions about physical laws.
—Max Planck (1858–1947), German physicist, letter to the American physicist Robert W. Wood

The breaking of a wave cannot explain the whole sea.
—Vladimir Nabokov (1899–1977), Russian-born American author, *The Real Life of Sebastian Knight*

The universe has its own constants, in the form of unvarying quantities that endlessly reappear in nature and in mathematics, and whose exact numerical values are of signal importance to the pursuit of science....Whenever a repeating pattern of cause and effect shows up in the universe, there's probably a [numerical] constant at work.
—Neil deGrasse Tyson, "The Importance of Being Constant," *Natural History* magazine

Max Karl Ernst Ludwig Planck is one of the most conservative of Germany's scientists. He is a physics professor—thoughtful and courtly—who wears formal, stiff, high collars and suits with vests. Everyone who knows him seems to agree: Max Planck is a fine human being. But there doesn't seem to be anything revolutionary or daring or far-out in his thinking. He is very much an "Establishment" person—part of the German cultural elite—someone not likely to make *waves*.

But that is exactly what Planck does in 1900, at age 42, when he solves a major scientific puzzle. No one can figure out why light from a furnace goes from red to orange to yellow (meaning shorter

Max Planck's eldest son was killed in World War I fighting for Germany. Two of Planck's daughters died in childbirth. "I could not hold back my tears when I visited him," wrote Einstein, after one daughter's death. "He behaves remarkably bravely... but one can see that he is eaten up by grief."

and shorter wavelengths) as it gets hotter. That isn't what classical mechanics predicts. Max Planck finds an answer when he uses a formula that assumes that some heated objects emit and absorb light in small, discrete units of energy. Planck calls those energy units *quanta*, after the Greek and Latin words for "how much." (*Quantum* is the singular form.)

What does color have to do with temperature? In this limestone kiln, yellow is the hottest, and dark red is the least hot.

He doesn't believe those quanta actually exist. No reputable scientist would. Everyone "knows" that light and other forms of EM (electromagnetic) radiation travel as waves in a continuous thread. Thomas Young settled that argument in 1803 with a famous experiment proving that light is made up of waves. Then James Clerk Maxwell came up with equations confirming that light is waves.

How can light be both continuous waves and also quantized bits of energy *at the same time?* That idea is impossible to accept, for Planck himself or for anyone brought up with classical physics—Isaac Newton's physics— where you can expect something to be one thing or another but not both. So, Planck doesn't buy what his results tell him: that electromagnetic energy comes in "packets" of *only* a certain magnitude (those quantum units). He is sure that other amounts of energy also exist (to create the continuousness of a wave), but he can't find them. *The light energy he has tracked is in quantized lumps.* Planck knows that his formula works. He goes with it, reluctantly.

Actually, the story is bigger than it sounds, and it is appropriate that it begins in 1900, at the brink of a new century, because it will introduce a whole new kind of physics to the world. Only a few scientists understand that a new kind of physics is even needed. Most think that with Newton's laws of motion and Maxwell's equations on electromagnetism and the laws of thermodynamics, you have explanations for all the action of the universe.

DISCRETE is a science word meaning separate, one from another.

CLASSICAL PHYSICS means non-quantum physics. Newton's laws of classical physics are certain and clear. Cause is followed by effect. Quantum laws are baffling. We know the hows but not yet the whys.

Max Planck wasn't convinced that energy quanta existed. The quanta worked in equations; that seemed to be enough. Einstein believed particles were real and that they came in quantum amounts. He went on from there.

A new concept appeared in physics, the most important invention since Newton's time: the field. It needed great scientific imagination to realize that it is not the charges nor the particles but the field in the space between the charges and the particles that is essential for the description of physical phenomenon.

—Albert Einstein and Leopold Infeld, *The Evolution of Physics*

Ocean waves and sound waves are not objects; they are disturbances in a medium, like water or air. But EM waves are different. They can travel through the vacuum of space. An artist created this beautiful wave pattern on a computer by starting with a sine wave (a basic periodic wave) and then trapping it in a box so that it bounced off the sides and overlapped.

But some physicists are aware that at times Newton and Maxwell are incompatible. Some things just won't work out right. The disconnects seem small, but they are bothersome.

One of the problems is with something that physicists call "blackbody radiation." Some surfaces reflect radiation (visible light, for example); some absorb it. A mirror is a good reflector of light; that's why you see yourself in the mirror. A blackbody is close to a perfect absorber or emitter of radiation.

Picture a backyard barbecue. At room temperature, the charcoal emits almost no radiation (and what it does emit is not visible). At 800 kelvins, the coals are dark red, at 1,300 kelvins bright orange, and at 1,800 kelvins, they have burst into yellow flame. Now think of a kiln. As it heats up, its interior (observed through a small hole) goes from red (hot) to orange (hotter) to yellow (still hotter). The same colors appear at the same temperatures no matter what the kiln is lined with. Physicists in 1900 realize that the color of radiation coming out of a closed cavity must say something about the nature of radiation itself, not about the cavity or the material it is made of. They are baffled; they call the intensity and wavelengths emitted from a heated kiln a "blackbody spectrum." Why does a rise in temperature change its color and spread? No one (at that time) knows. This leads to a crisis in physics. No one has been able to derive an equation that links temperature and emitted light.

That's what Planck does. After many attempts, he finds an equation that correctly describes *how much* radiation of different colors comes from the blackbody. But in order to get this equation—to make it work—Planck has to presume that the energy radiation from the blackbody is *not* continuous (it's not wave-like); it comes in quanta. Planck doesn't believe what the equation is saying. Light comes in waves—he is sure of that.

He hates this result and tries again and again to explain blackbody radiation some other way, but he fails. Only

If you take the Celsius temperature scale and shift all the numbers so that 0 degrees Celsius (the freezing point of water) equals 273, you get the kelvin temperature scale. The boiling point of water (100 degrees Celsius) is 373 kelvins.

A Body Is Not a Hole

Don't confuse blackbodies and black holes. We'll deal with black holes, which are heavenly sinkholes, in chapter 42. As for blackbodies and blackbody radiation, back in the nineteenth century, German physicist Gustav Kirchoff (KEERK-hoff), a pioneer in spectroscopy, defined a blackbody as one that absorbs all EM radiation that comes its way and gives off a spectrum of radiation that depends only on its temperature. Kirchhoff realized that there is no such thing as a perfect blackbody. The Sun, a yellow star with a surface temperature of almost 6,000 kelvins, comes close: It emits a spectrum of blackbody radiation. In a laboratory, a closed container with blackened inner walls and a tiny hole serves as a blackbody.

Kirchhoff thought he could chart blackbody radiation patterns mathematically, but he couldn't. He assumed that radiation flows continuously—which is where he, and everyone else, went wrong. Max Planck's formula made it clear that radiation comes in exact quantum lumps; that seemed astonishing, but turned out to be right. Keep in mind: All blackbodies give off radiation at all temperatures, and so none is really black. A blackbody is a perfect absorber and perfect emitter of EM radiation. It doesn't *reflect* light.

An erupting volcano is not a blackbody because it doesn't absorb all the radiation that reaches it, but you can gauge the temperature of its lava flow by its color. The red liquid rock at the feet of the famous French volcanologist Katia Krafft is cooler than the yellow-orange lava in the background. Likewise, by studying a star's spectrum, we can determine its surface temperature. Reddish stars (upper left corner at right) are comparatively cool (about 2,000–3,000 kelvins on the surface). Blue stars are hotter than 12,000 kelvins. The hotter the star, the shorter its life span.

A Word with Properties

"What is a quantum?" I ask a physicist friend. He winces. "You can't talk of 'a quantum' as if it were a thing. Instead, think of it as an adjective that describes Nature in the small. You would never say, 'a beautiful' but rather 'a beautiful picture.' There are no beautifuls and there are no quanta."

No quanta? No wonder I'm confused. The physicist, who is usually an easygoing fellow, takes a deep breath.

First, he tells me what the word *quanta* and its singular form, *quantum*, are not.

"Quantum is not a thing."

"So what is it?" I ask (still somewhat baffled). He informs me that quantum describes **properties.** "It starts with the idea that when an atom changes from one energy state to another, it gives off or absorbs energy in (discrete) bundles."

"Aha, that's what Max Planck discovered," I say, trying to look wise. The physicist continues, "Some people call photons 'light quanta,' but I don't really like it. The energy of a photon *only* comes in quantized units. Once a photon is emitted, it will be absorbed as one chunk of energy or not at all. No half photons allowed."

Now, that's like saying that I can only heat the temperature in this room to 60 or 70 degrees Fahrenheit but never an in-between 65 degrees. It seems like a strange rule to me, but I'm not an electron, and that's not the point. We're trying to define the word *quantum*.

"Charge is quantized. It comes in set amounts, and that's that," says the physicist. "Angular momentum is quantized."

"Whole classes of particles base their identities on the quantized value of their angular momenta. **In Quantumland you're known by your properties."**

(Just to remind myself, I look up angular momentum in a science dictionary and find that for a body in a circular orbit, its angular momentum is the product of the body's mass, radius, and speed.)

I want to be clear about this, and the physicist is helping. "Don't call an atom a quantum," he says. "Atoms are things; their properties are quantized. And subatomic particles? Their properties are also quantized. Get the language straight and you'll understand."

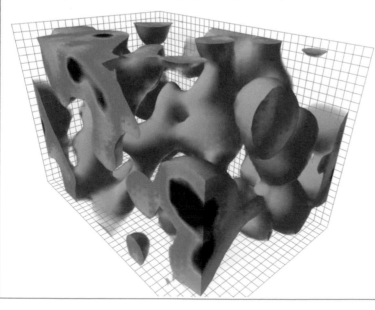

You're looking at a still from a computer animation of Quantum Chromodynamics (QCD). *Chromo-* means "color" and *dynamic* involves motion, which explains the lava-lamp look. Add *quantum* to the title, and you're talking about the weird world of atomic and subatomic particles. The box dimensions are programmed to be just big enough to hold a couple of protons. According to the Australian researchers who created it, QCD "describes the interactions between quarks and gluons as they compose particles such as the proton or neutron."

when he assumes that the energy comes in very tiny packets of an exact size—those quanta—does his equation explain blackbody radiation.

Planck deals with this by saying that it is "resonators" (by that he means atoms or electrons) inside the blackbody that absorb and emit energy in those discrete quantum units. Planck believes it is they, not the light itself, which come in quanta. He says it is the resonators oscillating at various frequencies that cause light to be emitted from a blackbody in individual packets, or quanta, which he believes then transition into waves of discrete energies.

The idea of resonators helps Planck accept what he has found. He's not ready to go further, nor is the scientific community. (Hold on—eventually they'll learn that the resonator idea isn't needed and, yes, light does come in packets, and it comes in waves, too.)

What size is h?
$h = 6.626 \times 10^{-27}$ erg-sec
Or,
.00000000000000000000000006626 erg-sec (an erg is a unit of *energy*). The important thing here, again, is to note how small it is.

His equation tells him something else that is unsettling, especially for someone as meticulous as Max Planck. Like many mathematicians, Planck loves music. He often compares light waves to the vibrations of a violin string. But imagine a violin on which you can only play whole notes (like B and G) and none of the tones between (like B-flat or G-sharp). That's what Planck has to deal with.

To get the formula right, Planck uncovers a constant. A constant is an unchanging number. (The universe seems to have a few unvarying quantities, numbers that keep appearing in nature. We describe them as constants.) Planck's constant—he calls it h—will become a basic of physics. It is a very, very small number—equal to the energy of one quantum of electromagnetic radiation divided by the frequency of its radiation. **The world is composed of atoms whose energy is quantized.** Once created, electromagnetic quanta can only be absorbed as a whole (not in chopped-off parts).

With his very useful formula, Max Planck was able to figure out a value for *h*, as well as improved values for Boltzmann's constant and Avogadro's number. (See *Newton at the Center* for details on those last two quests.) Only slowly and reluctantly did Planck come to realize that his formula made some classical ideas invalid.

You might be wondering: If Planck is so cautious, why did he write his big discovery on a postcard that any physics-minded mail carrier could read? In 1900, the physics-minded world consisted of a handful of people. Many scientists believed there was no future in physics—that everything of importance had already been discovered.

The young Spanish painter Pablo Picasso (1881–1973) arrived in Paris the same month that Max Planck wrote of his breakthrough quantum idea. Both men revolutionized their field, but Picasso did it with gusto. In *Man with Violin* (1911), created at the height of the cubist movement, he deconstructed a realistic subject into an exploded view with multiple facets almost to the point of abstraction.

Small but Powerful

With Planck's equation, $E = hf$, physicists can relate the energy (E) of one quantum unit of electromagnetic radiation to its classical frequency (f) by using Planck's constant (h), yielding the smallest possible "packet" of energy: hf for light of that frequency. The quantum hf cannot be further divided.

Planck's equation holds a surprise. It ties the particle nature of a quantum packet to its wave nature! Keep in mind, h is a very, very small number. It's because h is so astonishingly small that we are unaware of the wave-particle duality in our everyday world.

No one in the year 1900 understood this. It was a few decades before Planck and most other physicists grasped what his amazing equation is actually saying: Energy comes in the form of unbelievably small lumps!

Planck isn't ready to go where his calculations lead. He hopes someone will prove him wrong. He doesn't like the quantum idea at all, although he knows that his formula explains things, like blackbody radiation, that have been previously unexplainable.

An equation may not seem like much, but don't be fooled; Planck's formula describes the very nature of light. It wins Max Planck a Nobel Prize, and it sets the foundation for quantum physics and much of modern science.

On October 7, 1900, Planck sends a postcard to his friend Heinrich Rubens, telling him of his quantum idea. Planck knows that, if he is right, it is a very important discovery, but Planck is cautious by nature.

On December 14, Max Planck explains his theory to the members of the Berlin Physics Society. The physicists listen politely and yawn. (Some of

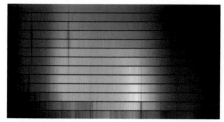

Each band represents the visible spectrum (the "rainbow colors") of a type of star, ranked in decreasing order by surface temperature. The top, hottest star appears blue to our eyes. The bottom star appears red.

Reading a Big-Deal Formula

The speed of light constant is represented in formulas by the lowercase letter c.

Planck's constant is symbolized by the lowercase letter h. It stands for the energy of a photon divided by its frequency.

Energy is uppercase E. (The charge of an electron is lowercase e.)

The symbol for frequency is f.

In this false-color image, a bright, young star cluster (in red) appears between the Milky Way's dense center (white) and closer stars (blue). The cluster is hidden behind clouds of dust, but the Hubble Space Telescope peered through the shroud by tuning into infrared frequencies, which are lower than visible light (and therefore of longer wavelength).

(Sometimes you will see it as the Greek letter υ, which is pronounced *nu* or "new," but I find that symbol confusing; it looks too much like the symbol for velocity, v.) Frequency means how many times per second something vibrates. **Electromagnetic waves of short wavelengths vibrate more rapidly, meaning they have higher frequencies. Longer waves have lower frequencies.**

If you want to figure the smallest amount of energy that can be turned into light, you can start with its color, which will lead you to its frequency (red and blue, for instance, have different frequencies), and then multiply by Planck's constant. **The key formula is $E = hf$. That formula is basic in quantum mechanics—the big twentieth-century science.** So don't be impatient when it gets repeated in the chapters to come—$E = hf$ is something to glue into your head.

Planck's previous theories have proved wrong.) They don't realize this is an historic occasion. It takes a few years before someone does recognize the significance of Planck's discovery. That someone is our young friend Albert Einstein.

Five Papers

Newton's mechanics had considered matter—the heavy stuff that falls when you drop it, sits still when you set it down, and rolls on forever if you hurl it through empty space. . . . [Einstein] saw that the most revolutionary mysteries had to do with light, heat, energy—all those immaterial enigmas that physicists lumped together as radiation.

—Edmund Blair Bolles, American science writer and author, *Einstein Defiant*

I sometimes ask myself how did it come that I was the one to develop the theory of relativity. The reason, I think, is that a normal adult never stops to think about problems of space and time. These are things that he thought of as a child. But my intellectual development was slowed, as a result of which I began to wonder about space and time only when I had already grown up. Naturally, I could go deeper into the problem than a child with normal abilities.

—Albert Einstein, quoted in *In Memoriam Albert Einstein*

Einstein, who is 21 in 1900, has been puzzling about visible light for five years. It doesn't always follow known rules. He hasn't yet been able to figure out what will happen if he travels on a beam of light. But he already knows—because of J. J. Thomson's electrons and the Curies' experiments with radioactivity—that there are particles smaller than atoms. Now, Max Planck's idea that a blackbody changes energy by discrete amounts during the emission process gives Einstein something else to consider. He will go way beyond the concept of

Forget that popular image of Einstein as an old man in a baggy sweater with a hairdo from a fright movie. In his most productive years, his big brown eyes and dark curly hair gave him a romantic allure. He seemed to care about the way he looked, and women were attracted to him.

blackbody radiation with his questioning mind. Can light really come in quanta? If so, maybe there are actual particles of light. Everyone thinks that light is continuous wave action, but why can't it be both particle and wave? According to classical physics, that is a preposterous idea. Respectable scientists don't believe that something can be two things at the same time. Still, Einstein wonders if it might be so. He has an unusual ability to visualize. So he pictures those light quanta in his head. He takes Max Planck's idea seriously.

The year of Max Planck's paper, 1900, is the year Einstein finishes his studies at the Swiss Federal Institute of Technology (later known as the ETH) in Zurich—but Einstein's university record is not very good. He has angered some of his teachers, he doesn't have a doctorate, and he can't get a teaching job. He lives in one small room and often goes hungry. Besides all that, he wants to get married.

Einstein sends letters to some scientists that he admires, asking for work, but none answer. So he puts an ad in the Bern, Switzerland, newspaper offering to teach physics to private students, with a free trial lesson thrown in as an added attraction. He ends up with two takers; each pays two Swiss francs a session (hardly anything) and gets far more than he expected. Calling themselves the "Olympia Academy" (they are mocking stuffy academies), the three

My dictionary says **PREPOSTEROUS** means utterly absurd and outrageous. Note: not just absurd but *utterly* absurd.

When Einstein ran this ad in the February 5, 1902, edition of a Bern newspaper, he was a month shy of 23 years old, the unmarried father of a newborn girl, and jobless. It offers private tutoring in math and physics, "trial lessons for free." Einstein landed a job at the patent office four months later.

Maurice Solovine (middle), a philosophy student, answered Einstein's ad (above). Soon joined by Conrad Habicht (left), who later became a mathematician, the trio formed the tongue-in-cheek "Olympia Academy" with Albert Einstein (right) as president.

What role did Einstein's wife Mileva (1875–1948), a physicist and mathematician, play in his work? Historians argue about that—sometimes passionately. We know the couple worked together and that Einstein called Mileva a "collaborator."

study as equals, not as teacher and students. They read philosophy, science, and literature—and share the excitement of serious brain stretching. Some of Einstein's neighbors are sure their merriment is liquor-induced, but the trio drinks nothing stronger than iced coffee. "Our means were frugal," says one of them later, "but our joy was boundless." Einstein tells them of his obsession: He wants to know what would happen if he could ride a light beam.

Finally, in June 1902, Einstein gets a job. He is hired as a technical expert, third class, at the patent office in Bern. Seven months later he marries Mileva Maric, and, before long, they have a son. (A daughter, born before they marry, may have been given up for adoption. Her fate is a mystery.) A second son will come later.

The patent office turns out to be a good place for the now 23-year-old thinker. He has a boss who Einstein says is strict but fair: "More severe than my father—he taught me to express myself correctly." Day after day, Einstein examines applications for patents on inventions. By law, each application must come with a model of the invention. His job is to decide, and quickly, if the invention is worthwhile.

Einstein wasn't only a thinker; he also loved to tinker and make things. When he was a professor in Berlin, he heard of a family killed by fumes from a faulty refrigerator. He and a young Hungarian physicist friend, Leo Szilard, invented a refrigerator (right) without moving parts—or noxious fumes. (It had a noisy magnetic pump, so their fridge never got produced.) Remember the name Szilard. While waiting for a streetlight to change, he got a powerful idea. (We'll get to it.)

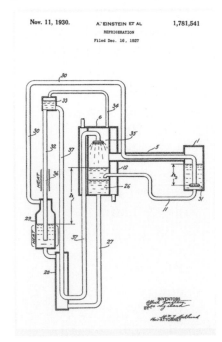

Nov. 11, 1930.   A. EINSTEIN ET AL   1,781,541
REFRIGERATION
Filed Dec. 16, 1927

Should it be given a patent? Then Einstein has to describe the invention and give the reason for his decision, all in as few words as possible. It is good mental training, especially as his boss will only accept precise, careful reports.

Einstein enjoys the work and does it well. He likes inventors and inventions; later he will look back at that time at the patent office as the happiest period of his life. It is certainly the most creative.

Albert Einstein has a wife, a child, a full-time job, and vibrant friends—yet somehow he seems to have plenty of time to think about the world around him, which is what he is really meant to do. He may think about travel on that light beam, about atoms, and about alpha, beta, and gamma rays. He thinks about the Michelson-Morley "failed" experiment (see chapter 4 if you've forgotten the M. and M. experiment); it is the talk of the scientific community. Of course he thinks about Maxwell's equations; he is a fan of Maxwell. Max Planck has given him bundles of energy to consider—along with a mathematical constant to use in his calculations. But thinking isn't enough for Einstein; he needs to talk out his ideas. He is lucky: He has Mileva, the Olympia Academy, and a close friend at the patent office as sounding boards.

In 1905, the unknown patent clerk publishes five scientific papers. Four appear in a physics journal, *Annalen der Physik*—three in the same issue. (Copies of that September issue, volume 17, are now rare and very valuable.)

Suppose you were interested in physics in 1905; would you pay attention to articles written by a young man who

Einstein's professor at Zurich Polytechnic, Hermann Minkowski, was one of the first to understand special relativity. "From now on," said Minkowski, "space and time separately have vanished into the merest shadows, and only a sort of combination of the two preserves any reality." Then, like a good professor, he set about correcting some of Einstein's math, taking it from algebra to geometry, which made it more visual. (More on special relativity in chapters 28 through 37.)

ANNALEN
DER
PHYSIK.

BEGRÜNDET UND FORTGEFÜHRT DURCH
F. A. C. GREN, L. W. GILBERT, J. C. POGGENDORFF, G. UND E. WIEDEMANN.

VIERTE FOLGE.

BAND 17.

DER GANZEN REIHE 322. BAND.

KURATORIUM:
F. KOHLRAUSCH, M. PLANCK, G. QUINCKE,
W. C. RÖNTGEN, E. WARBURG.

UNTER MITWIRKUNG
DER DEUTSCHEN PHYSIKALISCHEN GESELLSCHAFT
UND INSBESONDERE VON
M. PLANCK

HERAUSGEGEBEN VON
PAUL DRUDE.

MIT FÜNF FIGURENTAFELN.

LEIPZIG, 1905.
VERLAG VON JOHANN AMBROSIUS BARTH.

doesn't yet have his doctorate? Hardly anyone notices them. Then two people comment. Both are distinguished professors. One is Max Planck. The other is Hermann Minkowski, the teacher who had called Einstein a "lazy dog." Minkowski realizes at once that these articles are the work of a genius.

Classical science has an *annus mirabilis*, 1666, the year of Newton's greatest productivity. Modern

If you find yourself in Kansas City, Missouri, the Linda Hall Science Library there has one of the rare copies of this famous 1905 issue, volume 17, of *Annalen der Physik* (left), with Einstein's five groundbreaking articles.

science also has a miracle year. It is 1905, the year of those five articles.

• *The first article tackles the mysterious "photoelectric effect."* That effect has scientists puzzled. When light strikes a very clean metal surface, it knocks electrons out of that metal. But why? Wave theory is expected to explain it. The wave's varying electric field should shake electrons loose from the metal. But experiments turn out all wrong. The amplitude of the wave ought to determine the amount of shaking (and the number of ejected electrons): more amplitude, more shaking, and more electrons knocked out of the metal (*emitted* is the scientific word here). To everyone's surprise, experiments show that no electrons are emitted, no matter how large the wave's amplitude, when light is below a certain frequency (called the threshold frequency). At and above the threshold frequency, some electrons are ejected for all amplitudes, even if the amplitude is very small. This is baffling; no one can understand what's going on. Einstein makes sense of it by taking Max Planck's formula for energy ($E = hf$—the h is Planck's constant and the f is frequency) and applying it to light: when hf reaches the energy E needed to eject electrons from the metal, they fly out. Light quanta must be more than mathematical formulas; they must be real. Hardly anyone goes with him on this one. (It will take time, quite a bit of time, before physicists change their minds. For details, keep reading.)

• *Two of the articles are about atoms.* Einstein tells Conrad Habicht, his Olympia Academy buddy, "The second discusses the methods of determining the real dimensions of atoms.... The third proves that, according to the molecular theory of heat, bodies of dimensions of the order of $\frac{1}{1000}$ mm [millimeters] suspended in liquid experience apparent random movement." (That's Einstein's language translated from German. Be sure you understand that when he says, "bodies ... of the order of $\frac{1}{1000}$ mm ... experience apparent

---

Actually, There Are Five

Writing to his friend Conrad Habicht (a member of the Olympia Academy), Einstein says, "I promise you four papers...the first of which I might send you soon.... The paper deals with radiation and the energy properties of light and is very revolutionary."

Einstein is right: The concept he is developing, that light is a quantum duality (both particles and waves) was the hardest of all his ideas for most scientists to accept. It meant that light does not have to travel as James Clerk Maxwell said it does— through a medium. A patent examiner was questioning the great Maxwell. Was Einstein impudent, or brave, or astonishingly perceptive?

The molecular theory of heat states that heat is a form of kinetic energy— molecules moving around. Prior to this theory, many scientists thought of heat as a substance. A more scientific term for what most people call heat is *thermal energy*.

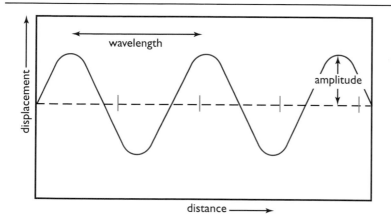

In wave motion, amplitude is the height of a crest or the depth of a depression. Wavelength is the distance from one crest to the next; higher frequency waves have closer crests.

random movement," he's talking about the particles floating in a liquid. He knows that atoms are much, much smaller than these particles. But the movement of the floating particles demonstrates that they are being bombarded by still smaller bodies.)

The technology to "see" atoms doesn't exist; Einstein has found a statistical way to show how H_2O (water) molecules *must* behave. He may not realize it, but he is laying groundwork that will make statistics an important part of modern science. (Later, he is never quite happy with the statistical approach.) Einstein's conclusions will be proved experimentally when scientists compare the predictions of his mathematical formulas with what actually happens in a container of water. **No question then: Atoms and molecules are real.**

• *"The fourth work…modifies the theory of space and time,"* writes Einstein to Habicht. Modifies space and time! This is no small concept. It will be known as **the Special Theory of Relativity**; this enormous idea has come from Einstein's attempt to hang on to a light beam.

Special relativity is a big deal but simple at its heart. It says that the laws of physics are exactly the same everywhere in the universe—on Earth, on another planet, or in an unpowered rocket ship. (Physicists call all those places inertial, or non-accelerating, reference frames.) Within those frames, special relativity describes the motion of objects moving uniformly or even accelerating.

Something else stays the same throughout the universe: **the speed of light (in a vacuum). It is the same as measured by everyone, in every inertial frame.** It is a fundamental constant. If you measure the speed of light while you are zooming in a rocket ship or while you are sitting in a chair on Earth, you will get the same result. This has surprising consequences: If the speed of light is the same for everyone, then other things have to vary, like time. Time is not the same for everyone everywhere.

We don't notice time differences in our slow-moving, Earth-bound lives. But people going very fast in a rocket and people hardly moving on Earth have clocks that tick at different rates. (The difference is very small, but it is measurable.) And your internal clock, which keeps track of your age, will tick more slowly than that of your Earth-bound brother if you are moving at a high speed.

Time and space are knotted together; neither stands alone, and neither is—as the clock on the wall tries to say—fixed and absolute. Only the speed of light has the same value for all.

This builds on James Clerk Maxwell, but it means Isaac Newton was wrong when he based laws on a "universal time"—time that is the same for all observers. (Who is this patent clerk? How dare he question the great Newton?)

• *The last paper, which is really an extension of the fourth, is published in November.* It says that "light carries mass with it," and that the "mass of a body is a measure of its energy content." That means mass and

These Are the Same for Everyone

To repeat, because it is important: Constants are numbers found in nature that have the same value as measured by everyone. Like the Ten Commandments given to Moses, they never change.

Some constants are minor players, important only in special circumstances, but there are three big ones *that seem to underlie the universe*—and you now know about all three of them. One is Newton's gravitational constant (G); the second is Planck's constant (h), which turned out to be the foundation of quantum mechanics; and the third is the speed of light in a vacuum (c). Modern science has been built on those three numbers.

energy are equivalent and, in some cases, interchangeable. Energy and mass are two faces of the same thing? Can this be? Einstein will soon rephrase that idea in his most famous equation. (Hold on if you don't know the equation.)

The patent clerk with "attitude" is challenging most of the scientific thinking that has come before him. He is offering a new world vision.

Before Einstein's miracle year, scientific learning was Earth-bound. Time was an absolute; it flowed forward at the same rate on all clocks. Matter was exact and indestructible. That made sense as long as we moved at Earth speeds. But attempting to catch a light beam gets Einstein moving at close to the speed of light, and he is soon beyond our planet's orbit. When he assumes that physics is the same for all observers everywhere in the universe, Einstein starts thinking in a new way; he begins to think as a citizen of the universe.

This is as revolutionary as Copernicus's moving the Earth so it goes around the Sun. Maybe more so. Writing 100 years after Einstein's *annus mirabilis* and comparing Albert Einstein with Isaac Newton, physicist Alan Lightman says, "These ideas are larger than scientific theories. They are philosophies, they are symphonic themes, they are different ways of being in the world."

But, in 1905, this is so unexpected, and it seems so counterintuitive, that most people just shrug their shoulders and leave special relativity to the experts. They are missing out. In the chapters to come, I plan to show you that you can understand special relativity and Einstein's other ideas. You will have to do some brainwork and, at times, you will have to throw out your common sense. Like our friend Albert, you will have to think of yourself as a citizen of the universe. It's an adventure worth the effort.

In 1905, with these five papers, Einstein has tackled, and solved, some of the deepest mysteries of the universe. What happens next? He goes back to work at the patent office. When he applies for a job as a university lecturer, he is turned down.

Here are more of Einstein's words from the letter he wrote to his friend Conrad Habicht:

One more consequence of the paper on electrodynamics has also occurred to me. The principle of relativity... requires that mass be a direct measure of the energy contained in a body.... The argument is amusing and seductive; but for all I know, the Lord might be laughing over it and leading me around by the nose.

In 1900, Einstein began work on his Ph.D. thesis. His first dissertation was rejected by the Zurich faculty. So were his next and his next (about relativity). Then he submitted a paper on the dimensions of molecules and atoms in liquids. His professor said it was too short. Einstein added one sentence. Finally, in 1905, the dissertation was accepted, and he became Doctor Einstein.

Having It Either Way

Thomas Young's double slit experiment with light is a turning point in the history of science. He sent light of a single frequency (color) through two slits in a panel. The split light beam produced an overlapping wave pattern, like two tossed stones creating ripples on a pond's surface (below).

The experiment proved that he was right: Light does travel in waves. What Young didn't know is that there is more to light's story. Light also has a particle nature.

Quantum scientists did another double slit experiment that proved that dual wave-particle identity. They did it again and again because they couldn't quite believe the results.

Shoot electrons or photons at a barrier containing one slit (or hole) and look at how they strike a parallel screen behind the barrier. You will see that the electrons fall randomly, like rain, on a single patch behind the slit or hole. With time, the patch is uniformly covered. Change the slit width, and electrons will still spread over the patch without a break.

Now, if you send electrons through a barrier with two slits, they will do something entirely different. The pattern on the screen changes dramatically from the single slit experiment. The electron hits build up a pattern of spots that interfere with one another with highs and lows and overlaps—just like waves. It is as if each electron goes through both slits at once, as a wave would.

How do electrons "know" to behave one way when there is one hole and another way when there are two? The scientists couldn't figure that out; they just knew that it happened.

When they set up experiments that send electrons one at a time through the slits, they got the same result. If there is one slit, an electron behaves sort of like a bullet; with two slits, it behaves as if it were a wave. When experimenters tried changing the number of slits while the electron is in motion, it immediately changed its response.

Electrons (and other elementary particles) seem to keep their options open as long as possible. They stay in a wave-particle double state until they hit the screen. Hard to imagine? You bet. But that's what happens.

Thomas Young pioneered the study of interference patterns, which are created when two or more waves collide, either enhancing each other (if they're moving in sync) or disrupting each other. Likewise, when two water waves meet, they either join crests to make a higher crest or join crest to valley to cancel each other out.

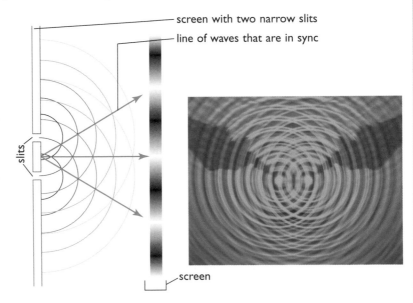

screen with two narrow slits

line of waves that are in sync

slits

screen

"and that is the reality and substantiality of the luminiferous ether."

When Einstein is 16, he joins the hunt and writes a paper on the ether. In college, he does a laboratory experiment to try and find the ether, but Einstein isn't much of an experimenter, and his effort fizzles. (Einstein understands the importance of experimental proof; he just isn't skilled in that area.)

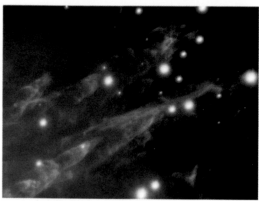

The fact that most of space is a vast void makes the discovery of any new object all the more fascinating. These orange tubular shapes, first spotted in 1983, are mysterious gas "bullets" shooting out of the Orion Nebula. Each one is roughly the size of our solar system and has a glowing blue tip made of iron gas. What's pulling the trigger? No one is sure.

Michelson and Morley are superb experimenters (see chapter 4), perhaps the best in the world (at that time), and yet no matter how they set their measuring devices, they cannot find any effects of the so-called ether—effects that (almost) everyone predicts should be there. It doesn't make sense. Might all the scientists be wrong? Albert Einstein keeps an open mind.

Einstein knows that if James Clerk Maxwell is right, **there is one thing in the universe—the speed of light in a vacuum—that has the same value for all observers, no matter how they move. And that means you can use light's speed in a vacuum as a measure to compare all other things.** Hardly anyone recognizes the usefulness of that information. But Einstein does.

Then he does something that geniuses (and other smart people) do: He makes a connection between two seemingly unrelated things. Newton made a connection between the orbit of the Moon in the sky and the fall of an apple from his mother's tree and came up with a theory of gravitation.

Einstein connects Maxwell's equation on the speed of light in a vacuum (c) to Max Planck's constant (h) dealing with energy quanta. (This is big; take note of it. Can you see a connection? At first, no one but Einstein could.)

Albert Einstein figures out that **light must come in particles with specific properties (quanta) that (in a vacuum) always travel at the same speed.** Einstein may feel as Archimedes did when he ran down the streets of Syracuse shouting "Eureka! I have it." Einstein now realizes that Max

Photons: Discovered and Named

Einstein thought of photons (he called them "light quanta") as "atoms of energy." Today, they are known as the basic unit of electromagnetic radiation (visible light, X rays, microwaves, etc.).

Light quanta were renamed "photons" by American chemist Gilbert Lewis (1875–1946) in 1926. We know that a lightbulb emits about 10^{20} photons every second (that's 100 billion billion). They travel so fast that one photon coming from that bulb lasts less than one-billionth of a second before it hits something or flies out the window. Photons lead fleeting lives; they come and go; they have no mass (although they do have momentum). You can't make molecules out of photons. They don't seem to be conserved in nature. But they are real. And they respond to experiments.

One more thing: Photons and electrons cannot be broken apart. They are both *quantum* particles. So you can't have half a photon or one-quarter of an electron.

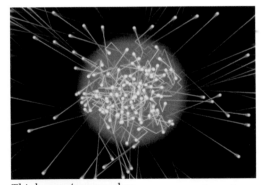

Think mass/energy when you consider electrons. Think just energy when you consider photons. Photons exist only as moving particles; absorb them and they disappear, giving up their energy. This luminous illustration (above), created at École Polytechnique in France, is the final frame of a series titled "The Random Walk of Photons Escaping from the Sun."

Planck's quanta are not just a mathematical device. Planck has quantized the source of radiation, says Einstein, which means the radiation (light) itself must be quantized as real packets of energy that have no mass.

So light is particles. And light is also waves. Light is both wave and particle, but never both at the same time. Young's experiment told only part of the story.

What does light wave? Nothing. It travels in a vacuum in fields that it creates. A changing magnetic field creates a changing electrical field, which creates a magnetic field, and so on. Energy is handed back and forth, which is what makes light wave.

Can there be particles without mass? Yes.

When you look at a light beam, you're seeing pure, massless energy.

When light travels through space, it has wave properties; when it interacts with matter, it behaves like a particle. Light can travel through a vacuum. It doesn't need an

Scientists were breaking the rules of the past, so why shouldn't artists? At left is Claude Monet's impressionist painting *Haystacks at Sunset, Frosty Weather* (1891). "The principal person in a picture is light," said artist Edouard Manet (1832–1883), in describing the impressionist style. Was it a coincidence that the artists who focused on light—the impressionists— arrived on the scene at the end of the nineteenth century?

ether—which means Einstein has solved the ether problem.

This is all so startling that, at first, most people have a hard time believing it. **There is no ether! Space is a vacuum!** Light is waves and also particles!

Albert Einstein is going way beyond Max Planck. Planck doesn't believe that those particles/quanta actually exist outside of mathematics. Einstein's genius is to assume that they do. He is sure light quanta are real.

When Einstein writes a paper about light's particle nature, most scientists think he has gone too far. He is aware of their skepticism. His 1905 paper on light quanta is the only

Turn Your Head and Think Both Ways!

Perched between December and January, the Roman god Janus looks two ways: forward and backward. Light does more than look two ways; it has two identities. It is both wave *and* particle (though never both at the same time). *And so is everything else!* An electron has wave properties. In principle, so does a golf ball, but the greater the mass, the harder it is to detect those waves.

Light is a different story. Modern technology makes it easy to discover light's wave-particle, two-faced nature, but those two faces do complicate predictions. You can do one set of experiments, and light will act at a specific point as particles. You can do another set of experiments, and it's waves all the way.

Like Janus, light has a two-faced identity.

Einstein's first wife, Mileva Maric, studied with Philipp Lenard (1862–1947) and almost certainly brought the photoelectric problem to her husband's attention. As for Lenard, in the 1930s he became a rabid Nazi and a passionate enemy of Einstein and all other Jewish scientists.

one he describes as "revolutionary." Planck writes, in a recommendation letter for Einstein to the Prussian Academy, "That [Einstein] may have occasionally missed the mark in his speculations, as, for example, with his hypothesis of light quanta, ought not to be held too much against him."

But Einstein is right. Light quanta (photons) do exist. They exist as particles and as waves. That's what he writes in his soon-to-be-famous paper explaining photoelectricity.

Photoelectricity has puzzled scientists. J. J. Thomson in England and Philipp Lenard in Germany found that when light, traveling at certain frequencies, hits a piece of metal, it knocks electrons out of that metal. That is the photoelectric effect.

Einstein Argues for Chunks

The scientific symbol for Planck's constant is h. The symbol for the frequency of electromagnetic waves is f. Planck multiplied them—got hf—and the outcome surprised him. When you multiply h by f, you have something continuous, like a wave, but made up of discrete particles. Bizarre! (Planck's actual words, translated from German: "[It was] a discrete quantity composed of an integral number of finite equal parts.")

Einstein was able to take abstract ideas and see them as real, but even he was astonished by this surprise. Matter—things like rocks—clearly comes in chunks; non-matter—like energy—seems to flow continuously. But if you multiply h by the factor f you get tiny chunks of energy (hf) with a flow and frequency like a wave.

Are matter and energy connected? Einstein introduced that idea in his paper explaining the puzzling features of photoelectricity.

He used hf as a symbol for the energy of his light quanta. He thought of those quantized particles as actual things. Few scientists agreed.

Albert Einstein cautioned, "Concern for man and his fate must always form the chief interest of all technical endeavors.... Never forget this in the midst of your diagrams and equations."

Twenty years later, they were still disagreeing. Most, like a young Danish physicist, Niels Bohr (1885–1962), would use hf to analyze light and solve equations but never quite believe in its reality. Einstein and Bohr would have a long-running battle about hf: Did it represent a real thing—a particle with a wave frequency—as Einstein believed? Or was light a wave and hf just a helpful mathematical device, as Bohr believed? Who do you think won the argument? (Keep reading for details.)

But there is something that seems inconsistent about it. Some light rays cause electrons to be released from metals, and others don't. That's the baffling part of the photoelectric effect. When Einstein realizes that light is particles (photons), he understands what is going on.

Einstein figures out that low-frequency radiation (like infrared) acts like low-energy photons. These low-energy photons can't release electrons from metal. When those photons strike the metal, they are absorbed as heat. (You can feel the heat on the metal's surface.) But as the frequency of light radiation gets higher, so does the photon energy. High-energy photons (like ultraviolet) easily knock electrons out of the metal (one photon can release one electron). The higher the energy of the photon, the more energetic the ejected electron can be. That solves the mystery of the photoelectric effect.

Einstein works all this out in formulas, and it's those formulas that impress his fellow scientists. As for there being *actual* photons, "particles of light," most of the scientific community thinks this is hogwash. For a long time Einstein is mostly alone in this belief.

Robert Millikan (1868–1953) is so annoyed by Einstein's theory that he decides to conduct experiments to prove him wrong. Millikan is a skilled American experimenter who spends 10 years trying to show that the photon idea is nonsense. But his experiments prove that Einstein is right. "The equation of Einstein seems to us to predict accurately all the facts which have been observed," he writes in 1916. Yet Millikan still resists this idea: "Despite then the apparently complete success of the Einstein equation, the physical theory... it was designed to [express] ... is found so untenable [impossible] that Einstein himself, I believe, no longer holds to it."

Hindsight Helps

In 1949 (note the year), Robert Millikan wrote: "I spent ten years of my life testing that 1905 equation of Einstein's, and, contrary to all my expectations, I was compelled in 1915 to assert its unambiguous verification [clear proof] in spite of its unreasonableness."

When Einstein finally got a Nobel Prize (in 1921), it was for his work on photoelectricity, not relativity. Millikan got one, too (in 1923), for his experiment that proved photons exist. But someone else had to prove it again, before Millikan or others really believed what the experiments told them: Photons are particles of light that also exist as waves.

A Letter from India

Bose worked with Einstein, Marie Curie, and de Broglie—becoming a highly respected physicist and mathematician.

An Indian physicist, Satyendranath Bose, showed that Planck's equations work if you treat light as particles and don't worry about waves. But he realized you can have it either way in the math world —waves or particles— just as you can in the "real" quantum world.

Bose was born in 1894 in Calcutta, the eldest of seven children and the only boy. As a student at the Hindu High School in Calcutta, he established a new record for math achievement. Then he went on to the Presidency College, a Calcutta school that produced a number of renowned scientists.

In 1924, when he was a teacher at Dacca University, Bose wrote an article in English on Max Planck's law. He sent the article to Albert Einstein. Einstein got lots of papers from would-be scholars. He tried to read them all, even though most were a waste of time. But Einstein saw that Bose was a real scholar. He was so impressed with Bose's work that he translated the article into German himself. Then he got it published in Germany (then the center of the scientific world). The article was about a statistical approach to science, today known as Bose-Einstein statistics.

Bose came to Germany and met Einstein and other leading scientists. He became an important physicist and mathematician himself. A family of subatomic particles, the bosons, is named for him. Bose returned to India in 1926 and became head of the physics department at Dacca University (now in Bangladesh and called Dhaka).

Millikan is wrong again. Einstein never loses faith in his light quanta. In a letter to his friend Michele Besso, Einstein writes, "The light quanta is practically certain."

Millikan just won't accept what his experiment tells him. He says, "Experiment has outrun theory.... [I]t has discovered relationships which seem to be of the greatest interest and importance, but the reasons for them are as yet not at all understood."

Robert Millikan has proved that photons do, indeed, exist—but he doesn't quite believe it. Hardly anyone does. By the 1920s, many physicists think that Einstein has given them a concept of light particles that works in mathematical formulas, and that allows for technology using the photoelectric effect (like today's automatic doors). But most scientists are not persuaded that light can really be both waves and particles.

Einstein doesn't doubt his ideas. He writes, "I have already attempted earlier to show that our current foundations of the radiation theory have to be abandoned." He calls for a "fusion of wave and particle theories." He says, "[The] wave structure and [the] quantum structure . . . are not . . . incompatible."

It doesn't matter what he writes; that particle-wave duality notion just seems wacky. (Would you have believed this idea from a young maverick when almost all of the distinguished professors aren't buying it?) For more than 15 years after Einstein announces it, the dual nature of light remains too much for most scientists to swallow.

You use the photoelectric effect every time you check out at the supermarket. A beam of high-frequency light shines across the conveyor belt and strikes a metal plate on the opposite side. That gets the plate to emit electrons. The emitted electrons turn on an electric current, which makes the belt move. If something gets in the way of the light beam—maybe a box of cereal—it stops the belt. Just to be sure you understand: Light is both wave and particle (but not at the same time). Einstein was right.

Why Small Is Big

We all know that big things can come in small packets. And most people are aware of the significance of atoms. But few really understand the importance of the discovery of tiny photons and other inhabitants of the quantum world. Let me tell you: You wouldn't have TV without quantum theory—how's that for important? And that's just a bit of it.

Quantum theory (about subatomic particles) is one of the most important ideas of modern science—maybe of all time. Understanding those tiny particle-waves, and being able to manipulate and use them, has changed the world we live in. More about this is coming, lots more, but right now we're heading back to 1905 and another of Einstein's papers. Some people still don't believe that atoms even exist. Einstein has to prove them wrong.

Blue Skies Smiling at Us

John Strutt (Lord Rayleigh) was so frail as a child that no one thought he'd survive, but he lived to become a famous physicist. (One of his students was J. J. Thomson.) A crater on the Moon is named Rayleigh in honor of him.

Have you wondered why the sky is blue? Well, so did Albert Einstein. In October 1910, he came up with an answer in a paper using statistical physics. He developed his ideas by studying the work of England's Lord Rayleigh (RAY-lee), a handsome, rich baron who was born with the name John William Strutt (1842–1919). Rayleigh built a scientific laboratory in his family's grand mansion in Essex, but he was a peripatetic fellow (that means he liked to travel around) and was living on a houseboat on the Nile River when he wrote a famous book about sound. Then he agreed to succeed James Clerk Maxwell at Cambridge University (but only for five years, during which time the number of physics students went from 6 to 70). Rayleigh won a Nobel Prize for his studies of gas densities, especially of the gas argon.

But most people associate Lord Rayleigh with blue skies. In 1871, he explained why we see blue when we look heavenward.

The visible light from the Sun includes the spectrum of colors of the

A rainbow (below left) is a spectrum of colors that appear in order of frequency. You can memorize the order by invoking the mnemonic name Roy G. Biv (red, orange, yellow, green, blue, indigo, violet). Note that some scientists insist that indigo (a deep blue) isn't a separate or distinct color and that the spectrum goes directly from blue to violet (a shade of purple).

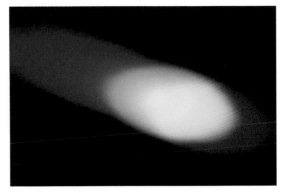

rainbow: red, orange, yellow, green, blue, indigo, and violet. When those colors are equally balanced, we see the color white. In outer space, where there is no atmosphere, an astronaut looking at the Sun (through a dark filter to save his or her eyes) sees a white orb.

But sunlight heading for Earth travels through Earth's atmosphere, where it bumps into things: mainly nitrogen and oxygen molecules. The charged electrons of those molecules act like small masses on springs, bouncing around in the light waves. When a light wave sweeps over a molecule, the molecule's electrons vibrate like charges sparking an antenna, and EM wave-particles are spewed out. The process of absorbing the light from one direction and then radiating it out in all directions is called scattering. This particular kind of scattering is known as "Rayleigh scattering." It's light interacting with matter.

As you know, light waves are not all created equal. In the visible-light portion of the spectrum, each color has a different wavelength. The blues have shorter wavelengths than the reds, but that means higher frequencies. The higher the frequency, the harder the shaking and the greater the light scattering. Colors with short wavelengths (the blues) keep molecules dancing and scattering. During daylight, the scattered blue dominates our view overhead. At sunset, the light passes through a greater thickness of air; the blue is scattered out by smoke, dust, and pollutants—giving us a red horizon.

Still, it might be hard to see that scattered blue light without a dark background. As it happens, the blackness of outer space provides that background.

Lord Rayleigh figured this out, but our friend Albert put in the numbers, adding quantitative confirmation to the explanation.

Molecules Move

> It is one thing to say, as some nineteenth-century scientists did, that Brownian motion might be explained…as the result of the impact of molecules on the suspended particles. It is quite another to calculate, as Einstein did, the precise statistical nature of the impact of very large numbers of molecules with suspended particles, and to use that calculation to predict the precise nature of the zigzagging Brownian motion that would result. A good scientific theory always has to make quantifiable predictions that can be tested by measurements and experiments.
>
> —Michael White and John Gribbin, *Einstein: A Life in Science*

> In those days many chemists and a few physicists still considered atoms to be theoretical fictions. Einstein's proof of their reality built on statistical ideas developed by a thick-bearded Austrian named Ludwig Boltzmann….Einstein sometimes wondered why [Boltzmann] had not found the proof first. It was a modest man's question.
>
> —Edmund Blair Bolles, science writer and author, *Einstein Defiant*

> What we call physics comprises that group of natural sciences which base their concepts on measurements…[that] lend themselves to mathematical formulation.
>
> —Albert Einstein, *Science* 91 (May 24, 1940)

This early photograph shows the Smithsonian Institution, nicknamed "The Castle," before a fire destroyed much of the main building in 1865.

It is 1827, and the United States has a president, John Quincy Adams, who is fascinated by science. He believes the government should sponsor scientific research and build a national observatory. But there is little support for his ideas. A scholarly man, the son of a former president, Adams has a formidable political enemy: Andrew Jackson. Besides that, Adams is clearly antislavery, and that issue is dividing the nation. John Quincy will be defeated when he runs for reelection. Later, as a congressman, he will have success when he champions the idea of a national organization to further science: the Smithsonian Institution.

Meanwhile, in Scotland, in this year of 1827, a quiet plant scientist, a botanist named Robert Brown, looks through a microscope at tiny bits of pollen floating in water and notices something puzzling. The pollen is dancing around, even though the water seems still. What makes it move? Can the pollen be alive? Brown doesn't think so, but he isn't sure. Being a careful scientist, he decides to float some other particles in water. He uses powdered tar, ground-up arsenic dust, and other things he knows have no life. The particles all move actively. The moves are like jitterbug or break dancing, with jumps here and there. Brown calls it a "tarantella," which is an Italian folk dance. What causes the movement? No one knows.

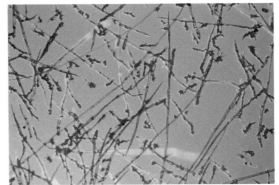

Botanist Robert Brown described the "dance" of pollen grains floating in water as a tarantella, which is both a fast, whirling circle dance and a lively instrumental song (below). A sycamore tree produced the pollen above.

Scientists will spend years puzzling over those random moves—which come to be called "Brownian motion." In 1865, the year the American Civil War ends, a group of scientists float particles in a liquid, seal the liquid under glass, and then watch for a whole year. The particles never stop moving. No one can fathom why.

They don't realize that if they keep watching and watching, the dance will go on and on. If they had been able to preserve the sample in the glass jar, we could see the very same Brownian motion they saw (and so could our grandchildren). It is ceaseless. Why? What makes it happen?

Seventy-eight years after Robert Brown first watches the bizarre action, Albert Einstein solves the problem of Brownian motion. In one of his 1905 papers, he explains that a glass jar filled with water holds billions and billions

Artist Antoine Coypel (1661–1722) imagined Democritus as a jolly philosopher.

At the end of the nineteenth century, physicists were divided into two camps: One refused to believe in atoms; the other said that atoms might never actually be seen, but the idea made sense and it provided a hypothesis that could lead to useful experiments. That second group called the smallest unit of an element an "atom," and they called two or more atoms bonded together "molecules" (as we do today).

Ernst Mach never conceded that atoms are anything but a convenient fiction.

of H_2O (hydrogen/oxygen) molecules. They are all moving very fast, randomly bumping and banging the pollen.

No one can see the water molecules; they are much too small for the microscopes of the time. Einstein figures this out in his head—but not all by himself. The atomic idea goes back to Democritus (di-MAHK-ri-tuhs), who lived in Greece about 2,500 years ago. Democritus said nature must have basic building blocks. He called them "atoms," and then he said that they are in constant motion, even in a substance that seems at rest.

Einstein knows of the ancient atomic theories (Democritus wasn't the only Greek who believed in atoms), and he knows of John Dalton and Ludwig Boltzmann and the nineteenth-century scientists who believed in atoms. He also knows that some scientists of his day don't take the atomic idea seriously.

How can you be sure of something you can't nail down? Many scientists think molecules and atoms are fictional devices that are helpful in working out formulas—but that it is unscientific to believe in something you can't actually find. The skeptic Ernst Mach keeps asking, "Have you ever seen one?"

Einstein admires Mach (who, among other things, wrote a famous science text). But Einstein ignores Mach's question. Instead, he figures out mathematically the way the water molecules would behave if they do exist.

Einstein devises a formula that says the distance the pollen particles move increases by the square root of the time considered. In other words, in four seconds, the particles will move (on average) twice as far as they do (on average) in one second. You would expect that the average change of position would be four times as much in four seconds as it is in one second. Not so!

He is right: Trillions of unseen but active water molecules are moving the visible particles of pollen, just as Einstein's math says they do. Keep in mind, the atoms of water are not hitting the pollen grains equally from all sides. If they did, there would be no movement. The pollen moves when a

Probable? Average? Oh, No! Where Will This Lead?

Note that phrase "on average" in the text—it's an important part of understanding atoms. Scientists are not going to be able to figure out what any one or even any ten atoms will do. They'll be forced to look for averages. They'll have to rely on statistics that deal with lots of atoms. They won't enjoy this. But once they master the averages, they can express them as probabilities (based on statistics) that will allow them to predict how gazillions of atoms are likely to act together.

This will be tough for scientists who are used to precise measurement of individual units. Albert Einstein started it all with his 1905 paper on Brownian motion, but he won't be happy with where it leads. When others take the idea of probabilities (based on statistics and averages) still further and show that probability is the only way to describe a single electron in a single atom, Einstein will fuss. But, like it or not, it seems to be the way to deal with the atomic and quantum worlds.

 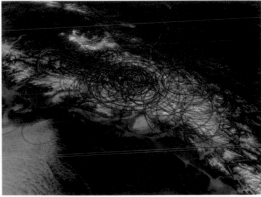

Salvador Dali titled this 1937 painting (left) *Average Pagan Landscape*. What does he mean by "average"? You get to decide—and that's just fine in art. But not in science and math. The word *average* demands a precise answer to questions like: How often does a strong earthquake hit southern Alaska (above) on average? The circles mark the location and size of strikes over a period of years. By counting the strong quakes and dividing by the years, you get an average. The longer the time period and the greater the data (photo on right), the more precise the average.

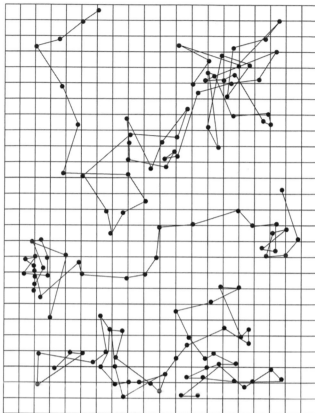

In 1908, Jean-Baptiste Perrin peered through a microscope and painstakingly noted and tabulated the movement of individual particles of gamboge (a kind of pigment). In 2006, his Brownian motion experiment was recreated at Harvard University, this time using a CCD camera (which collects light on an electronic silicon wafer instead of on film). A computer analyzed the movements of the gamboge. The results were the same as Perrin's. The graph at left plots the random, zigzag movements of three granules every 30 seconds.

Why doesn't a particle move four times as much in four seconds as it does in one second (on average)? Because the motion is random. After a particle has moved one step to the right, it is equally likely to take the next step back to the left or farther to the right.

bunch of molecules randomly hits it more from one side than the other side. That's what causes the herky-jerky, tarantella motion. But no one can be sure of that until Einstein's theory is tested experimentally.

"By 1908, the French experimental physicist J. B. Perrin had tested and confirmed Einstein's formula," says physicist/writer Jeremy Bernstein. "Moreover, by actually observing the distance that the Brownian particles traveled, [Perrin] was able to deduce approximately the number of molecules per cubic centimeter in the liquid through which they were traveling."

"The objective reality of molecules and atoms which was doubted twenty years ago, can today be accepted as a principle," announced Perrin in 1926, when he was awarded a Nobel Prize for his work.

Einstein's reasoning hasn't just answered the question of Brownian motion; it offers statistical proof that atoms and molecules exist. And, says science writer Edmund Blair Bolles, "He made his theory so precise that a person holding a stopwatch and measuring the jigs could calculate the number of molecules in the water."

Einstein has shown that statistics can be taken seriously in the creation of scientific theories. His explanation and the follow-up tests finally convince the skeptical scientists: Atoms are real. It is a sweet victory for atomic theory.

"Nevertheless," Perrin says, "it would...be a great step forward...if we could perceive directly these molecules."

It will happen.

We think we live in a world where most objects have smooth surfaces, but that's because our eyesight is limited. We can't see atomic structures. If we could, we'd realize that the world is a grainy place. Glass, silk, and an eyelash (magnified by an SEM, at left) are actually collections of zillions of separate particles (grains) of matter.

Big Numbers Have Powers

Once physicists began computing the numbers of atoms and molecules in any substance, they found those numbers staggeringly large. How large?

Well, let's start by imagining a very big number: 1 billion. It is $10 \times 10 \times 10 \times 10 \times 10 \times 10 \times 10 \times 10 \times 10$. Written in mathematical notation, with an exponent, that's 10^9.

Now consider the total molecules in a teaspoonful of water: It's approximately 1.67×10^{23} molecules. That's a number so big it confounds your head to attempt to picture it, but let's try. Imagine this: If all the world's people (roughly 6.5 billion or 6.5×10^9) could be turned into water molecules and put in that teaspoonful of water, they would be so outnumbered by the other 10^{14} molecules that they would be invisible. So, if it takes gazillions of molecules to fill a teaspoon, each molecule must be very, very, very tiny.

In 1905, no one knew quite what to make of that information.

In this paper, it will be shown that according to the molecular-kinetic theory of heat, bodies of microscopically visible size suspended in a liquid will perform movements of such magnitude that they can easily be observed with a microscope, on account of the molecular motions of heat.
—Albert Einstein, 1905 paper on Brownian motion

Getting the Picture Right

> Newton's laws are wrong—in the world of atoms...[because] things on a small scale behave nothing like things on a large scale. That is what makes physics difficult—and very interesting.
>
> —Richard P. Feynman (1918–1988), American physicist, *Six Easy Pieces*

A bar of gold, though it looks solid, is composed almost entirely of empty space: The nucleus of each of its atoms is so small that if one atom were enlarged a million billion times, until its outer electron shell was as big as greater Los Angeles, its nucleus would still be only about the size of a compact car parked downtown.

—Timothy Ferris, American science writer and author, *Coming of Age in the Milky Way*

When walking about on the Earth, we often get the impression that a force is some kind of invisible entity that pushes and pulls us around. But on the level of atoms, physicists prefer to describe a force as a kind of tennis game: A force between two particles arises from their continually exchanging another, identifiable particle—a sort of subatomic tennis ball. In electromagnetic interactions...the tennis ball is the photon.

—Marcia Bartusiak, American science writer and author, *Through a Universe Darkly*

Just what is it like inside an atom? John Dalton's hard, solid model no longer makes sense. Atoms seem to have a structure and parts. Are they a doughy mass with electrically charged specks (electrons), spread out like soft raisins in a plum pudding (right), as J. J. Thomson has suggested?

Christmas with J. J. Thomson and Plum Pudding

An English plum pudding is a traditional Christmas dessert. It rarely has plums in it. Actually, it's a kind of bread pudding with a soft texture; it usually has raisins, currants, and sometimes other fruit bits mixed throughout. Here's a recipe from *Miss Leslie's Directions for Cookery*, a popular 1851 English cookbook that's likely to have been in the Thomson household:

Recipe for a Baked Plum Pudding

Grate all the crumbs of a stale six cent loaf; boil a quart of rich milk, and pour it boiling hot over the grated bread; cover it, and let it steep for an hour; then set it out to cool. In the meantime prepare half a pound of currants, picked, washed, and dried; half a pound of raisins, stoned and cut in half; and a quarter of a pound of citron [lemon] cut in large slips; also, two nutmegs beaten to a powder; and a table-spoonful of mace and cinnamon powdered and mixed together. Crush with a rolling-pin half a pound of sugar, and cut up half a pound of butter. When the bread and milk is uncovered to cool, mix with it the butter, sugar, spice and citron; adding a glass of brandy, and a glass of white wine. Beat eight eggs very light, and when the milk is quite cold, stir them gradually into the

Here's what you need to bake a plum pudding, according to a 1961 pictorial recipe by Eliot Hodgkin. It's chemical reactions all the way. What happens to those eggs? You won't notice an egg when you eat plum pudding. How about milk and butter? The atoms and molecules of those ingredients do some bonding. Chemistry can be yummy.

mixture. Then add, by degrees, the raisins and currants (which must be previously dredged with flour) and stir the whole very hard. Put it into a buttered dish, and bake it two hours. Send it to table warm, and eat it with wine sauce, or with wine and sugar only.

No one knows. But Thomson's first research student, Ernest Rutherford (who has a walrus mustache and a lot of self-confidence), is searching for a key to unlock the puzzle.

Back in 1895, 24-year-old Rutherford was digging potatoes on his father's New Zealand farm when he got the news that he had won a scholarship to study at the University of Cambridge. "That's the last potato I'll dig," he said and then borrowed money to pay for his boat fare to England.

After training with J. J. Thomson, Ernest Rutherford heads for Canada to teach at McGill University. There, he

When Rutherford first arrived in England, he wrote to his fiancée in New Zealand describing J. J. Thomson: "I had a long talk with him. He's very pleasant in conversation and is not fossilized at all. As regards appearance he is a medium-sized man, dark and quite youthful still: shaves, very badly, and wears his hair rather long."

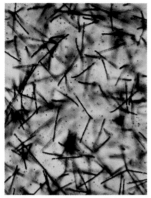

A photographic emulsion impregnated with a radioactive thorium salt reveals the tracks of alpha particles when developed. As Ernest Rutherford and Frederick Soddy found in 1900–1902, the decay of thorium initiates a chain of decays. First one radioactive element is formed, then that decays to another, and so on. This is why the emulsion picture shows the tracks of two or more alpha particles emerging from the same spot: They are the result of successive stages in the decay of a single nucleus.

When we *can't* know something exactly, knowing the probability or likelihood of something—by using statistics—can be almost as useful. Albert Einstein didn't like using statistics to help understand the composition of atoms, but he was flexible and worked with what he had.

teams up with chemist Frederick Soddy; they decide to study radioactive decay. Aside from the Curies, most scientists aren't bothering with that subject. But Rutherford and Soddy are fascinated by the way radioactive elements—like radium and uranium—emit alpha or beta particles or gamma rays. Since schools teach that all elements are fixed forever and unchanging, an explanation is needed.

Rutherford and Soddy are looking for one. They try to manipulate the radioactive process in their experiments. But they can't do it. You can heat atoms, you can freeze them, you can shake them, and you can bake them. It doesn't matter; the rate of radioactivity does not change. Radioactive atoms do their own thing, in their own way, on their own timetable. The process is random.

> Until the interpretation of radioactivity, the atoms had been [thought of], as [James] Clerk Maxwell put it in 1873, as the permanent foundation stones of the universe. A new world was opened up when it was discovered that the two heaviest and most complex of the atoms were spontaneously breaking up.... Discoveries, such as the discovery of the production of helium from radium and the production of radium from uranium, any one of which would have startled chemists of any earlier century, now followed from the theory of atomic disintegration as a matter of course, being predicted beforehand, much as a skilled billiard player will declare his stroke before he plays it.
>
> —Frederick Soddy, English physicist and Ernest Rutherford's lab partner

However, the Rutherford-Soddy experiments do show that **half of the atoms in a radioactive element will decay in a fixed amount of time.** Problem: No one knows which half. So no one can pick out the specific atoms that are going to decay. Since Rutherford and Soddy can't know exactly which atoms will decay, they have to use statistics.

They do what insurance companies do when they make charts of probabilities and predictions called "actuarial tables." Those insurers understand that, in a given population, a certain number of people will die at age 50, and another number will die at 60, and so on. They don't know which individuals will die. They can write insurance policies because they have statistics that tell them the probable percentages.

PROBABILITY

J. Harris

IF YOU HAVE 5 DOGS, 3 WILL BE ASLEEP

Lest men suspect your tale untrue, Keep probability in view.
—John Gay (1685–1732), *The Fables, Volume 1*, "The Painter Who Pleased Nobody and Everybody"

Rutherford and Soddy help bring that method into science. They believe what they are doing is temporary and that eventually someone will be able to pinpoint atoms and predict their decay. (But it hasn't happened. We still have no idea which atom will decay at which time.) This is a whole new way of looking at the physical world. Statistics will become a significant part of science, even though many scientists wish it weren't so.

Rutherford's time in Canada is productive, but he is soon back in England, this time at the University of Manchester. That's where he discovers something that brings him world fame.

It is 1909, and this is what Rutherford does: He shoots alpha particles at gold. The gold is in foil sheets $\frac{1}{50,000}$ of an inch thick, and it's actually Rutherford's assistants Ernest Marsden and Hans Geiger who do the experiment.

The alpha particles Rutherford used were ejected naturally from radioactive elements, like uranium and radium. Pierre and Marie Curie gave Rutherford samples of their precious radium. They were friendly rivals.

Remember (from chapter 8): An alpha particle is the nucleus of an atom of helium-4—two protons and two neutrons held together by the strong nuclear force; it is unusually stable and behaves like a single particle. Even before Rutherford figured that out, he used alpha particles to probe atoms. Alpha particles emitted from a radioactive source move at about 16,000 kilometers per second (10,000 miles per second). Wham them into an atom, and you might break apart the nucleus.

The alpha particles go right through the gold foil, hitting a metal screen, where they give off tiny flashes that can be detected and marked. That is just what Rutherford expects. Then Marsden and Geiger do some further checking. They don't think they will find anything. But they have been trained to be good scientists, and good scientists check. To their astonishment they find that a few particles have *not* gone through the foil (about 1 in every 8,000 particles). Some have bounced back.

That is amazing. "It was...as if you had fired a 15-inch shell at a piece of tissue-paper and it came back and hit you," says Rutherford. (Imagine shooting BB pellets at a soft cake and having a few of the BBs smack the cake and turn around.) What do the particles hit that makes them bounce off the gold foil? It can't be a tiny electron (at that time the only thing known to be inside the atom); it has to be something very hard and dense—with a charge so strong that alpha particles are turned back when they get near it.

The atom Rutherford has been picturing is like a plum pudding dotted with electrons. He knows alpha particles would pass right through that atom, scattering the

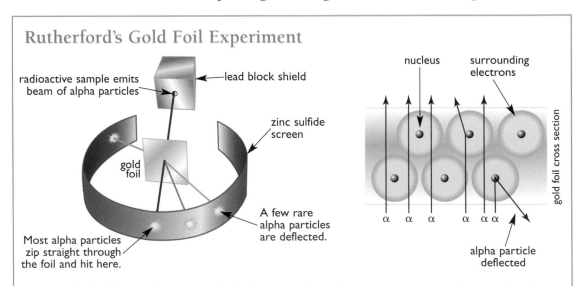

Rutherford's Gold Foil Experiment

radioactive sample emits beam of alpha particles

lead block shield

zinc sulfide screen

gold foil

Most alpha particles zip straight through the foil and hit here.

A few rare alpha particles are deflected.

nucleus

surrounding electrons

gold foil cross section

α   α   α   α   α α

alpha particle deflected

When Rutherford bombarded a thin foil of gold with alpha radiation, most of the particles zipped right through, as expected. Every once in a rare while, a particle veered off track at an unexpectedly sharp angle. Rutherford concluded that atoms must have very tiny but dense nuclei (right).

The Boss Gets the Credit

Ernest Marsden and Hans Geiger (left) did the first gold foil experiment at Cambridge. But it was Rutherford (right) who suggested the experiment and knew what to do with the results.

"One day," Marsden recalled later, "Rutherford came into the room where we were counting the alpha particles [and said], "See if you can get some effect of alpha particles directly reflected from a metal surface."

Working for Ernest Rutherford could be a life-changing experience. More than a dozen of his aides went on to become Nobel Prize–winners. One of them, James Chadwick, wrote: "He had the most astonishing insight into physical processes, and in a few remarks he would illuminate a whole subject.... To work with him was a continual joy and wonder.... He was, in my opinion, the greatest experimental physicist since [Michael] Faraday."

lightweight electron "plums." But an alpha particle can get very close to a massive, tiny, highly charged nucleus and be reflected right back the way it came, which is exactly what happens sometimes. So the atom must have a tiny nucleus with a positive charge to balance the negative electrons and make the atom electrically neutral. This nucleus takes up very little room in an atom.

Rutherford has taken a new snapshot of the atom!

Almost all (but not quite all) of the atom's mass is in that tiny nucleus. The rest of the space in the atom is just about empty. Since you and I and everything else we know about is made of atoms, that means most of us and most of everything is empty space. Electrons exist in that mostly empty arena.

Tall, hearty Ernest Rutherford has come up with the first close-to-accurate picture of an atom. Atoms aren't hard balls, and they aren't blobs of doughy plum-pudding cake. They are a whole lot more interesting than anyone has suspected. **In Rutherford's model each atom has a nucleus orbited by fast-whizzing electrons.** When the atom of a radioactive element decays, its nucleus may spit out a gamma ray—a high-energy photon—which means it drops in energy. Or it may spit out an alpha particle (a helium nucleus) or a beta particle (an electron), which changes the

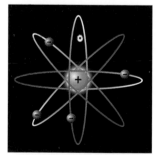

Rutherford's now-outdated "solar system" model of an atom began with a single, dense nucleus that is positively charged (see below) and later expanded to include newly discovered neutrons along with protons (see chapter 15, page 129).

Why the Alchemists Never Had a Chance

This 1771 painting by Joseph Wright is titled *The Alchymist in Search of the Philosopher's Stone.*

The ancient and medieval alchemists were doomed to fail. They were trying to change elements—lead into gold, for example—by chemical means. Although they didn't know it, they were rearranging electrons. (That's what happens in chemical changes.) You can't transform one element into another element that way.

Nuclear scientists know that protons will do it. Each element is set apart by the number of protons in its nuclei. Change the number of protons, and you change (transmute) one element into another. Sometimes that happens naturally.

A few elements are radioactive—over time they change themselves into other elements. How does that happen? Some radioactive elements eject alpha particles, which are helium nuclei (two protons and two neutrons apiece). Getting rid of an alpha particle leaves the parent element with two fewer protons in its nucleus, and that makes it a different element.

In beta decay (another form of radioactivity), a neutron turns into a proton, emitting an electron—the beta particle—along with another particle now called an electron antineutrino. In this case, the number of protons increases by one, which changes the atom into a new element.

Neutrons weren't identified until 1932, but in 1903, Rutherford and Soddy realized that radioactivity releases energy and that energy is "maybe a million times as great as the energy of any [chemical] change." Rutherford joked that if that nuclear energy could be unleashed, "some fool in a laboratory might blow up the universe unawares." Soddy said, "The man who put his hand on the lever... of this store of energy would possess a weapon by which he could destroy the earth if he chose."

A nuclear physicist is a modern "alchemist," able to change one element into another. Here, a researcher at the Gran Sasso Laboratory in Italy is studying the decay of tellurium-130 into xenon-130.

atom from one element to another. Stable elements don't spit.

Figuring all this out isn't easy.

Later, Rutherford will write about the early struggle to understand the atom: "While the vaguest ideas were held as to the possible structure of atoms, there was a general belief among the more philosophically minded that the atoms could not be regarded as simple, unconnected units. For the clarifying of these somewhat vague ideas, the proof in 1897 of the independent existence of the electron as a mobile electrified unit of mass, minute compared with that of the lightest atom, was of extraordinary importance."

Rutherford is saying that J. J. Thomson's discovery of the electron got atomic science moving. Then, the Curies' discovery that radioactive atoms are unstable—that they change—led Rutherford and Soddy to delve further. "Our whole conception of the atom was revolutionized by the study of radioactivity," says Rutherford. His discovery of the nucleus gives the atom a new identity. But there is still much to learn. Rutherford is about to encourage one of his students to study the hydrogen atom.

"There came into the room a slight-looking boy," Rutherford's biographer, Arthur Stewart Eve, will write about the first time

A colleague of Rutherford described him as "a force of nature." His scientific achievements earned him a knighthood (making him a British lord and giving him the right to officially put "Sir" in front of his name). Like J. J. Thomson, Ernest Rutherford, too, won a Nobel Prize.

A Poet Finds a Metaphor

English playwright Tom Stoppard, in his drama *Hapgood*, compares a hydrogen atom to the whole of St. Paul's Cathedral in London (right):

If your fist is as big as the nucleus of one atom, then the atom is as big as St. Paul's, and if it happens to be a hydrogen atom, then it has a single electron flitting about like a moth in the empty cathedral, now by the dome, now by the altar.... Every atom is a cathedral.

Tom Stoppard has used a fine metaphor, but keep in mind: Your fist (representing a nucleus) would weigh several thousand times more than the cathedral (representing the rest of the atom)!

We now know the nucleus of a hydrogen atom takes up only about $\frac{1}{100,000}$ of its diameter. Since almost all the atom's mass is in that nucleus, the nucleus is incredibly dense: A pea made out of nothing but atomic nuclei would weigh about 133,000,000 tons! No wonder alpha particles bounced off the atomic nuclei in Rutherford's gold foil.

Niels Bohr (right) poses next to his inseparable brother, Harald, in 1902. "[Bohr] saw life as a whole and was immune to the scholarly delusion that brain power is superior to muscle power," said science writer Timothy Ferris.

In a letter to his future wife, Niels Bohr wrote, "Now I am going to read a little more about electrons, and afterwards if there is time I shall read a little of *David Copperfield*." He especially loved this line from the Dickens book: "I think I simply told, gravely, what I had to tell." Bohr read the novel for two reasons: It's a great story, but it also helped him learn English. The 1924 lithograph (above) is *David Copperfield and Little Emily*.

Rutherford sees Niels Bohr (pronounced "neels bore"). But "slight" conveys the wrong image. Bohr is an athlete. It is his manner that may have seemed hesitant or tentative.

Bohr has come to England from Copenhagen, Denmark, where he and his brother have been national heroes on the soccer field. Bohr skis, bikes, sails, and is unbeatable at Ping-Pong. When he climbs stairs, it is almost always two at a time. He has a mop of dark, unruly, combed-back hair, a huge head, big hands, a long face, and a prominent jaw.

Bohr struggles with dyslexia—reading and writing are difficult for him—but he doesn't let that stop him. He is fascinated by physics and awed by J. J. Thomson's discovery of the electron. So Bohr applies to the Cavendish Lab at Cambridge University, where Thomson is in charge. When he gets there, one of the first things the 25-year-old graduate student does is to try to point out to Thomson that his plum pudding model of the atom doesn't work. Bohr's English isn't very good—he doesn't always make himself clear—and on top of that he talks softly, you have to strain to hear him, and he is very intense. Thomson begins avoiding him.

Then big, friendly Ernest Rutherford appears. He and Bohr connect. Bohr is asking questions that intrigue Rutherford; he invites Bohr to join him at the University of Manchester to see if they can work out a still better model of the atom. Because the electron's path is labeled an orbit, atoms are being described as miniature solar systems. It is a picture that is not accurate. Unlike the planets, electrons can change their orbits. (Imagine Venus suddenly leaping into Earth's orbit.) Rutherford cautions, "The atom has often been likened to a solar system where the sun corresponds to the nucleus and the planets to the electrons.... The analogy, however, must not be pressed too far." (What Rutherford doesn't know is that electrons will later come to be seen as probability waves that fill the space inside an atom like resonant sound waves fill an organ pipe. Even though the orbiting particle image isn't quite right, in its time, the solar system model is useful for certain predictions, and it's a big

Hunting Atoms

When Rutherford returned to the Cavendish Laboratory to take over as J. J. Thomson's successor, the physicists there wrote and sang a song to him (to a Gilbert and Sullivan tune). Here is part of it:

What's in an atom,
The innermost substratum?
That's the problem he is working at today.
He lately did discover
How to shoot them down like plover,
And the poor things can't get away.
He uses as munitions
On his hunting expeditions
Alpha particles which out of radium spring.
It's really most surprising,
And it needed some devising,
How to shoot down an atom on the wing.

Chorus:
He's the successor
Of his great predecessor
And their wondrous deeds can never be ignored:
Since they're birds of a feather,
We link them together,
J. J. and Rutherford.

Rutherford's nickname at Cambridge was "The Crocodile," which his colleague Peter Kapitza immortalized by carving this image on a laboratory wall.

step ahead of the plum pudding model.)

According to the soon-to-be-developed Bohr model, electrons travel around the nucleus in one or another orbit determined by quantum rules. (Bohr pays attention to Planck's formula.) When left alone, the electron stays in the lowest-energy orbit. When disturbed, it can jump to a higher-energy orbit. It's a bit like hopping up or down the steps of a ladder. You can't land between steps; electrons are never found between orbits.

When they jump, electrons do something that can't happen in the macro world. They leap from one orbit to another without crossing over the space in between. They just blink off in one orbit and on in another. (Some people call this a "quantum leap" or an example of "quantum weirdness.") What are the rules that guide the quantum world? Niels Bohr intends to find out.

Today the phrase **QUANTUM LEAP** is often used non-scientifically. To take a quantum leap is to try something new, something way beyond current practice.

Getting Atom

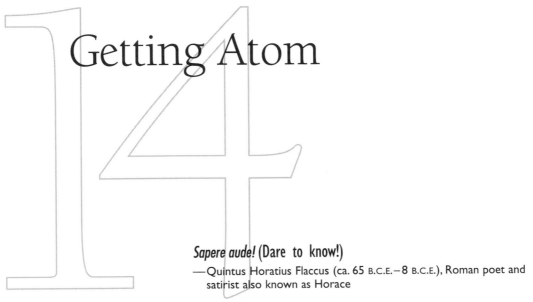

Niels Bohr in 1913 devised a theory that was spectacularly successful....He forced a connection between the Einstein relation $E = hf$ for photons and the Newtonian picture of electrons in orbits around the atomic nucleus.
—A. P. French and Edwin F. Taylor, *An Introduction to Quantum Physics*

The opposite of a correct statement is a false statement. But the opposite of a profound truth may well be another profound truth.
—Niels Bohr (1885–1962), Danish physicist

R utherford and Bohr turn out to be a terrific team. Thoughtful, intense Niels Bohr is the better theoretician (idea person). High-spirited, energetic Ernest Rutherford is the better experimenter. "Science walks forward on two feet, namely theory and experiment," says American scientist Robert Millikan, and no one disagrees.

Bohr is trying to figure out what is keeping atoms stable. Why don't atoms collapse when they give off energy? If they follow the rules of classical (non-quantum) physics, they should collapse. According to those classical laws, orbiting electrons must spiral closer and closer to the nucleus as they radiate energy until they hit the nucleus and the atom is destroyed. But that doesn't happen. Something unknown

Electrons are all the same whether they are in an electric wire or an atom. (Electric current is like a river of free-flowing electrons.)

Bohr and Rutherford Have a Problem

Galileo's law of *inertia* says anything in orbit (like the Moon) must keep accelerating, or it will fly off in a straight line. Maxwell's laws of electromagnetism say accelerating charges must radiate energy. That's what is happening when a broadcasting antenna emits signals for TV, radio, or cell phones; free-flowing electrons are radiating energy—just as expected. But electrons *inside* atoms *don't follow either of those classical rules of physics.*

Imagine those electrons inside atoms radiating energy in a classical manner. As they lose energy they move into orbits of lesser radius and greater speed. The orbits get faster and faster and the radii get smaller and smaller until the electrons smash into the nucleus. If the quantum world worked that way, what would happen when electrons crashed into nuclei? In a very small part of a second—good-bye us and the universe as we know it.

is going on. Bohr has an idea that he may have to enter a different world to understand. That different world is newly discovered Quantumland.

Many physicists haven't yet accepted the quantum concept that there are particles smaller than atoms and that they follow their own rules. According to those rules, most things in Quantumland come in tiny lumps (even though they may seem continuous to us). Max Planck discovered the size of the smallest lump. Einstein applied that discovery to light.

Bohr takes the quantum idea seriously. He suspects that there is a tie between the energy in Einstein's tiny light quanta and the energies of electron orbits inside the atom. Is that a coincidence, or can these amounts be related?

Bohr is working on this in 1912 when he goes back to Copenhagen to get married. He and his fiancée, Margrethe, have planned a long honeymoon in Norway. But now Niels is deep into experiments and calculations. What should he do? Bohr manages to convince his new wife to cancel the honeymoon. They head for Manchester, England, where he gets back to the problem of the atom. (It will be a long and happy marriage. Margrethe Bohr often helps her husband by reading to him and writing for him.)

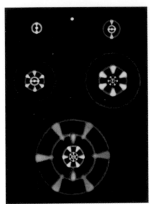

Do you recognize these atoms by their (artistically interpreted) electron clouds? The blip at the top is easy—that's hydrogen, with one electron. Carbon and silicon make up the second row, iron and silver the third. And the big, colorful atomic posy? The artist chose to showcase europium, a rare earth element.

An electron that falls into a lower-energy-level shell gives off a photon. Where does the photon come from? It is created on the spot! Surprising? Only to us outside Quantumland.

In Rutherford's model, the atom has a positively charged nucleus surrounded by tiny, fast-moving, negatively charged electrons that circle the nucleus like planets orbiting a star. Bohr and Rutherford both know there is something wrong about that solar system model. It's the action of the electrons that is bothersome.

Bohr has a hunch that **electrons don't all orbit a nucleus at the same energy level**—and he has the skills to find out for sure. It takes mathematics—mostly algebra—along with an understanding of electric charges, of Newton's laws of motion, of Maxwell's equations, and of Einstein's formula for photons. Equally important is Bohr's ability to visualize, to imagine himself inside an atom.

Bohr figures out—this is a breakthrough discovery—that the **electrons in an atom stay inside three-dimensional corridors**; he calls them **shells**. But don't picture hard shells. Shell is just a name for the layer where electrons are found. There, until it is interfered with, an electron is stretched out and fuzzy, like a cloud.

Some shells demand high-energy electrons; some take lower-energy electrons. What is puzzling is that electrons seem to jump back and forth between shells.

Bohr realizes that, just as a car needs fuel, an electron needs to absorb energy in order to jump to a shell where the energy is greater. When it drops into a shell where the energy level is less, it must get rid of that extra energy.

An electron doesn't lose energy gradually; rather, it does so suddenly, spitting out a quantized unit equal to the energy difference of the two levels between which it jumps. That difference in energy is emitted as a photon (a quantum of light).

The process works two ways. When a photon of the same energy, coming from outside, is absorbed by the atom, it can boost the electron from a lower to a higher energy level. So photons are the key to both kinds of leaps.

Einstein's formula for light quanta ($E = hf$) helps Bohr make sense of this. With it he can explain the dance of the electrons inside the atom (changing energy-level shells

Electrons in a Copper Atom

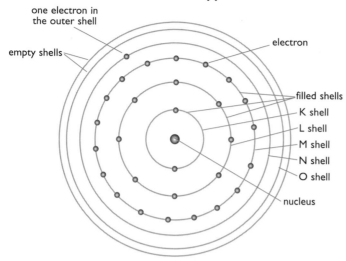

one electron in the outer shell

empty shells

electron

filled shells

K shell

L shell

M shell

N shell

O shell

nucleus

On this chart (not to scale) of a copper atom's electron configuration, the innermost shell (K) is stable—always filled with two electrons. L and M have no room at the inn, either, with a maximum occupancy of 8 and 18 electrons, respectively. The outer resident, a lone electron in the N shell, just might leap to a higher shell when excited by, say, an incoming photon.

rather than spiraling down into the nucleus). It helps him explain the energy that is radiating from atoms. And if electrons are confined to discrete (separate) shells, this explains why atoms don't collapse. Once he understands all that, Bohr can search for details.

Here are some:

When electrons leap from one shell to another, it's a clean leap. No in-between shells are allowed. The quantum world has fixed quantities; that's a basic quantum idea.

Each of the shells around the nucleus has room for only a certain number of electrons. If a shell is filled up, then no more electrons are allowed in. The shells don't all have the same capacity. The innermost shell holds 2 electrons, the next shell 8 electrons, the one after that 18, and the next 32.

Now here is another key piece of information: If an electron is in the innermost—or stable—shell, it *cannot* give off photons of energy; it can only absorb them. So it will *not* go further and fall into the nucleus and expire. **In its lowest stable state the electron does not give off energy.** An electron can stay in that state forever! (In the macro world that we know so well, nothing is forever. Death is a part of life. Not so in the quantum world, where some particles seem to be immortal.)

This is important: When an electron changes atomic orbits to emit a photon, it is the same atom, the same element; the electron has just rearranged itself around the nucleus. In contrast, when a radioactive nucleus changes by emitting an alpha particle (helium nucleus) or beta particle (electron), that atom does not remain stable. It changes its nature, becoming a different element. Atomic electron jumps and nuclear radioactive decays are different processes. Don't confuse them.

ANGULAR MOMENTUM
(*mvr*) is the momentum of
a body orbiting or rotating
in a circle. If one value is
altered (like the radius, *r*),
another value (like the
velocity, *v*) must change to
keep the value of the angular
momentum the same.
Regular momentum—the
momentum of a car racing
straight down a highway—
is *mv* (mass times velocity).

There's something else about Quantumland. It has to do with angular momentum. Imagine a weight at the end of a rope; now swing it in a circle around your head. The weight is *m* for mass, the circle has a radius of *r*, and the weight is twirling at a velocity, *v*. The angular momentum is *mvr* (mass times velocity times radius). In our everyday world, angular momentum can have *any* value (just change the weight, or the speed of the twirl, or the length of the rope). But Bohr figures out that in the quantum world angular momentum can only exist as a multiple of a basic unit.

That's the big thing—well, the little thing—about the quantum world. Things there come in lumps, but only lumps of certain sizes and angular momenta are allowed.

Bohr begins to understand that a new scientific rule book is needed to explain Quantumland; the rules of classical physics don't work there. He starts writing that book. He takes Einstein's formula (which includes Max Planck's constant) and applies it to the structure of the atom. Because of Bohr, we soon have a model of a mostly empty atom with fast-moving electrons in specific energy-level shells.

To summarize: Bohr discovers that electrons can and do change orbits. Each electron spins on its axis as it zooms in

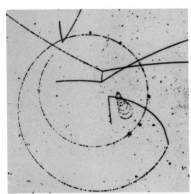

That neat, round spiral is the path of a low-energy electron. The lines and squiggles mark where subatomic particles met, decayed into something else, and veered off.

A computer model of Bohr's atom is mostly empty space with electrons (dark blue balls) orbiting a nucleus of protons and neutrons. Today, we don't use the word *orbit* to describe the motion of atoms. That turned out to be an inaccurate picture. (More on this to come.)

elliptical paths around a nucleus. (Electron spin will be discovered later.) These are not ordinary particles—an electron can stretch out, like a cloud. Given added energy—in the form of photons or collisions—electrons leap from one shell to a higher-energy shell. They can go back again, giving off photons. In an innermost or base shell, they are stable and cannot release photons.

In Bohr's picture, the electron jumps instantaneously from one energy shell to another without existing in an intermediate state. This is called a quantum leap. It defies common sense. It's bizarre.

In 1913, Bohr produces the mathematics to express all this numerically. He publishes three papers that describe the hydrogen atom; they catapult him into the first rank of physicists. "There is hardly any other paper in the literature of physics from which grew so many new theories and discoveries," writes Austrian physicist Victor F. Weisskopf.

Technology follows pure science. Spectroscopes—able to measure electron action by examining the light each element emits—begin to confirm theory. When light is confined by passing through a slit and is then split into different frequencies—meaning different colors—the colors form images of the slit that are like the lines in a bar code.

It turns out that each element has its own pattern of colored lines. Each color (which means each frequency) corresponds to the difference in energy level of the two energy levels that the electron jumps between. The pattern of colors identifies the emitting atom by tracking Einstein's photons ($E = hf$)! (See "In the Elemental Grocery Store," on page 126.)

About the same time Bohr published his papers on the hydrogen atom, English writer H. G. Wells (1866–1946) was working on his visionary science-fiction novel *The World Set Free* (1914), which influenced a generation of thinkers. Wells imagined a great war fought by the major powers in which the world's leading cities are destroyed by atomic bombs. Science caught up with fiction within Wells's lifetime, with the 1945 bombing of two Japanese cities. That's a test bomb, above.

A spectroscope (ca. 1900s) reveals that the element in the cylinder is sodium, based on its spectrum (upper right).

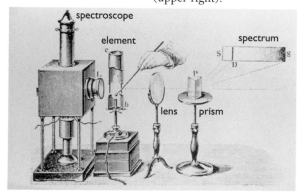

spectroscope
element
spectrum
lens   prism

Soon, astronomers are using that information to read the light coming from distant stars, which tells them the specific elements in a star. (They are startled to find that all stars are not alike; some contain different elements from others.) Astronomy is stretched. So is chemistry.

Add a prism to a telescope, and all the little dots of light in the night sky reveal their true colors. Look closely at these rainbow profiles, called spectra, of stars in the Hyades Cluster. Each one is unique. Alan MacRobert explains in *Sky & Telescope*, "The spectral code tells at a glance just what kind of object the star really is: its color, size, and luminosity compared to the Sun and stars of all other types; its peculiarities, its history, and its future."

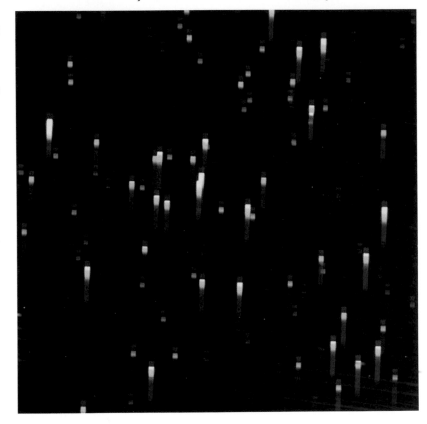

What has knowledge of electrons done for us? Understanding the way those negatively charged entities circle the nucleus of an atom revolutionized the chemical industry— it gave us the ability to make detergents, plastics, and synthetics. Knowing electronics allowed us to design microwaves, computers, lasers, fax machines, and all the other electronic devices that have transformed our homes and offices and have made us information-age people.

Why are there no intermediate atoms between elements? For example, why is there nothing halfway between gold and mercury, which are adjacent elements in the periodic table? Bohr's model of the atom doesn't explain why, but it does describe the atom as no one has done before.

Bohr makes it clear that quantum rules determine the number of electrons in an atom's orbital shells. And that explains the periodic table of elements. It now makes sense. Elements behave the way they do chemically because of the arrangement of electrons in their outer shells. Shared electrons can lead to interesting molecules, including a group of carbon molecules that help make life possible. It's

Solid State?

Einstein and Bohr were both helped in their thinking by *hf*, which is Planck's constant times the frequency of the corresponding classical wave. Einstein used it in 1907 to calculate the amount of energy it takes to raise a substance's temperature and came up with a formula for energy (*E*) that would become the basis of condensed matter physics: *E = hf*.

Condensed matter? It used to be known as "solid state."

Take quantum theory and apply it to solids, like transistors and silicon chips that are made of one piece of material, and you can get some amazing electronic devices. They are small and fast compared to older devices that joined components by wires or on circuit boards.

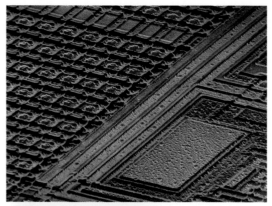

You can store a short novel (about 1 megabyte) on a computer chip (above) that's smaller than an ant (left)!

the electrons that control chemical reactions. Bohr has linked physics and chemistry and biology.

Einstein, at first, is skeptical. Then a fellow scientist, George de Hevesy, explains the details to him, and "Einstein's big eyes appeared even bigger, and he said, 'Then this is one of the greatest discoveries. . . . Bohr's theory must be right.'" Later, Einstein adds that "this is the highest form of musicality in the sphere of thought."

Niels Bohr has found his way through a scientific tunnel into a tiny world where electrons zoom, jump, spit out energy, do not die, and give off information. Think of Columbus's 1492 voyage when you consider Bohr's discovery. He has found a new world. Soon a bevy of scientists begins serious exploration. Some focus on quantum physics, some on nuclear physics, others begin an electronic revolution. In the minute arena of the atom, there are vast territories to explore. Bohr's achievement is dazzling. He will be at the center of quantum physics.

As for Albert Einstein, he and Bohr become good friends.

Bohr knew that he couldn't think about physics the way he was trained to do it: the Isaac Newton way (the physics of our large-scale world). He realized that other rules are at work in the atom.

Atoms Go from Weight to Number

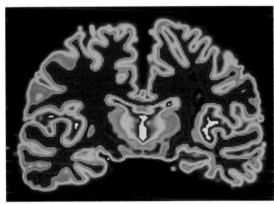

What's in a human brain? This computer image shows a slice of white matter (falsely colored deep pink) and gray matter (yellow). But all matter, including your brain, is made of the elements of the periodic table (opposite page). Your thoughts, feelings, opinions, memories, and sensations boil down to these atoms, from most to least, by mass: oxygen (O), carbon (C), hydrogen (H), nitrogen (N), calcium (Ca), phosphorus (P), potassium (K), sulfur (S), sodium (Na), and traces of others.

Bohr's picture of the atom turned out to be very important to chemistry as well as physics. Chemistry, which is concerned with the composition of matter, begins with elements—those basic substances from which everything else is made. "We stand on the elements, we eat the elements, we are the elements. Because our brains are made up of elements, even our opinions are, in a sense, properties of the elements," writes English chemist P. W. Atkins (in a fine small book called *The Periodic Kingdom*).

By the nineteenth century, chemists had become sophisticated enough to identify a passel of elements. Then Dmitri I. Mendeleyev (1834–1907) came along and found an intriguing order in nature's basic elements: Elements line themselves up by chemical properties and atomic weight. That observation let Mendeleyev (men-duh-LAY-ef) make the first periodic table of elements. (For details, see chapter 29 of book two in this series, *Newton at the Center*.) He used pure chemistry to find the patterns—measuring and weighing

The periodic table of elements (below) isn't finished yet. In 2006, a Russian/American collaboration announced that they had produced three atoms of a new element, number 118, by sending a beam of calcium ions into californium atoms. The atomic numbers of calcium (20) and californium (98) add up to—guess what?—118. The element 118 atoms lasted 1 millisecond before decaying into element 116, then 114, and then 112. Then they each fissioned (split) into two daughter particles. How did the scientists know they had produced a new element? Very heavy elements decay quickly, releasing alpha particles in a unique sequence that can be detected.

each of the elements and observing and listing their properties.

Niels Bohr, being a physicist, looked at elements differently. He considered the structure of their atoms. Bohr found that the number of electrons in an atom determines that atom's potential for combining with other atoms. When he counted electrons, he got almost the same order of elements as Mendeleyev did using weight. (Electrons hadn't been discovered when Mendeleyev was alive.) Bohr's electron count confirmed Mendeleyev's periodic table of elements!

Mendeleyev had taken an outside look; Bohr's was an inside job. Their results were almost identical. Atomic number (the number of protons, balanced by an equal number of electrons) is the method scientists use to identify elements today. (Atomic weight, used by Mendeleyev, can be misleading because the number of neutrons may vary. Mendeleyev didn't know that.)

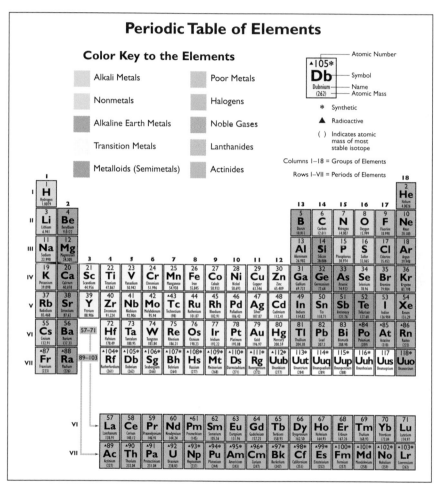

Otto Dix (1891–1969), a young German portrait painter, eagerly volunteered to fight in World War I. What he experienced turned him into an angry antiwar activist who was especially cynical about the way Germany treated its wounded veterans. Dix was part of the *Neue Sachlichkeit* (New Objectivity) movement, made up of artists whose startling works questioned the values of those with political power. In this painting, legless veterans march in front of a *Schuhmacherei*, a shoemaker's store!

But, for 30 years, they will battle over ideas—mostly about the reality of photons and about the place of probability in quantum physics.

Einstein and Bohr meet for the first time in Berlin, in 1920. Germany is ravaged by inflation, and people are going hungry. Bohr arrives from calm, productive Denmark bringing a package filled with butter, cheeses, ham, and sweets. They hit it off at once and walk across the city, talking intensely. Neither cares about appearances or ceremony; physics is their shared passion (so is sailing). Both are theorists rather than experimenters, and both are great talkers. Einstein is 5 feet 9 inches tall, better than average height for his time; Bohr is just a bit shorter, six years younger, with the powerful body of an athlete.

Bohr is a pragmatist: For him, what works is what is important. For Einstein, that it works is not enough. He has to understand, and the result has to be beautiful to him. Bohr's quantum leap is an example of their difference. Einstein feels there must be an explanation for it. Bohr thinks it is unexplainable. What's crucial to him is that he can predict it statistically, use that information, and go from there.

For Einstein, an equation that doesn't exist in physical terms has no meaning. He "sees" light waves that are made up of a flow of tiny grains (photons) with space between them. For him, they are actual grains of energy that also behave as waves.

Bohr doesn't quite agree. It is one thing to think of matter in terms of definite, tangible particles, but energy? Bohr doesn't believe those massless energy grains actually exist; he just knows that if you pretend they exist, the mathematical formulas work. (See chapters 16 and 17 to find out where this argument goes.)

Bohr and Einstein have an even bigger disagreement, and most of the scientific world comes down on one side or the other. Bohr thinks quantum theory stands on its own.

He isn't concerned about fitting that theory into a grand universal scheme that unifies the everyday macro world and the quantum world. Actually, he feels it can't be done.

Einstein is convinced that quantum theory is not the whole picture, although he says it "undoubtedly contains part of the ultimate truth." He doesn't like its quirky quality—that the action of an individual particle can't be predicted. In Bohr's quantum world there's an element of probability that Einstein can't accept. Einstein is looking for a way to put quantum theory into an overarching explanation of the cosmos. He wants to bring that theory of the small into harmony with theories of time and space and the greater universe. He is searching for a **theory of everything**.

This clash of ideas is between men who respect and admire each other. American physicist John Wheeler describes it as "the greatest debate in intellectual history that I know about.... I never heard of a debate between two greater men over a longer period of time on a deeper issue with deeper consequences for understanding this strange world of ours."

And so Bohr and Einstein talk and argue and talk and argue, each trying to convince the other. It never happens, but their disagreements seed the scientific world, which leads to some surprising sprouts.

Mix and Match, and You Get New Worlds

Everything in our complex universe comes from a relatively few elements (92 natural ones), which get arranged and rearranged. Some of the very light elements (hydrogen, helium, and lithium) were created in the hot cosmic soup that existed when the universe was first born. But most elements are cooked in stars through nuclear fusion. Then they get spewed into space. Huge, hot star explosions known as supernovas cook heavy elements like gold and silver. Elements are the raw materials of solar systems and—eventually—of you.

In addition to the 92 natural elements, many others have been created in particle accelerators. These heavy elements (with a high number of protons and electrons) are very unstable and decay quickly into lighter elements.

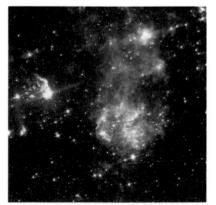

The remnant of a supernova explosion contains bits and pieces of a star.

In the Elemental Grocery Store

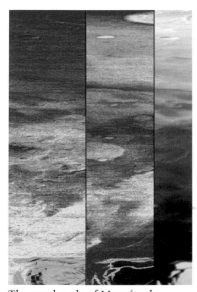

The south pole of Mars (at the bottom of this triple image) is capped with ice; that's plain to see. In visible light (right panel), the ice cap appears turquoise and was thought to be made of carbon dioxide (CO_2). And it is. In 2004, a spectrometer on the European Space Agency's Mars Express measured the reflected light of the polar region at infrared frequencies and found the spectral "fingerprint" of carbon. The large CO_2 ice cap is colored light blue in the left panel and purple in the center panel. But notice the fringes! The areas that are dark blue (left) and bright green (center) are water ice—H_2O! It's mixed with the soil as permafrost and so was hidden from ordinary view.

In the nineteenth century, scientists discovered that elements give off EM energy at specific frequencies. The atoms of each element create a pattern of lines, spaces, and colors that is unique and can be used as a "fingerprint" to identify that element (see illustration at right). A single element's atoms emit only a few colors, not the whole rainbow of light. Those same atoms absorb light of specific frequencies, not light of all frequencies.

The emitted pattern looks a bit like the bar codes that identify grocery items. An element's spectral lines can be figured out mathematically. Niels Bohr made the connection between the EM spectrum and the spectrum identifying elements and put it all into equations for others to follow.

Using Planck's formula for action, Bohr figured out that if an electron switches to a different-energy-level shell, it must get energy from a photon (to go into a higher-energy shell) or it must expel a photon (to go into a lower-energy shell). Bohr realized that the

Atoms give off only a few colors, not the whole rainbow of light. And atoms absorb light of specific frequencies, not all light. The result is a unique spectral "fingerprint" for each type of atom—that is, for each element on the periodic table. The visible light portion of the absorption spectrum of hydrogen, shown in the center image below, mirrors its emission spectrum (bottom image). The pattern of lines never changes, no matter what the temperature of the hydrogen or whether it's on Earth or in the stars.

The Spectral "Fingerprint" of Hydrogen

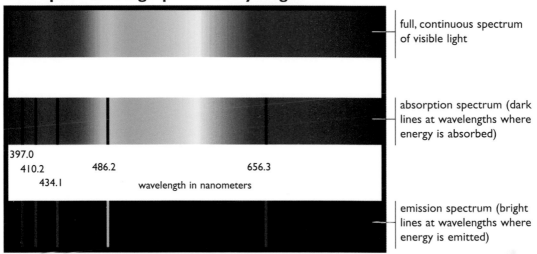

full, continuous spectrum of visible light

absorption spectrum (dark lines at wavelengths where energy is absorbed)

397.0
410.2
434.1
486.2
656.3
wavelength in nanometers

emission spectrum (bright lines at wavelengths where energy is emitted)

spectral lines—the bar codes—that elements give off are a record of their electrons absorbing or expelling photons. Those spectral lines are proof that electrons jump from one orbit to another.

The same effect is at work outside the visible spectrum. Radio wavelengths record an element's EM peaks and valleys beyond visible light. Radio waves reveal atoms and molecules in clouds of dust and gas in space that are invisible to the eye. The study of spectral lines is called spectroscopy. In the twentieth century, spectroscopy, improved photography, and new kinds of telescopes were used together. Astrophysics soared.

Still Shooting Alpha Particles

In 1916, while much of Europe is at war, Niels Bohr is wooed home to Denmark. The Royal Danish Academy of Sciences and Letters agrees to build a physics institute in Copenhagen and to give grant money for fellowships to young physicists. The Carlsberg Brewery will eventually provide most of the funds. (Beer and physics? Strange? Perhaps, but no stranger than Alfred Nobel using his fortune from inventing dynamite to create his famous peace prize.)

The Institute for Theoretical Physics (*Institut for Teoretisk Fysik* in Danish) is built on Blegdamsvej (BLEYE-dams-veye) Street in Copenhagen. In the 1920s and early 1930s, it becomes the world's center for quantum research and the place to be if you want to deal with atoms. (It's still there if you'd like to visit, now called the Niels Bohr Institute.)

In this 1921 building, Niels Bohr and his colleagues changed theoretical physics. Today, it's part of the Niels Bohr Institute, 10 buildings connected by tunnels at the University of Copenhagen.

Scores of scientists come to the institute to talk, listen, collaborate, and accomplish things. "Bohr could not think creatively without human company," writes William Cropper in a chapter in *Great Physicists* called "Science by Conversation."

Meanwhile, Ernest Rutherford, who remains a lifetime friend of Niels Bohr, is in England, still firing fast-moving alpha particles at things. In 1919, when J. J. Thomson retires as head of the now-famous Cavendish Laboratory at the University of Cambridge, Rutherford takes over. It is there that he shoots some of those alpha particles down a glass tube filled with *nitrogen* gas. He doesn't think much will happen. Something does.

Hydrogen nuclei appear. That is amazing. Where have they come from? There was no hydrogen in the tube, only nitrogen. Nitrogen has an atomic number of 7 (seven protons orbited by seven electrons). Hydrogen has an atomic number of 1 (one proton, one electron). Rutherford realizes he has knocked nitrogen nuclei apart and gotten hydrogen nuclei (and other nuclear fragments)!

This is a big story. Newspapers trumpet the news: Ernest Rutherford has "split the atom." Rutherford, not given to modesty, says, "If, as I have reason to believe, I have disintegrated the nucleus of the atom, this is of greater significance than the war." (He's talking about World War I.)

Rutherford figures out that **the hydrogen nucleus is a basic particle** found in *every* atom's nucleus. He has discovered the proton. (The most common hydrogen nucleus is a proton—more on this later.)

He soon realizes that the number of protons in the nucleus is an exact match to the number of electrons orbiting the nucleus (in a neutrally charged atom). Finding protons is like finding another missing piece—a key piece— in a jigsaw puzzle.

You already know (from chapters 8 and 13) that **ALPHA PARTICLES** (α) are helium nuclei: two protons and two neutrons. They're held together tightly by the glue of the strong nuclear force (more to come on this) and act like a single particle.

Until transistors came along in 1947, electronic devices used an evacuated (airless) glass tube, like the 1926 model advertised at left. Inside this vacuum tube, electrons are boiled off a hot wire (like the filament of an old-fashioned lightbulb) and flow under the control of charged grids and electrodes. The tube amplifies electric current or voltage.

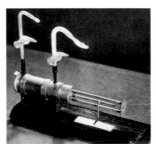

Here's Rutherford's first apparatus for shooting alpha particles at nitrogen nuclei, causing them to "disintegrate" (his word) into hydrogen artificially.

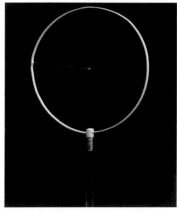

In 1919, newspapers reported that Rutherford had "split the atom." (Actually, he had split the nucleus of the atom.) What made this big news? Marie and Pierre Curie had worked on radioactive— or unstable—nuclei that decay and transmute into other elements naturally. Rutherford's bombardment broke apart the *stable* nuclei of nitrogen into hydrogen nuclei. Above is the model of a hydrogen atom that emerged after the experiment: a nucleus of one proton orbited by a single electron.

The atomic scientists have been looking for answers to two questions:

Why are atoms electrically neutral when electrons have a negative charge?

And what makes the tiny nucleus so massive (heavy)?

Protons answer both questions. **Protons carry a positive electrical charge that exactly balances the negative electrical charge of electrons.**

And a single proton weighs almost as much as 2,000 electrons, which is why most of the mass of the atom is found in the nucleus. (The mass of the proton is actually 1,836 times that of the electron.)

Is that all there is in an atom—electrons circling a nucleus of protons? Rutherford isn't sure. But his experiment has opened a way for the nucleus to be studied.

In France, Irène Joliot-Curie (Marie and Pierre's daughter) and her husband, Frédéric, find something else in the nucleus—something that has no electrical charge at all, but when it bumps into one of the massive protons, the proton goes flying at thousands of miles per second. That seems incredible. Rutherford's colleague James Chadwick sets to work to figure out what this heavyweight is.

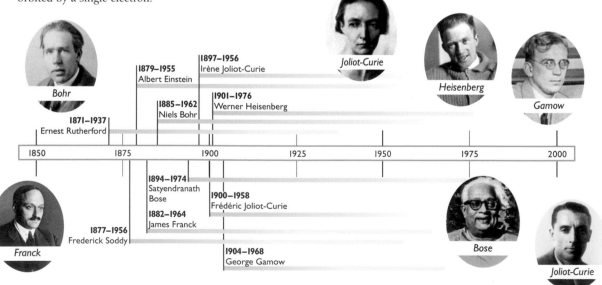

Protons First

Rutherford's favorite experimental tool was the alpha particle (α), a helium atom with the two electrons stripped off, which he called his "strong right arm." Alpha particles are fast, powerful, and can tear apart other particles. They led Rutherford to discover the nucleus, so he kept using them.

"Rutherford's α particles were very energetic—they carried the energy they would have had if they were accelerated through millions of volts—and on an atomic scale they were massive," says William H. Cropper, in an excellent book titled *Great Physicists*.

By 1914, Rutherford had figured out that the hydrogen nucleus is the smallest of all positively charged particles. Later, he called the hydrogen nucleus a *proton*, from the Greek word for "first."

In 1919, Rutherford was shooting alpha particles into *nitrogen* gas when his detectors showed *hydrogen* nuclei (protons) were present. He realized that "hydrogen which is liberated formed a constituent part of the nitrogen nucleus." In other words, there are hydrogen nuclei—protons—in each nitrogen nucleus. It was a short step from there to figure out that those **protons are in every atom's nucleus**.

The most common hydrogen nucleus is just one proton; hydrogen has an atomic number of 1 and is first on the periodic table of the elements (see page 123). Helium, number 2, has two protons. Lithium, number 3, has three protons. Each time you add a proton, you get a new element. Uranium is number 92 and has 92 positive protons and 92 negative electrons.

Do you want to picture a proton? It isn't easy. Writer Bill Bryson says, "Protons are so small that a little dib of ink like the dot on this i can hold something in the region of 500,000,000,000 of them, rather more than the number of seconds contained in half a million years."

Today we know that protons are made of three quarks held tightly by particles called gluons and that protons are kept in the nucleus by the strong nuclear force (more on quarks and gluons coming up).

Rutherford and Chadwick began by bombarding nitrogen with alpha particles, whose tracks are shown in this 1940 photo, but then they probed boron, sodium, and phosphorus. These atoms all have lightweight nuclei that are easily blown apart.

When Rutherford and Chadwick tried to split the nuclei of heavy atoms, they found the nuclei couldn't be penetrated. The solution seemed clear: Alpha particles would have to be accelerated to high velocities. Particle accelerators—which are really big-time peashooters—were needed. At left is one of the first such machines to do the job. It's a Van de Graaff generator built by its namesake, Robert J. Van de Graaff, in the 1930s. Today, you can watch the high-voltage devices produce spectacular displays of electric charge in science museums and classrooms.

The Joliot-Curies took this photo in 1932, shortly after Chadwick's discovery of the neutron. It proves that a neutron (unseen) entered the cloud chamber and hit a piece of wax (horizontal white line), knocking off a proton (vertical white track).

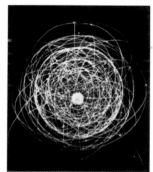

It's tough to visualize something you can't see, as evidenced by the variety of atom models in this book. Compare this concept of a uranium atom to Rutherford's description (text at right).

Chadwick does some experimenting and realizes that the Joliot-Curie duo has discovered another particle inside the nucleus. This one is slightly more massive than the proton. Because it is electrically neutral, Chadwick, in 1932, calls it a neutron. He finds that there are neutrons, along with protons, in each atom's nucleus—except in the number-one atom, hydrogen: The most common hydrogen atom has no neutrons (see box, page 134).

More pieces of the puzzle are fitting together. Three components of the atom are now known: Neutrons and protons are in the nucleus, and electrons are speeding around that nucleus. An atom is definitely *not* a plum pudding.

Every neutral atom—an atom with no net charge—has an equal number of electrons and protons. The elements found in nature range from hydrogen with one electron circling its nucleus—and an atomic number of 1—to uranium with 92—and an atomic number of 92.

Here is Rutherford's description of an atom of uranium:

At the center of the [uranium] atom is a minute nucleus surrounded by a swirling group of 92 electrons, all in motion in definite orbits, and occupying but by no means filling a volume very large compared with that of the nucleus.... [T]he electrons of one group may penetrate deeply into the region mainly occupied by another group.... The maximum speed of any electron depends on the closeness of the approach to the nucleus, but the outermost electron will have a minimum speed of more than 600 miles per second, while the innermost K electrons have an average speed of more than 90,000 miles per second, or half the speed of light.

Read Rutherford's words a few times. Then picture the action right now in every atom in your toes, your hair, and the clothes you wear. Think about this: You and your atoms are mostly composed of empty space. Particles inside your atoms are in constant motion. To give you an idea of their size, keep in mind: **It takes 4 billion atoms to cover the**

On the Air: It Is February 2, 1939

Columbia Broadcasting System (CBS) presents "Adventures in Science," a radio dialogue between Enrico Fermi, 1938 Nobel Prize–winner in physics, and Watson Davis, journalist.

Davis: Professor Fermi, when did modern atomic transmutation of elements really begin?

Fermi: The first and most important step toward its solution was made by the late Lord Rutherford. It was only in 1919 that he pioneered the method of nuclear bombardments. He showed that when the nucleus of a light element is struck by a fast-moving alpha particle, it is transformed into the nucleus of a different kind of atom.

Davis: Rutherford blasted the H out of matter, the H in this case standing for hydrogen.

Fermi: The new atomic nucleus.... formed by the bombardment, may be the same as one of the known stable nuclei. But this is not necessarily so. Sometimes the nucleus formed is different from all ordinary nuclei. It is not stable. It disintegrates spontaneously with emission of electrons. This phenomenon is called artificial radioactivity. Frédéric Joliot and Irène Curie of Paris discovered it in 1933. They produced their first three kinds of artificial radioactivity by bombarding boron, magnesium, and aluminum with alpha particles....

Enrico Fermi's home was Rome, which named a street after him. This photo was taken in the mid-1930s.

Immediately after these discoveries it occurred to me that alpha particles very likely did not represent the only type of bombarding projectiles that would produce artificial radioactivity. I decided to try bombardment with neutrons.

Neutron sources are much less intense than those of alpha particles. But the fact that they have no electric charge makes it easier for them to reach the nucleus of the bombarded substance. They do not have to overcome the repulsion of the electric field that surrounds the atomic nucleus.

Davis: What did you find, Professor Fermi?

Fermi: Beginning with the very first experiments, I proved that most of the elements tested—37 out of 63—became radioactive under the effect of neutron bombardment.

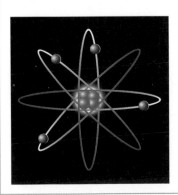
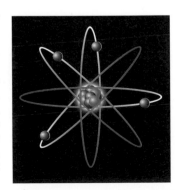

In 1932, the solar system model of atoms changed from a nucleus with protons (left) to one that included the newly discovered neutrons, too (right). With four protons (and the same number of neutrons and electrons), this atom represents atomic number 4, beryllium (Be). Fermi realized that he could bombard nuclei with neutrons to make elements artificially radioactive. (Nuclei are shown larger than correct scale.)

period at the end of this sentence. Now consider the astonishing structure of each of those minute atoms and further consider that human brains figured out what atoms

Isotopes? Almost the Same, but Not Quite!

In 1919, Francis Aston (J. J. Thomson's assistant) invented a spectrograph that sorted atoms by their mass. Surprise! He found that all the atoms of an element are not alike. Each atom of a given element has the same number of protons. (It's the protons that give an element its chemical identity.) Sometimes (especially for lighter elements), neutrons match the number of protons, but not always. Neutron variations of atoms are called isotopes. Isotopes are *not* rare; uranium has more than 200. All elements can have synthetic isotopes. Some isotopes are stable; some are radioactive. Think of isotopes as not-quite-the-same twins.

Chemists identify an isotope by writing its atomic mass (the total of neutrons and protons) as a small superscript before the element's symbol. Neon has 10 protons (its

atomic number) and 10 neutrons. So it is ^{20}Ne. An isotope of neon with 12 neutrons is ^{22}Ne. Carbon (atomic number 6) has a handful. A famous one (see box, next page) is carbon-14 (^{14}C), with six protons and eight neutrons.

In New Mexico, birthplace of the atomic bomb, there's a minor league baseball team called the Albuquerque Isotopes. Here are their nuclear counterparts, all the isotopes for the first 12 elements, from hydrogen (H) to magnesium (Mg). Hydrogen (bottom row) is the simplest element with one proton and one electron. It has three isotopes: The most common hydrogen atom has no neutron, stable deuterium has one neutron, and radioactive tritium has two neutrons. Helium (He) and lithium (Li) each have five aliases, beryllium (Be) has seven, boron (B) has six, and so on.

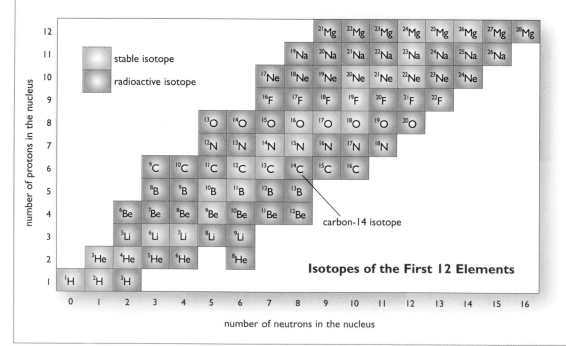

Isotopes of the First 12 Elements

carbon-14 isotope

Dating? It's Not Just for Girlfriends and Boyfriends

The air you breathe contains carbon dioxide—including a very small amount of a radioactive carbon isotope, carbon-14. Carbon-14 is continuously added to the upper atmosphere as cosmic rays interact with nitrogen molecules.

Plants, growing through photosynthesis, take carbon dioxide from the air and incorporate the carbon—including that small amount of carbon-14—into their tissue. When animals eat plants, they get their share of carbon-14. The carbon-14 in living plants and animals stays in balance with the carbon-14 in the atmosphere until the organic matter stops growing (the tree is chopped down or the rabbit expires). Then the carbon dioxide isn't replaced. Meanwhile, the carbon-14 atoms already in the rabbit and the tree decay at a known rate called a half-life. We can't tell which carbon-14 isotopes will decay at a given time, but we do know—statistically—that half of them will decay every 5,730 years.

California physicist Willard Libby, working in the 1940s, figured out that if you measure the carbon-14 in a dead object (up to 50,000 years old) and compare it to the carbon-14 in living matter, you can tell when the organism died. Radiocarbon dating is very useful to archaeologists, geologists, and detectives, too. So, if you have any old skeletons hanging around, you can date them if you check their carbon-14 content.

This human thighbone was thought to be medieval. A carbon-14 test of a tiny sample is about to provide proof.

were like long before they could be seen with any kind of microscope.

Ernest Rutherford says with justified pride and pleasure, "We are living in the heroic age of physics!" His colleagues seem to agree. One of them is C. P. Snow, a physicist at the Cavendish Laboratory at the University of Cambridge. Snow, who will later become a novelist, writes, "Week after week I went away through the raw nights, with east winds howling from the fens down the old streets, full of a glow that I had seen and heard and been close to the leaders of the greatest movement in the world." Snow, Rutherford, and their fellow scientists realize that understanding the atom is one of the outstanding achievements of the human mind. But none of them knows how many heroics are still ahead.

16

Bohr Taking Quantum Leaps

Anyone who is not shocked by the quantum theory does not understand it.
—Niels Bohr (1885–1962), Danish physicist

In his forties, Bohr was a jovial father figure to scores of students, most in their twenties, from all over the world. He relaxed by watching cowboy films, but he always needed a couple of students to go with him to explain the complicated plots.
—Denis Brian, American journalist and biographer, *Einstein: A Life*

[Bohr's] theoretical mind showed even in these movie expeditions. He developed a theory to explain why although the villain always draws first, the hero is faster and manages to kill him....We disagreed with this theory, and...I went to a toy store and bought two guns in Western holsters. We shot it out with Bohr, he playing the hero, and he "killed" all the students.
—George Gamow (1904–1968), Russian physicist, *Thirty Years That Shook Physics*

Niels Bohr is a skinny graduate student struggling to learn English when Rutherford recognizes his genius. It isn't long before Bohr becomes a dynamic mentor to a generation of physicists. In the 1920s and 1930s, they flock to Copenhagen, Denmark, to work with him. That city becomes the center of a *charged* field, the field of quantum mechanics, attracting an international band of physicists.

To a scientist, mechanics is motion. **Quantum mechanics is the science that deals with the *motion* of electrons, photons, and other tiny particles in the subatomic world.** This is new territory for science; Isaac Newton's familiar laws of motion (that guide our everyday world) just don't work inside atoms. Niels Bohr is *the* great early explorer in the realm of atoms and molecules. He becomes a hero and a daddy figure in the field.

Bohr has an intensity that often wears out those around him. He thinks aloud—and talks and works and works and talks—until he figures out a problem. When he relaxes, it is with the same gusto he brings to physics. Once, while Bohr is walking along a Copenhagen street with a physicist who is a mountain climber, they both decide to climb up the outside of a building. It happens to be a bank. The police are called, but when they see who the "bank thief" is, they just shrug their shoulders; they are used to the professor and his escapades.

A reproduction of a mosaic by Danish artist Ejnar Nielsen (1872–1956) depicts Niels Bohr surrounded by four scientific disciples. The mosaic is part of a ceiling decoration in the Staerekassen (Royal Theater) in Copenhagen, a building filled with Art Deco ornamentation and provocative artworks.

Bohr and his team are doing things that, to the traditionalists, seem as disturbing as bank robbery. **The Copenhagen picture of the subatomic world will replace certainties with probabilities, which means the future is not exactly predictable from past events.** (Einstein started all this, but he never does agree that, at the fundamental level, probabilities are all we have to go on.)

Bohr and his young physicists are caught up in the exhilaration that comes with mind work that is going somewhere. Like detectives searching for clues, they explore the quantum world, but they also find time to ski, go to cowboy movies, sail, party, and play music. Quantum mechanics is a small enough field that everyone knows everyone else and information is shared. Europe is not the only place producing first-rate scientists; they come to Copenhagen from Japan, India, the United States, and

James Franck (1882–1964), a physics professor who won the Nobel Prize in 1925, left Germany in 1933 to protest the rising Nazi Party. As a naturalized U.S. citizen and professor at the University of Chicago, Franck worked on the atomic bomb and then headed a committee that urged that the bomb not be used.

elsewhere. To communicate, most speak German as well as the language of mathematics.

James Franck, who has the good luck to be included, says:

> *One could correctly describe Bohr's house as resembling a Greek academy. The conversations and discussions there were by no means limited to physics and natural science; they encompassed philosophy, history, fine arts, religious history, ethical questions, politics, current events and many other topics.... [Bohr] read a great deal, had a good memory, and thought about everything he read and experienced.*

The physicists' work isn't easy, and money is never abundant, which makes their achievements all the more remarkable.

"The Institute buzzed with young theoretical physicists and new ideas about atoms, atomic nuclei, and the quantum theory in general," writes George Gamow (GAM-off), remembering his sojourn there (from 1928 to 1929).

Gamow, a Russian student, gets a fellowship for a summer of study in Germany. On his way home he has just enough money to stop for two days in Copenhagen. He

The oil painting *Copenhagen by Gaslight*, by W. Behrens, captures a vibrant evening about to begin in the heart of the Danish capital, ca. 1920s. The large public square is called Rådhuspladsen, after the City Hall building (right) that dominates it.

Good Physicist, Poor Sailor

George Gamow was a one-of-a-kind maverick. He thought for himself, did what he believed was right, or fun, and wouldn't let anyone stand in his way. Born in Odessa, Russia (now in Ukraine), in 1904, Gamow was a scientific standout at an early age. After he made some discoveries in nuclear physics, a Russian newspaper proclaimed that Gamow "has shown the West that Russian soil can produce her own Platos and sharp-witted Newtons."

The Gamows aren't escaping in this photo. They're just out for a kayak joyride.

But Platos and Newtons don't blossom in an unfree atmosphere. A revolution in 1917 left Russia with a dictatorial Communist government. Freethinking was not allowed. Gamow wanted to escape. In 1932, he and his wife, Lyubov Vokhminzeva (also a physicist), decided to paddle across the Black Sea to Turkey. Gamow recalled: "We hardboiled [some eggs] and saved them for the trip. We also managed to get several bricks of hard cooking chocolate, and two bottles of brandy, which turned out to be very handy when we were wet and cold at sea." Being a mathematician, he quickly figured out "that it was rational to take turns in paddling, rather than paddling together, since in the latter case the speed of the boat did not increase by a factor of two." But, after 36 hours of paddling, the weather changed, pushing the kayak backward, and they had no choice but to turn around.

The Gamows tried again, this time across the Arctic Sea from Murmansk to Norway. Again it was an amateurish gamble that didn't work. Then, in 1933, Gamow was invited to the prestigious Solvay Conference in Brussels. (Ernest Rutherford, Niels Bohr, and Marie Curie were among those who attended that year.) He got government permission (not easy) to bring his wife along. They didn't go back home. The Gamows headed for the United States, where George did important work in physics and contributed to the theory of the Big Bang (coming in chapter 43). In his spare time, Gamow wrote some good books (see illustration) and seemed to have fun wherever he went.

His books have become classics. *Thirty Days That Shook Physics* is an insider's look at the quantum theory revolution. In *Mr. Tompkins Explores the Atom* (1945), the title character is a bank clerk who shrinks to the size of an electron and goes on adventures in the particle world. Here's an action-packed paragraph:

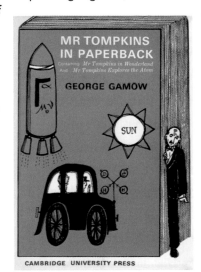

Mr. Tompkins the bank clerk invites readers to join him on a wild atomic ride.

"Hold on tight!" yelled one of his companions, "or you will be thrown out by photo-effect forces!" But it was already too late. Mr. Tompkins was snatched away from his companions and hurled into space at a terrifying speed, as neatly as if he had been seized by a pair of powerful fingers. Breathlessly he hurtled further and further through space, tearing past all kinds of different atoms so fast he could hardly distinguish the separate electrons. Suddenly a large atom loomed up right in front of him, and he knew that a collision was unavoidable.

Zeno's Paradoxes

This serious fellow (left) is Zeno of Ilea, a famous Greek who lived in southern Italy circa 495 to 430 B.C.E. and came up with paradoxes that drove some people crazy.

In the flying arrow paradox, Zeno proposed that time and space can be neither continuous nor come in discrete particles. Aristotle tried to prove him wrong, but nineteenth-century mathematicians Lewis Carroll and Bertrand Russell thought Zeno had raised serious issues—and Einstein almost certainly knew about his paradoxes.

Zeno's Arrow (1964), by the surrealist René Magritte (1898–1967), was inspired by a paradox that claims: At any given instant in time and point in space, a moving object is actually at rest. Therefore, motion is impossible.

photoelectricity. The momentous 1905 paper that he wrote on that subject described light as particles (quantum chunks of energy). That concept helped quantum theory take off.

Almost everyone expected Einstein to win the Nobel Prize for his papers on relativity, but that subject is still too controversial for the Nobel Committee. Einstein is asked not to talk about it in his acceptance lecture. But Sweden's King Gustav V, who is part of the audience of 2,000, has other ideas. The king sits in the front row and says he wants to learn about relativity. Einstein obliges. Still, it is quantum mechanics that is producing new technology and practical wonders. Einstein now seems to have been left behind by much of that.

After giving his Nobel Prize speech, Einstein takes a ferry to Copenhagen. It is his first trip to Denmark, and Bohr meets him at the dock; they board a tram heading for Bohr's home and are soon deep in discussion. What do they say? We don't know, but it is a good guess that they are arguing about those light quanta (soon to be called photons). Are they real or not?

Each time you turn on a light, you put billions of photons (light quanta) on the move. You're converting electrical energy into light energy. But you're not creating energy. Remember: The first law of thermodynamics says that *energy can neither be created nor destroyed.*

Bohr, who doesn't think they are, uses light quanta as mathematical abstractions; they work wonderfully well in equations. This disagreement may seem minor, but it is not to these two intellectual giants. Do photons really exist? "Yes," says Einstein. "It doesn't matter," says Bohr.

When they look up from their seats on the tram, they realize they have gone way beyond their stop. Sheepishly they get out, cross the tracks, take another train, and head the other way. But, still talking, they miss their stop again and don't discover it until they are almost back at their starting point. These two geniuses—more intelligent than almost anyone in the world—can't seem to focus on a train ride. Their intensity is a shared trait. But Bohr works best as a teacher in collaboration with others, while Einstein mostly stays inside his own head.

Physicist Paul Ehrenfest (1880–1933) snapped this candid photo of Bohr (left) and Einstein relaxing at Ehrenfest's home in Leiden, the Netherlands, in December 1925. "Relaxing," to these great minds, meant debating quantum theory.

They are on the same track—although they sometimes don't know it. The scientific train they are on is laying new lines; no one is sure where those lines will lead. The old Newtonian universe was based on Earth-centered absolutes. Every effect had a cause. Our pal Bohr and his intense young colleagues shatter that sureness when they bring Einstein's probability concept to quantum theory—even though Einstein hates the idea of probabilities on the fundamental level. The Copenhagen gang won't go along with Einstein on this (and they are right); they are introducing a whole new realm to scientific scrutiny. For his part, Einstein is changing the scientific focus from the Newtonian system that seems to work on Earth to one that works for the whole universe. It takes some getting used to.

As for Bohr and Einstein's dispute over light quanta? That argument has been settled by an experiment done in the United States, as they will soon learn.

17
An American Tracks Photons; a Frenchman Nails Matter

Are not the Rays of Light very small Bodies emitted from shining Substances?
—Isaac Newton (1642–1727), Query 29, in *Opticks*

Einstein's intuition and insights guided him to the view that, despite the fact that light is waves, light must also be particles: indeed Nature is preposterous!
—Roger Penrose, British physicist, "Newton, Quantum Theory and Reality"

We are...confronted with the dilemma of having before us a convincing evidence that radiation consists of waves, and at the same time that it consists of corpuscles. It would seem that this dilemma is being solved by the new wave mechanics. De Broglie has assumed that associated with every particle of matter in motion there is a wave....
—Arthur Compton (1892–1962), American physicist, 1927 Nobel lecture

Arthur Compton decides to settle the Bohr-Einstein argument on light quanta. Compton is a physics professor at Washington University in St. Louis, Missouri, and a superb experimenter. He wants to prove that Einstein's "light quanta" are mathematically useful but not real. As it turns out, he will win a Nobel Prize in 1927 for what he proves. It is unexpected.

Here's the story as it unfolds:

In 1923, Compton tells himself that *if* corpuscles of light actually exist, he should be able to detect them. He is an expert on X-ray scattering, so that's the tool he chooses for his experiments.

As with J. J. Thomson's corpuscles of electricity (chapter 5), **CORPUSCLES** of light are tiny bits or particles. Isaac Newton introduced the idea of light corpuscles in the early 1700s. (See chapter 15 of *Newton at the Center*.)

Scattering? When X rays hit the human body, they go right through the skin. But when they hit certain atoms, they bounce and scatter. Why? That puzzle was first tackled in 1912 when Max von Laue (1879–1960) passed X rays through a crystal of zinc sulfide; the

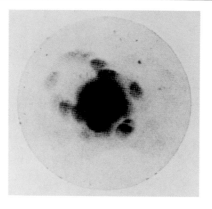

Max von Laue's first X-ray diffraction photograph, made in 1912, produced a pattern of dots that proved his hypothesis that X rays are short-wavelength electromagnetic waves.

rays bent, and he recorded the pattern on a photographic plate. Von Laue realized that some solids contain an internal arrangement of atoms, molecules, or ions geometrically ordered in three-dimensional layers. Called crystal lattices, they can diffract, reflect, and scatter X rays. That becomes important information for those trying to analyze atoms.

The British father-and-son team of William and Lawrence Bragg take scattering further. They find that the layers of atoms in metals are arranged in crystal patterns that act as mirrors reflecting X rays. When the Braggs diffract those rays onto a film, they see a pattern of spots that can be analyzed and measured. The Braggs chart and map whole crystal structures inside some atoms. In the United States, Arthur Compton is soon following their lead.

In popular language, a **CRYSTAL** can be almost any transparent object, like ice or glass or the crystal balls used by fortune-tellers. Scientifically, a crystal is a solid with a regular, repeating pattern of atoms, ions, or molecules. The surfaces of the particles are flat faces that meet at regular and definite angles (or lie in planes that meet at regular, definite angles).

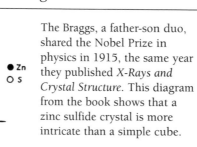

The Braggs, a father-son duo, shared the Nobel Prize in physics in 1915, the same year they published *X-Rays and Crystal Structure*. This diagram from the book shows that a zinc sulfide crystal is more intricate than a simple cube.

Compton beams X rays (high-frequency EM waves) at an assortment of metals. He can detect electrons rebounding. (That's not a surprise: Electrons are particles of *matter*; when whammed by a wave, they react.) What's amazing is that whether the X rays hit copper, tin, or gold, the scattering of electrons is always the same. It's like spraying a hose at a

Keep in mind: X rays, radio waves, gamma rays, microwaves, infrared light, visible light, and ultraviolet light are all electromagnetic (EM) wave-particles—also known as photons or light quanta.

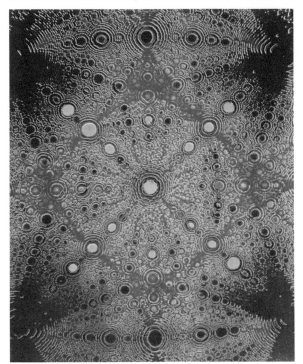

The geometric patterns of a platinum crystal, falsely colored to bring out the details, could easily be mistaken for abstract art. The image is a field-ion micrograph, the first technique invented (in the 1950s) to record the presence of atoms on a nanoscale.

Is the wave-particle duality a paradox (as it is often termed)? Not really. You can be both a school student and a son or daughter. Those are two aspects of your identity, and you become one or the other according to where you are and to whom you are talking.

mud wall and at a piece of rock and getting the same outcome. What's going on? Compton asks himself.

He figures out that the electrons he scatters are outer electrons not tightly bound to their atoms. They are essentially free electrons. That's important information. Compton decides to probe further.

Next, he finds that when X rays hit metal at one frequency, they bounce back at another frequency. This is really astonishing. Imagine that you are wearing a green scarf, but, in a mirror, the reflection of your scarf is pink (remember: Frequency determines color). This doesn't make sense: Energy and frequency seem to be linked. In Maxwell's wave theory, energy and frequency are *not* linked.

Compton realizes that when an incoming EM wave encounters a slow-moving electron in a metal, there's a transfer of energy. The electron (which is almost at rest in the metal) must pick up speed from the very fast EM wave.

According to the *known laws of waves*, as the electron accelerates, the scattered waves should consist of a *smear* of changing frequencies—like the downward slide in pitch you hear when a car blows its horn as it passes you. (Musicians call that slide down the scale a glissando.)

And there *is* a smear in Compton's experiment, but perched on top of the smear is a peak at one discrete lower frequency. **This peak can only be explained if particles are hitting particles, which is just what Einstein predicted.** Yes, particles of electromagnetism must be hitting and scattering electrons. (The smear says, "I'm a wave"; the peak says, "I'm a particle.")

Electrons are carrying away energy lost by the EM (light) particles. That explains why those scattered photons end up with less energy and longer wavelengths (meaning lower frequencies).

Compton has been trying to prove that light can't be particles. Instead he has proved exactly the opposite, just as Einstein said. That means that Isaac Newton's idea of light as corpuscles is right, but so is Maxwell's idea that light is waves. **Electromagnetism has a dual particle-wave nature.** (Imagine two photographs of a face. One looks smooth; the other is made of a zillion tiny dots. It's the same face.)

When the news of Compton's proof (called the Compton effect) reaches the small world of physics, finally the existence of photons (light quanta) can no longer be doubted. Compton has surprised himself, settled a long-running argument between Bohr and Einstein, and opened a new door into quantum physics. Right away, others cross that threshold.

One of the first is Louis-Victor de Broglie (brah-GLEE). In 1923, he is trying to decide on a subject for his doctoral thesis when he learns of Arthur Compton's experiment.

De Broglie (1892–1987) is a French aristocrat (a real prince in both senses of the word) who graduated from college with a degree in history. Then, during World War I, he served in the French army, where he was assigned to the very new field of radio communications. While many of his friends were in the trenches, de Broglie was high in the Eiffel Tower in Paris, sending and receiving telegrams. He also installed the first radio on a French warship—which may explain why he switched from history to science and why he had waves on his mind.

When de Broglie hears of Compton's work he asks: If light is both particles and waves, how about matter? Could matter also have a dual nature? **Might there be matter waves? Might electrons be waves?**

"I was drawn towards the problems of atomic physics . . . [because of] the mystery which surrounded that famous Planck's constant, h, which measures the quantum of action; it was the disturbing and badly-defined character of the dualism of waves and corpuscles which appeared to assert itself more and more," de Broglie will write later in a book called *Physics and Metaphysics*. "After long reflection in solitude and meditation, I suddenly had the idea, during

There's no question now: Photons are real. Einstein has won a much-argued battle with Bohr. But chalk one up for Bohr: Those photons (along with other quantum particles) live in a state of uncertain duality. They don't follow the same rules that work in the large-scale, everyday world.

Louis-Victor de Broglie's official title was Louis, seventh *duc* (duke) de Broglie. The younger son in a wealthy aristocratic French family, Louis wasn't expected to be a serious scholar, but kids often surprise their parents. De Broglie became a theoretical physicist.

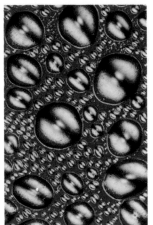

Nickel oxide molecules bead up on a surface whose atoms are of incompatible size and spacing. On a more fitting substrate, they form a mirror-smooth film.

the year 1923, that the discovery made by Einstein in 1905 should be generalized by extending it to all material particles and notably to electrons."

De Broglie imagines electrons in atoms as standing waves (like pure notes plucked on a guitar string).

In his Ph.D. thesis, presented at the Sorbonne in Paris, de Broglie takes Einstein's formula (relating mass and energy) and Planck's formula (relating frequency and energy) and shows mathematically that **electrons—and all particles of matter—must have wave properties.** Matter, like energy, must also have a dual particle-wave nature. Bohr's Copenhagen gang doesn't believe it, but Einstein does. He says that de Broglie has "lifted a corner of the great veil." Can this be proved?

At the Institute for Theoretical Physics in Göttingen, where Max Born and a cluster of great minds ponder

What about people? Do we have wavelengths, too? Yes, but far too small to measure. The "fuzziness" of our size introduced by our wave nature is truly minuscule.
—Kenneth W. Ford, American physicist, educator, and author, *The Quantum World*

quantum mechanics, a graduate student, Walter Elsasser, reads Einstein's paper and de Broglie's, too. Then Elsasser writes his own paper suggesting an experiment to test the wave aspect of electrons. In the United States, Clinton Davisson and Lester Germer are doing an experiment at Bell Labs. They have an accident and unexpectedly produce large crystals. When electrons are reflected off those crystals, the results are puzzling. Elsasser's analysis explains that they are seeing wave action in electrons.

The term *photon* was coined in 1926 by the American physical chemist Gilbert Lewis.

Meanwhile, in England, G. P. (George Paget) Thomson (J. J.'s son) reads Elsasser's paper. He experiments and understands what he sees: Electrons exhibit wave diffraction patterns. His father showed that electrons are particles. G.P. shows that they are also waves. So, in 1927, there is experimental proof of electron particle-waves. Like photons, electrons aren't just particles; they also have wave properties.

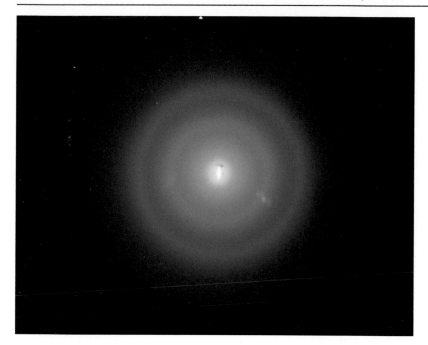

A gun that fires electron "bullets" made this glowing bull's-eye. It's a diffraction pattern, similar in technique to shooting X rays through crystals, and proves that the "bullets" (particles) can alternately act like waves. The opening through which the electrons passed is close in size to their wavelength.

In 1929, de Broglie wins the Nobel Prize. A Swedish physicist introducing him at the ceremony in Stockholm says, "When quite young you threw yourself into the controversy raging round the most profound problem in physics. You had the boldness to assert, without the support of any known fact, that matter had not only a corpuscular [particle] nature, but also a wave nature. Experiment came later and established the correctness of your view."

Quantum mechanics, once fully developed, showed that the electron wave in the atom is three-dimensional, spread out over space, not just stretched along an orbit, and that the electron itself must therefore be viewed as spread out, not executing a specific orbit.
—Kenneth W. Ford, *The Quantum World*

What's Uncertain? Everything, Says Heisenberg

I had Heisenberg here during the winter....He is easily as gifted as [Wolfgang] Pauli but has a more pleasing personality. He also plays the piano very well.
—Max Born (1882–1970), German physicist, letter to Albert Einstein in 1922

I felt so ill with hay fever that I had to ask Born for fourteen days' leave of absence. I made straight for Helgoland [a small island in the North Sea], where I hoped to recover quickly in the bracing sea air, far from blossoms and meadows....Apart from daily walks and long swims, there was nothing to distract me from my problem, and so I made swifter progress than I would have done in Göttingen.
—Werner Heisenberg (1901–1976), German physicist, *Physics and Beyond*

The more Niels Bohr and the quantum physicists dig into the smaller-than-atoms world, the more surprised (and even dismayed) they become. Behavior in Quantumland is unlike anything we experience in our macro world. For instance, atoms and other elementary particles are never still. Never. Newton's laws of motion deal with the state of rest. But **according to quantum rules of motion— which we can observe but still not fully understand—rest is not allowed.** Brownian motion goes on forever.

Inside the atom, Bohr and his quantum mechanics find an incredible, elfin world of peewee particles all jumping around like frantic frogs. (That's the scene inside you, and the chair you're sitting on, and the book you're reading—right now!)

Making this scientific voyage takes imagination; the physicists can't actually see the quantum world—but they can see the results of experiments. They can see what those particles do.

Well, they can see up to a point, and then a barrier gets in the way. When they attempt to make measurements in the quantum world, they run into problems.

Big-world physics is all about measurements and formulas. Classical physicists dreamed of measuring with perfect precision all the physical quantities (momentum, position, energy, etc.) in the universe; they expected to put those measurements into mathematical equations. Quantum

What Physicists Want Is Not What They Get

For classical physicists, making accurate measurements is a cornerstone of experimental science. Why?

One: to make predictions. Two: to verify existing laws of physics. Three: to develop new laws of physics. Four: to describe things and apply the results for practical purposes.

Classical physicists have it easy. Their kind of measurement is certain.

But today's physicists have to deal with the Uncertainty Principle. It makes measurement a challenge. Why?

To begin with, the Uncertainty Principle says particles/waves can't be pinpointed in *both* time and space. That means you can never know both the location of a given particle and its momentum (mass times velocity). The same is true of other complementary (paired) properties, like time and energy.

You can make accurate *single* observations and measurements of a quantum object. You can know exactly where an electron is. You can know its exact momentum. What's the hitch? You can't know *both* properties for the same electron. So what do you do?

Eric Heller programmed a computer to model the random path of quantum wave-particles, starting at the top, as if descending a ramp and "lapping up against a shoreline." The pink line is a wavelength threshold beyond which the wave-particles can't pass.

You experiment on a huge number of electrons. Keep in mind: **Every electron is exactly alike.** So is every photon of the same energy, but they don't all act in exactly the same way.

That's why you have to be satisfied when experiments with identical hydrogen atoms give you **a spread of results for each complementary property.** Those spreads are continuous and limited by a maximum and minimum result. The smaller the spread in one property (like location), the larger the spread in its corresponding property (momentum). With the spread of results, you can make accurate predictions, called probabilities, for the behavior of a large number of hydrogen atoms.

If you do the same experiment over and over again on *identical* systems, you will begin to see a big picture with a predictable spread in outcomes (but not for any one particle). You can never pin down *all* the properties of any individual particle, but you can accurately predict the probable attributes of a quantity of identical particles.

That's August Heisenberg (center) in his military uniform just before he left to fight for Germany during World War I. You can understand why his two sons, Erwin (left) and Werner were proud, patriotic Germans.

theory squashes that dream with a measurement catch. There seems to be no way to escape the problem. Then along comes a dashing, 24-year-old German physicist, Werner Heisenberg. He does some very difficult math to help deal with the measurement crisis.

Heisenberg, the son of a professor of ancient languages, is a *Wunderkind*, which is the German word for a very bright boy. As a youngster he reads Greek literature and philosophy and is fascinated by Greek ideas on atoms and mathematics. By the time he is 19, Heisenberg is studying theoretical physics at the University of Göttingen with some of the world's leading physicists. One of his teachers takes him to a lecture on the new atomic theory. Niels Bohr is the speaker. Later Heisenberg will write,

> *I shall never forget the first lecture. The hall was filled to capacity. The great Danish physicist . . . stood on the platform . . . a friendly but somewhat embarrassed smile on his lips. . . . [E]ach one of his carefully chosen sentences revealed a long chain of underlying thoughts, of philosophical reflections, hinted at but never fully expressed. I found this approach highly exciting.*

After the lecture, Bohr talks to the students, and young Heisenberg corrects him on a point. Bohr is impressed. "At the end of the discussion he came over to me and asked me to join him that afternoon on a walk over the Hain Mountain," Heisenberg recalled later.

On that walk, Bohr invites Heisenberg to Copenhagen. The next night at a university dinner spoof, two German policemen arrest Bohr for "kidnapping." The policeman are graduate students, and the kidnapping charge is all about Heisenberg. Bohr won't capture him immediately; Heisenberg has college studies to complete. Two years later, he will arrive in Copenhagen and become one of the brightest stars at Bohr's institute.

A few years after that, Heisenberg is teaching at Göttingen when he suffers a horrendous hay fever attack. Heisenberg is red, swollen, and miserable. So he goes off to an isolated island in the North Sea where there is little pollen and lots of quiet. There he concentrates on the measurement problem in quantum theory. And it's there, in June 1925, that Heisenberg comes up with **one of the most important theories in modern physics: the Uncertainty Principle.**

What it says is simple: **You can't measure or know all aspects of a quantum system at the same time.** If you measure the position of a particle, you can't measure its momentum, and vice versa.

In other words, in the world of atoms and subatomic particles, **you can't measure *every* property of any one particle-wave.** The closer you come to measuring one thing with perfect precision, the less precise or certain your measurement of *something else* becomes. Picture a mother

Position and momentum are complementary properties. In classical physics, you can tell both where something is (position) and its momentum (mass times velocity). Not so in Quantumland. This is more than a measurement problem. A narrower spread of position measurements means a wider spread of momentum measurements. This is a key to understanding the Copenhagen idea of complementarity.

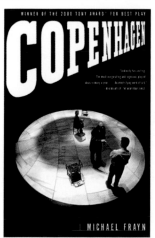

A Playwright Speaks

From *Copenhagen*, a 1998 play by Michael Frayn:

BOHR: Then, here in Copenhagen in those three years in the mid-twenties we discover that there is no precisely determinable objective universe. That the universe exists only as a series of approximations [probabilities]. Only within the limits determined by our relationship with it. Only through the understanding lodged inside the human head.

We know that Werner Heisenberg visited Niels Bohr during World War II. What did they talk about? The atomic bomb? No one knows. The play *Copenhagen* imagines that moment, weaving our uncertainty and their science into the dialogue.

One-Time-Only Systems

To a physicist, the hydrogen atom is an example of a system. After you experiment on a system, it's no longer in the same state, which means it can't be described by the same quantum numbers, so it can't be used for other measurements. Scientists are used to grasping, analyzing, and measuring things—again and again. But there is no way of holding on to a particle or a wave without changing it.

So atoms, or subatomic particles, get experimented on once, and then they are discarded. It means that cause and effect can never be measured in a single system (like an atom) in the quantum world.

Picture a Kleenex. Finicky people don't use one twice. If you bought a box of Kleenex, you'd be upset if each sheet had been sneezed on, so tissues get thrown away. Think Kleenex when you experiment in the subatomic world. You can't do the same experiment again on the exact same object. In Quantumland you need a supply of identical things, like hydrogen atoms. You do the same experiment on each of these—knowing they are all alike.

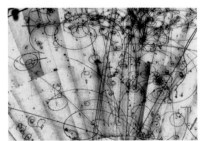

The story behind these subatomic tracks is: "Photon gives birth to electron-positron pair in a magnetic field; they split in opposite directions."

with two rambunctious kids; whenever she gets one child under control, the other acts up.

The Uncertainty Principle tells us that we can't know everything about the physical world. It's a big-deal theory. And it isn't conjecture; it has passed all the tests. Experiment after experiment has proved it right.

Scientists will come to learn that uncertainty is everywhere—in the macro world as well as the quantum world—but its effects are so very small in our everyday world that we ignore them.

Newton's theories suggest that everything follows exact patterns that are preordained (set in advance). Heisenberg's Uncertainty Principle says that is not so. The universe has an element of flexibility at its core; that leaves some freedom and fluidity in our world.

Usually, if you set up an experiment—carefully—you expect to get the same results each time you try it. But not in Quantumland. When an electron, for instance, has a choice of paths to travel, no one can tell—with certainty—which path it will pick. One electron may do one thing, the next one something else. The best you can hope for is to predict the *probability* that any one electron will go a certain way. You can calculate the average—and the probable spread of results around that average—and that is all. You can do that if you base your calculations on past observations. But knowing in advance what any *one* of those pesky critters will do or where it will go? Electrons (and photons, too) are quirky and independent—they won't tell.

Actually, "critters" is the wrong word to use. Particles exist in a state of potential: They have the potential to be a wave undulating through space or a stream of bullet-like particles. Bohr calls this dual potential "complementarity." Some experiments reveal waves, some particles—but never both at the same time. Until they are observed, those quantum wave-particles can only be described as in a fuzzy state of potential. Weird? You bet.

You can imagine how difficult it is for the classical thinkers, trained on the logic of Newton's physics, to accept quantum mechanics, which has a whole different set of rules.

This new kind of physics is very disturbing to most twentieth-century scientists. It makes them rely on statistics and probabilities.

This is something they have to live with.

Now Einstein and Bohr have something else to argue about. It's the Uncertainty Principle. Einstein isn't happy with it. And he doesn't like the idea that the subatomic world can only be understood statistically, even though his paper on Brownian motion helped make statistics acceptable in scientific research. He writes to his friend Max Born (in a much-quoted letter): "Quantum mechanics is certainly imposing. But an inner voice tells me that it is not

Is it a vase or a face-to-face encounter? This popular optical illusion is both, but our brains can focus on only one or the other at a time. Likewise, an electron can be a wave or a particle but not both at once—a property called "complementarity."

Why Uncertain Truth Is Not an Oxymoron

Do you want to measure the *position* of the electron in an atom? No problem. You can use gamma rays. Their wavelengths can be much smaller than the size of the atom, so they can find electrons. But hitting an atomic electron with a gamma ray will whack the electron to kingdom come, ruining that particular atom for any later

The Gammasphere collects gamma rays from bombarded, deformed nuclei for physicists to analyze.

observations. This means there's no way to use the same atom again. If you want to repeat the experiment, you'll need a brand-new hydrogen atom. Using a bunch of hydrogen atoms will give you a spread of locations.

You can also measure the momentum of an electron in an atom as accurately as you want. But each time you do that experiment, you disturb the atom in its ground state, making it useless for further experiments. Again, by measuring many electrons, you'll end up with a spread of momentum results. The uncertainty relation tells you the smallest possible product of these two spreads (location and momentum)—and that's certain!

Raphael (1483–1520) imagined God, whom Einstein called "the old one," in this famous creation painting, a fresco at the Vatican Palace in Rome.

yet the real thing. The theory says a lot, but does not really bring us any closer to the secret of the 'old one.' [He means God.] I, at any rate, am convinced that He is not playing dice."

Einstein is looking for a grand theory that will bring the quantum and macro worlds together: a theory of everything, something that is not uncertain.

But, like it or not, the Copenhagen picture of the atom works; the "why" doesn't concern Bohr and his team. They realize that, for centuries, no one knew why a compass needle points north, and that didn't make the compass less useful. Bohr believes the truth of an idea should be judged by its

Unfair? You Decide

Werner Heisenberg was struggling with the idea of uncertainty. So he went to Einstein's friend Max Born, head of the Institute for Theoretical Physics at the University of Göttingen. Born and a student, Pascual Jordan (1902–1980), guided him through something known as "matrix mechanics." Today their work is usually called the "three-man paper." But it was Heisenberg alone who was awarded the Nobel Prize in 1932. Here's a letter he wrote to Born (who had fled Germany because he was Jewish):

> Dear Mr. Born:
> If I have not written to you for such a long time, and have not thanked you for your congratulations, it was partly because of my rather bad conscience with respect to you. The fact that I am to receive the Nobel Prize alone, for work done in Göttingen in collaboration—you, Jordan and I—this fact depresses me and I hardly know what to write to you. I am, of course, glad that our common efforts are now appreciated, and I enjoy the recollection of the beautiful time of our collaboration. I also believe that all good physicists know how great was your and Jordan's contribution to the structure of quantum mechanics—and this remains unchanged by a wrong decision from outside. Yet I myself can do nothing but thank you again for all the fine collaboration and feel a little ashamed.
> With kind regards,
> Yours,
> W. Heisenberg

Twenty-two years later, in 1954, Max Born was awarded a Nobel Prize "for his fundamental research in quantum mechanics." Pascual Jordan never received the prize.

Speaking of Particles, We Can't. Or, When Language Fails, Try Math

You could say some quantum particles are thick. But, of course, they are so tiny that that word is all wrong. Dense? No, no—that won't quite do either. Some are massless, and that is almost impossible to picture. So how do we explain them in words? When it comes to normal spoken language, we have a big problem describing particles.

There is no way to explain their actions and make "common sense."

Maybe the trouble is with our languages. "Language evolved to help people get around on earth, not down inside atoms," says science writer George Johnson. Our spoken languages—all of them—are too limited for the realities of modern science. Try as we might, our words just aren't fluid and expansive enough to completely capture ideas like relativity and quantum theory. And, beware, sometimes the words we use in daily life lead us astray when used to describe the quantum world.

Those aren't planets or suns or even galaxies in this illustration of space. They're entire *universes*. Is it true? Is our universe part of a so-called *multi*verse? It's one interesting possibility stemming from the manifold outcomes of quantum theory.

The Uncertainty Principle means that nature, on its smallest level, can't be predicted. Therefore, we have to be content with statistical probabilities. That fact drove Einstein nuts. As for Bohr, he seemed to say that all he was doing was predicting results of experiments, and those results kept proving that uncertainty exists in the quantum world.

Our everyday languages don't convey that idea easily. But there is a language that does explain all this—beautifully and understandably—and that is the language of mathematics. So here's a bit of it.

The Uncertainty Principle for momentum and position can be written this way:

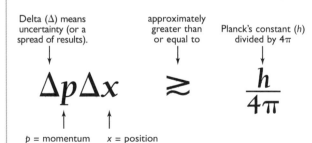

Delta (Δ) means uncertainty (or a spread of results).

approximately greater than or equal to

Planck's constant (h) divided by 4π

$$\Delta p \Delta x \gtrsim \frac{h}{4\pi}$$

p = momentum   x = position

The equation says that the spread of momentum (Δp) times the spread of position (Δx) can be approximately equal to or greater than the size of h (Planck's constant) divided by 4π. The value of h divided by 4π is so small it can be called zero, so the uncertainties of position and momentum are so tiny that they are almost meaningless.

practical workings. He doesn't worry about deep, underlying meaning. The strength of the Copenhagen School (as Bohr and his associates come to be called) is its ability to discover practical formulas that lead to technological marvels. The founders of quantum mechanics become heroes in an atomic and technological age. Niels Bohr is their mentor and poet.

A Cat, Quarks, and Other Quantum Critters

Schrödinger: "If all this damned quantum jumping were really here to stay then I should be sorry I ever got involved with quantum theory."

Bohr: "But the rest of us are extremely grateful that you did; your wave mechanics has contributed so much to mathematical clarity and simplicity that it represents a gigantic advance over all previous forms of quantum mechanics."
—as quoted by Werner Heisenberg (1901–1976), German physicist, *Physics and Beyond*

The task is...not so much to see what no one has yet seen; but to think what nobody has yet thought, about that which everybody sees.
—Erwin Schrödinger (1887–1961), Austrian physicist, *Problems of Life*

E rwin Schrödinger wears hiking boots and carries a backpack long before it is commonplace to do so. "When he went to the Solvay Conference in Brussels, he would walk from the station to the hotel...carrying all his luggage in a rucksack and looking so like a tramp that it needed a great deal of argument at the reception desk before he could claim [a] room," writes fellow physicist Paul Dirac. Max Born says, "He was a most lovable person, independent, amusing, temperamental, kind and generous, and he had a most perfect and efficient brain."

Schrödinger is Austrian. As a student at the University of Vienna, he studies painting, languages, and philosophy—along with mathematics and physics. He's not quite sure what he wants to do with his life. Then, as a soldier in World War I, he is stationed at a quiet outpost where he has free time, so he

Schrödinger thought in terms of unity. Like Einstein, he worked much of his life trying to unify gravity and electromagnetism.

Four More Famous Papers

In 1926, Schrödinger published four famous papers in *Annalen der Physik*, the same journal that published Einstein's 1905 papers. When Einstein read Schrödinger's papers, he wrote, "The idea of your work springs from true genius."

In the first of his papers, Schrödinger wrote, "Bohr's rules for quantization can be replaced by another requirement, in which the mention of 'whole numbers' no longer occurs. Instead the integers occur in the same natural way as the integers specifying the number of nodes in a vibrating string." That vibrating string will inspire a new generation of physicists.

A fixed string vibrates at a fundamental wavelength, or unit of frequency—a musical note, for instance. That's one wave in the middle, above. Harmonics are "quantized" integer multiples of that frequency.

reads more physics on his own. By 1920, Schrödinger is a physics professor at the University of Stuttgart in Germany. There, he reads a footnote to one of Einstein's papers. It deals with Louis-Victor de Broglie's assumption that electrons (matter) have wave-like properties. Einstein, discovering de Broglie's work, understands at once its importance.

"Einstein had shown that light, long thought to be a wave, was like a particle. De Broglie had brought the argument round full circle by suggesting that matter, long thought to consist of particles, must be accompanied by waves and thus partake of their nature," Banesh Hoffman will write later.

When the Copenhagen physicists describe electrons mathematically, they treat them as particles and don't take waves into account. **Werner Heisenberg has produced something called matrix mechanics; its equations only deal with the electron as a particle.** The matrix method uses tables of numbers (matrices) and follows rules of algebra. The equations are very difficult; Schrödinger thinks he can do better. He, too, isn't comfortable with a dual description of the electron—as wave and particle; he thinks it complicates things. Inspired by de Broglie, he decides to set up a theory of waves only. His theory is rooted in calculus.

So **Schrödinger does the mathematics to describe electrons as waves.** He wants his wave function equation to replace matrix mechanics. His math is easier to use than the

Solving Things at Solvay

Back in 1861, Belgian chemist Ernest Solvay (1838–1922), shown at left, figured out a way to produce sodium carbonate (also called washing soda or soda ash) from sodium chloride (salt), ammonia, and calcium carbonate (limestone). Sodium carbonate, Na_2CO_3, is an important industrial chemical used to make paper, glass, soap, porcelain, and bleach. It is also an ingredient in cleaning compounds and water softeners. Today more than 5 million metric tons of Na_2CO_3 are produced each year using Solvay's method. In the nineteenth century, sodium carbonate helped accelerate the Industrial Revolution. Solvay took out a patent on his chemical process, and it made him very rich.

He used some of his riches to found institutes of physics, chemistry, and sociology. He also paid the bills for a series of conferences. Beginning in 1911, the world's leading physicists got together in classy surroundings in Brussels to talk—and argue—about their favorite subject. Einstein, Bohr, Heisenberg, and Schrödinger were among those who met at the Solvay Conferences. Hendrik Lorentz, a much-admired Dutch physicist, was the first director of these international gatherings. "Lorentz is a marvel of intelligence and exquisite tact. A living work of art!" Einstein wrote to a friend, after attending the first Solvay Conference. "In my opinion he was the most intelligent of the theoreticians present." (Some of Lorentz's ideas helped lead Einstein to the theory of relativity.)

Here's one of the most famous photos in physics history. Seventeen of the 29 scientists at the 1927 Solvay Conference in Brussels were or became Nobel Prize–winners.

P. Debye · A. Piccard · E. Henriot · P. Ehrenfest · E. Herzen · T. de Donder · E. Schrödinger · E. Verschaffelt · W. Pauli · W. Heisenberg · R. H. Fowler · L. Brillouin

I. Langmuir · M. Knudsen · M. Planck · W. L. Bragg · M. Curie · H. A. Kramers · H. A. Lorentz · P. A. M. Dirac · A. Einstein · A. H. Compton · P. Langevin · L.-V. de Broglie · C. E. Guye · M. Born · C. T. R. Wilson · N. Bohr · O. W. Richardson

matrix approach, and it does help chemists and others work out formulas that illustrate the way atoms interact to form molecules. It explains why electrons can extend in a cloud-like wave.

His wave equation is a big success. It is especially popular among the older physicists. Max Planck calls it "an epoch-making work." After reading Schrödinger, one physicist chastises Heisenberg: "We are now finished with all that nonsense about quantum jumps." (Presumably, waves don't jump.) But that wave equation turns out to be the equivalent of Heisenberg's matrix math. Called wave mechanics, it is just another way of stating the same thing. You can do matrix math or wave math and get the same results.

Schrödinger fails in his attempt to dump the particle-wave duality and fully explain electrons as waves. He can't get rid of the weirdness of quantum physics and those quantum leaps. Electrons are both waves and particles—the duality can't be eliminated. **It's no more true to say that an electron is a wave than it is to say it is a particle. Electrons are both.** Schrödinger hoped to eliminate quantum probabilities, but his waves turn out to describe packets of probabilities, too.

Finding the Wave

Schrödinger was lecturing at the ETH (Einstein's alma mater) when Peter Debye, a professor friend, suggested that he give a talk on de Broglie's ideas about waves. Einstein had called attention to the concept, so there was much interest among physicists.

Schrödinger did some research and gave a lecture. It was a big success. Then Debye said to him, "This is a childish way to do physics. If there is a wave then there should be a wave equation."

A few weeks later, Schrödinger gave another lecture. He started this one by saying, "My colleague Debye suggested that one should have a wave equation; well, I have found one!"

Schrödinger compared understanding the quantum world to reading *Don Quixote* by the Spanish novelist Cervantes. "Like Cervantes's tale of Sancho Panza, who loses his donkey in one chapter but a few chapters later, thanks to the forgetfulness of the author, is riding the dear little animal again, our story has contradictions."

To the average American, the rules of chess written in Russian will seem to have nothing in common with the same rules written in Chinese. But let him see a Russian and a Chinese [person] play an actual game and the connection becomes immediately obvious.... Dirac discovered what was the prime reality beneath the confusion of theories in the new quantum mechanics; the basic rules of the new game the physicists were playing.

—Banesh Hoffman, *The Strange Story of the Quantum*

That's one psychedelic cat, not two (below). It's a quantum cat, both alive and dead, featured on the cover of the May 1994 issue of *Scientific American* about a new take on Schrödinger's famous thought experiment.

Still, what Schrödinger has done is impressive and very useful. It wins him a shared Nobel Prize in 1933 and also Max Planck's job at the University of Berlin (Planck has retired). But Schrödinger has to accept quantum theory with its weirdness.

He's not happy about that. Schrödinger says of his own theory, "I don't like it, and I wish I'd never had anything to do with it." To demonstrate the strangeness of quantum theory and especially of the Uncertainty Principle, he comes up with a thought experiment that becomes very famous. It focuses on an imaginary cat. Like an electron or photon, Schrödinger's cat is in a dual state. Instead of being both particle and wave, this cat is both dead and alive. (Note: If you want to explain something, using an analogy—even a wild one—will help.)

This is the story of Schrödinger's imaginary cat. It's fiction—but meant to prove a point. In the story, a live cat is put in a box along with a radioactive device. If the device emits one gamma ray, it will cause a hammer to fall and crush a vial of cyanide gas. Cyanide is poison; it will kill the cat. But there is an equal probability that the radioactive device will not emit a gamma ray in the given time. Then the hammer won't fall, and the cat will live.

So, does the pussycat survive? Of course Schrödinger knows that the state of a cat can't be described in the same way as the quantum state of a particle. But, for the story, kitty has the same attributes as an electron or photon or other quantum entity. The cat exists in a state of probability. So,

"Honestly, Erwin. Can't you just flip a coin?"

Words Can't Explain the Weirdness

Do elementary particles behave like waves? Yes.

Do elementary particles behave like particles? Yes.

So which are they? Both.

Physicist Erwin Schrödinger said, "Both the particle picture and the wave picture have truth value, and we cannot give up either one or the other. But we do not know how to combine them."

Physicist Richard Feynman put it differently, and more accurately, when he said that elementary particles "behave in their own inimitable way, which technically could be called a quantum mechanical way. They behave in a way that is like nothing that you have ever seen before."

As if that weren't hard enough to grasp, elementary particles do bizarre things, like being in one spot and then another *without actually jumping from one to the other.* It's that quantum leap.

Hold on. It gets even more fantastic than that. Can a particle be in two places at the same time? Common sense says, Of course not. Yet it does seem to happen. (See chapter 45 on entanglement.)

As for motion, elementary particles never, ever stop moving. While you might think that you are at rest, your atoms are in constant motion (and so are you on this moving globe that we inhabit).

You can measure the momentum of a particle, or you can measure its position. The Uncertainty Principle (chapter 18) says you can never do both measurements at the same time—by measuring one, you change the other.

Did you think science fiction was way out? Well, try reality if you want to sample the bizarre. But is it really bizarre? Or is it just that we live on a different scale and find our words and our imaginations limited?

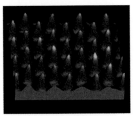

Chromium atoms on a silicon crystal "tuned into" the frequency of a laser, spacing themselves evenly at the light wave's nodes—the lowest-energy intervals.

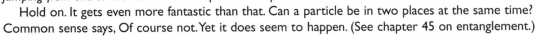

before the box is opened, he can be described as both dead *and* alive. Each state is equally probable.

In the world we know best, you are either dead or alive—no dual states allowed. With this meant-to-be-absurd experiment, Schrödinger makes clear the difference between the strange-to-us quantum world and the classical everyday world. As for Schrödinger's cat? He doesn't tell if it survives.

Schrödinger has made his point: The quantum world *is* weird. Duality exists. This doesn't seem to bother Niels Bohr, who says that quantum action can't be explained, it can only be described.

Then a young Englishman, Paul Adrien Maurice (P.A.M.) Dirac (1902–1984), takes Heisenberg's matrix mechanics and Schrödinger's wave equations and, building on both, comes up with an even more complete and useful quantum theory.

Dirac's Swiss father is a French teacher in a school in

Like a particle, Schrödinger's cat is described as in a superposition. A particle in a superposition can be both wave and particle.

What a Life!

In 1927, Max Planck asked Erwin Schrödinger to replace him as a professor of theoretical physics at the University of Berlin, where Einstein was also on the faculty. Most Berlin professors dressed formally and read notes when they lectured. But not Einstein or Schrödinger. These intellectual friends sailed together, walked in the woods, and talked informally to their students. Then Hitler came to power in 1933, and Germany's creative role in science soon ended.

Schrödinger left Germany for Oxford University in England. He was fluent enough to lecture in English. (He also spoke French, Spanish, and, of course, German.) But he was homesick for his native Austria, so in 1936 he took a job as a professor at the University of Graz. Then Hitler moved into Austria. In 1940, Schrödinger headed for Ireland where the Irish prime minister, Éamon De Valera, a math enthusiast, had created the Dublin Institute for Advanced Studies. Schrödinger stayed in Ireland for 17 productive

Follow Schrödinger from Berlin to Oxford to Graz to Dublin. Other physicists went farther— to the United States.

years. There he wrote *What Is Life?*, an influential look at molecular biology. Then he returned to his beloved Austria.

Schrödinger believed that narrow specialization endangered science and that if scientists "continue musing to each other in terms that are, at best, understood by a small group of close fellow travelers," then science "is bound to atrophy and ossify."

Here's another historic cloud chamber photo, this one showing the track of a positively charged particle slowing down as it passed through lead. It proved, in 1932, the existence of the positron (the antiparticle of the electron).

Bristol, England. He insists that his wife and children speak French at home. But only Paul speaks well enough to sit at the dinner table with his father. The rest of the family must eat in the kitchen. "My father made the rule that I should only talk to him in French," Dirac recalled later. "He thought it would be good for me to learn French in that way. Since I found that I couldn't express myself in French, it was better for me to stay silent than to talk in English. So I became very silent at that time."

Dirac grows up to be a famous physicist; among his colleagues he is famously taciturn. When Niels Bohr complains to Ernest Rutherford that he can't get Dirac to say anything, Rutherford tells Bohr the story of a pet store owner and a customer who returns a parrot that won't speak. The storekeeper says to the customer, "You didn't tell me you wanted a talker. I gave you a thinker."

No one complains that Dirac can't think. Later, Bohr will say that, of all the physicists he knows, "Dirac has the purest soul."

Dirac looks at matrix mechanics and wave mechanics and realizes they lead to the same answers. He doesn't think that quantum mechanics needs two voices. So he comes up with a single, graceful fusion of the two approaches. He is 23 when he publishes his first major paper. Max Born (Heisenberg's boss at Göttingen) is amazed. "The author appeared to be a youngster, yet everything was perfect in its way and admirable," he recalled later.

The following year, 1926, Dirac takes quantum mechanics to a mathematical plane that goes beyond both matrix and wave mechanics. He applies it to very fast particles and to the creation and annihilation of particles allowed by relativity. (Hold on, details on relativity are coming.)

Dirac's calculations show that electrons have a kind of **spin** and that **they behave like tiny magnets with north and south poles.** Dirac also says that the negative electron must have a positive twin (later called a positron). The positron will be known as an **antiparticle**. It turns out that all particles have antiparticles (though some, like the photon, are their own antiparticles).

Dirac uses the terms *beauty* and *ugliness* often in his writings. "Mathematical beauty is a quality that cannot be defined, any more than beauty in art can be defined," he writes, "but which people who study mathematics usually have no difficulty appreciating."

When the classical physicists looked at reality they saw solid molecules at its base. For the quantum physicists there is *nothing solid* underlying reality. Mostly they see the quantum world in terms of statistics. Paul Dirac finds great beauty in the math that expresses that astonishing form of reality.

Paul Dirac (on the left, standing next to Heisenberg) intended to be an electrical engineer, but when he couldn't get a job, he switched to mathematics. It was a good choice. He became Lucasian Professor of Mathematics at Cambridge (Isaac Newton's old position) and shared the 1933 Nobel Prize in physics with Heisenberg. Dirac figured out that every charged particle, like the negative electron, must have a twin, an antiparticle, with an opposite charge (like the positron). He was proved right. In 1950, he suggested that particles might be string-like rather than points. No one is sure if that is true or not, but some physicists are taking the idea seriously.

Up and Atom: A Review of Atomic Theory Basics

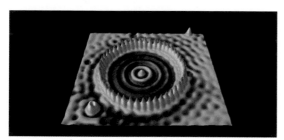

Einstein said, "It's the theory which decides what can be observed." When we finally "saw" atoms using new microscopes, it was just as theory had predicted. In 1981, Swiss scientists Gerd Binnig and Heinrich Rohrer invented the scanning tunneling microscope (STM), whose tip probes extremely close to a surface. When it encounters an atom, the flow of electric current changes. A computer records that change and plots the atom's position. In this 1993 image, an STM tip formed a "corral" of 48 iron atoms on a copper surface.

At the beginning of the twentieth century, some scientists were still saying atoms don't exist. By the end of the century, knowledge of the atom had become basic science, in all its fields. Here's a famous quote from America's great physicist Richard Feynman:

If, in some cataclysm, all of scientific knowledge were to be destroyed, and only one sentence passed on to the next generation of creatures, what statement could contain the most information in the fewest words? I believe it is the atomic hypothesis (or atomic fact, or whatever you wish to call it) that all things are made of atoms—little particles that move around in perpetual motion, attracting each other when they are a little distance apart, but repelling upon being squeezed into one another.

Bill Bryson said this about atoms in *A Short History of Nearly Everything*:

It is still a fairly astounding notion to consider that atoms are mostly empty space, and that the solidity we experience all around us is an illusion.... When you sit in a chair, you are not actually sitting there, but levitating above it at a height of one angstrom (a hundred millionth of a centimeter), your electrons and its electrons implacably opposed to any closer intimacy.

What Is an Atom?

All matter is made of atoms. An atom is mostly empty space.

Most of the mass of an atom is **in its small, dense nucleus.**

The nucleus is composed of **protons and neutrons**, which are each made up of three particles called quarks (see illustration). Protons carry a positive charge; neutrons have no electrical charge.

The nucleus is surrounded by an electron particle-wave cloud. In a neutral atom, the number of electrons equals the number of protons, which is also the atomic number. Hydrogen has one electron and one proton and is number 1 on the periodic table of elements; uranium has 92 electrons and 92 protons and is number 92. (Some atoms, called ions, are missing at least one electron or have at least one extra and so are not neutral.)

It's the number of protons that makes an element unique. The periodic table includes 92 natural elements followed by transuranic elements (number 93 and above) that are created in laboratories. Like us, each atom has its own personality. Atoms of gold don't look like atoms of mercury or atoms of oxygen or carbon. But each atom of an element is the same, with an important exception: Isotopes are atoms of the same element that have the same number of protons but a different number of neutrons (and so different atomic masses). Isotopes can be stable or radioactive.

Atoms never stop moving. They are in perpetual motion, even in a solid. At absolute zero, there is hardly any vibration, but *it is never zero vibration.* Those atoms don't give up!

The basic structure of an atom is an electron cloud surrounding a nucleus (top) made of protons and neutrons (center) that are in turn each made of three quarks (bottom).

Atoms attract one another when they are a little distance apart. Atoms repel one another when they are squeezed together.

Atoms are not the smallest particles known. There's a "zoo" of subatomic particles, including electrons, protons, photons, quarks and many others.

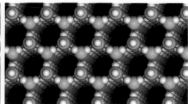

While all atoms are in perpetual motion, the motion differs in a gas, a liquid, and a solid: Atoms in a gas zoom freely (left), atoms in a liquid move within a structure (middle), and atoms in a solid vibrate in place. When water turns into a solid (ice), something interesting happens. The molecules make a hexagonal and symmetrical pattern called a crystalline array (right). Even so, don't think the jiggling of the atoms stops. They no longer move around at random, but they keep oscillating.

How Big and Fast Is an Atom?

The largest atom is about 0.0000005 millimeters across (5×10^{-7} mm). Here's a description to help your imagination:

If an apple could be enlarged until it was as big as the Earth, its atoms would be the size of the original apple. Think of that Earth-sized McIntosh or Rome Beauty filled with zillions of ordinary apples bouncing and jiggling at the same time and you'll have an idea of what you're biting the next time you chomp a Granny Smith.

"A man is about 10 billion times 'taller' than an oxygen atom," writes Brian L. Silver in *The Ascent of Science.*

As for speed, atoms and molecules cavort at about 1 to 10 times the speed of sound in the air, or about 330 to 3,300 meters per second. (Light travels almost 10,000 to 100,000 times faster.) Compared to quantum particles, like electrons, atoms don't seem fast.

Why Do Bonds Matter?

In this world with only a limited number of elements, how come there are tens of thousands of different forms of matter? Because *atoms can link themselves into compounds in almost endless combinations.* That process, called chemical bonding, happens when the electrons in different atoms rearrange themselves and form a link, or glue, that holds them together as compounds. Knowledge of that process is **the substance of modern chemistry**.

Once scientists figured out that it is electrons that create the ties (the bonds), they experimented and did their own bonding. Soon, they were making detergents, plastics, and synthetics. Think of chemical bonding as like baking bread. The yeast, flour, and water that go to make the dough get transformed in the oven. Hydrogen and oxygen, two gases, don't resemble liquid water. In the model below, two carbon atoms (green) bond with seven hydrogens (white) and one nitrogen (purple) to form ethylamine (C_2H_7N), an ammonia-like compound used in oil, rubber, and medicine. When atoms bond (think marriage), the offspring can amaze.

Picturing an Atom

When Werner Heisenberg was asked how to picture an atom, he said, "Don't try." He couldn't imagine that we humans would ever actually see an atom. But, in 1980, an atom was photographed in Heidelberg, Germany. A few years later, scientists at the National Institute of Standards and Technology, in Boulder, Colorado, were adding snapshots to the atomic scrapbook. They were using an "atom trap," a tiny, electrically charged ring suspended inside a glass vacuum. An atom hangs between the walls of the doughnut-shaped ring, while ultraviolet light shines through the trap. A light detector, connected to a monitoring screen, enlarges and pictures the atom.

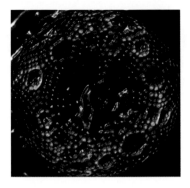

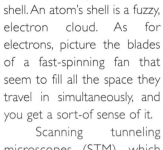

Here are two more amazing atom images, one of the earliest (below) and one of the latest. The silicon atoms below are enlarged 20 million times. The reptilian-looking surface above is the world's sharpest object (National Institute for Nanotechnology/University of Alberta, 2006). The nano-needle naturally tapers to a one-atom tip as nitrogen atoms in the air interact with tungsten atoms—those round pits. The red blobs are atoms that moved as the microscope's probe recorded their presence.

"Individual atoms can now be counted, and photographed, and kept in captivity; the surface roughness of materials can be magnified a millionfold to reveal its atomic character; and atoms can be combined one by one in the construction of synthetic materials," writes Hans Christian von Baeyer, describing this astonishing process in *Taming the Atom*.

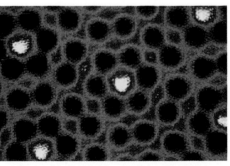

So what does an atom look like? Not like those solar system models you've seen. The atom takes up the space filled by its electron shell. Don't imagine a skin-like shell. An atom's shell is a fuzzy, electron cloud. As for electrons, picture the blades of a fast-spinning fan that seem to fill all the space they travel in simultaneously, and you get a sort-of sense of it.

Scanning tunneling microscopes (STM), which "see" atoms, don't use light, and they don't magnify things. An STM scans the surface of an object (such as the element gold) with electron beams, reading it atom by atom, almost as a blind person reads Braille.

Now that we can "see" atoms, is uncertainty still out there? To put this another way, was Bohr right (uncertainty rules) or Einstein right (there's a deeper meaning here)?

Quick answer: We're still working on that. But **no one has disproved Werner Heisenberg's Uncertainty Principle**, which says that you can't know both the position and momentum of an atom or a subatomic particle. And that means there's an element of mystery in the quantum world.

Yes, finally, we can see atoms, but they still don't act as predictably as things do in our everyday world. As Richard Feynman said, "[T]hings on a small scale behave *nothing like* things on a large scale."

Smashing Atoms

Niels Bohr in 1913 devised a theory that was spectacularly successful....He forced a connection between the Einstein relation $E = hf$ for photons and the Newtonian picture of electrons in orbits around the atomic nucleus.
—Kenneth W. Ford, American physicist, educator, and author,
The Quantum World

There is more to the universe than the mere vibrations of its particles, just as there is more to music than merely the vibrations of instruments.
—John Polkinghorne, Anglican priest and physicist, "Dialogue on Science, Ethics, and Religion" speech

Ernest Rutherford and his colleagues have been using "table-top physics" to establish the basic laws of nuclear physics. But then they can go no further with their experiments.

In 1927, Rutherford announces that, to move ahead, nuclear physics needs a machine that will send particles at higher energies than the helium nuclei naturally coming from radium. The available alpha particles (those helium nuclei) are too few and have too little energy. Rutherford says he needs "a copious supply" of high-energy particles to do more research. He imagines a machine that will propel particles at high

TABLE-TOP PHYSICS?
It means small-scale experiments that can be done in an old-fashioned laboratory.

At Fermi Laboratory in Illinois, a chain of smaller accelerators gets protons and antiprotons up to speed before they enter the giant, circular Tevatron accelerator (left). Traveling in opposite directions, the two types of subatomic particles nearly reach the speed of light. The Tevatron has a circumference of 6.28 kilometers (about 4 miles).

Tunnel Vision

In 1928, George Gamow came to the Cavendish Lab and talked to John Cockcroft about the possibility of low-energy alpha particles escaping on their own from atomic nuclei.

In classic physics, there is no way for a ball to go through a wall. In quantum mechanics, if a particle reaches a barrier, it has a small (but not zero) probability of passing through. Given a large number of impacts, odds are that a particle will tunnel through the barrier.

Gamow believed alpha particles could *tunnel* out of a nucleus. Cockcroft said, "If they can tunnel out, they can probably tunnel in." He was right.

speeds inside a vacuum tube, where there are no molecules to deflect the particles. That machine, called an accelerator (or a collider or voltage multiplier), would need to produce high voltages to give the particles high energy. The intent is to split atoms apart.

Two of Rutherford's associates start to work on such a machine. They aren't alone. It becomes a scientific race. In the United States, scientists work long hours and weekends, too, trying to build a particle accelerator. In Cambridge, England, Rutherford locks the Cavendish Laboratory at six p.m. every evening and on weekends, too. He believes his scientists should spend time with their families.

But it is John Cockcroft and Ernest Walton at the Cavendish Lab who, in 1932, win the race. Their device sends a stream of hydrogen nuclei (protons) down a long, straight tube called a linear accelerator. Rutherford tests it

Notice Ernest Walton (far left) inside a safe "hut." The Cockcroft-Walton voltage multiplier is in the background. Walton described what happened on April 13, 1932: "When the voltage and the current of protons reached a reasonably high value, I decide to have a look.... I saw scintillations on the screen.... I then phoned Cockcroft who came immediately.... He then rang up Rutherford... [and] we maneuvered him into the rather small hut.... [Rutherford said,] 'Those scintillations look mighty like alpha particles. I should know.'" From left to right, that's Walton, Rutherford, and Cockroft soon after the experiment.

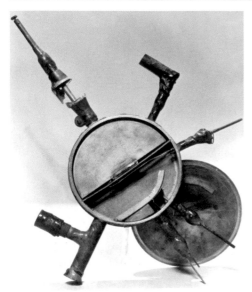

In a linear accelerator, electromagnets keep the beam of particles narrowly focused as it travels in a vacuum down a long, straight, copper tube. In a circular accelerator, like this first successful cyclotron (above), the particles travel around and around. The goal is to get them to bang against a barrier or hit another particle in a high-speed collision. Detectors record what happens: how many particles and how much radiation is released. This 1931 cyclotron accelerated hydrogen ions, the most energetic of which made 40 or more laps before falling into a collecting cup near the center.

by whamming protons into the metal lithium at an energy level of about one-half million volts. The lithium nuclei split! Each lithium nucleus becomes two alpha particles. (An alpha particle, a helium nucleus, has two protons and two neutrons.)

That's not all. Those alpha particles have *30 times the energy* of the original hydrogen nuclei. Rutherford is ecstatic. He now has a supply of high-energy particles.

Cockcroft and Walton have designed what will be called an "atom smasher." With it, they have split the nucleus of an atom, showing that it can be done in a laboratory (atom smashers will get larger and larger). And here is the really big news: **This experiment is the first to prove Einstein's formula: $E = mc^2$. Cockcroft and Walton have shown that mass and energy are equivalent.**

A few weeks later, Robert Van de Graaff at the Massachusetts Institute of Technology (MIT) has a working linear accelerator (some say it is better than the Cambridge version). And, right after that, Ernest O. Lawrence in Berkeley, California, has one that is still better (he started work on it in 1929). Lawrence has designed a circular accelerator named a cyclotron, which sends charged particles in a magnetic field moving in a spiral. That means the same boost in energy can be applied again and again each time a particle whizzes around a loop. The first cyclotron is about 11.5 centimeters (4.5 inches) in diameter, but it does the job of accelerating nuclei. This astonishing technology charges forward until the early 1940s when physics (and most of the world) goes to war.

After the war, in the 1950s and 1960s, quantum physicists finally get back to particle physics. Before long they have huge cyclotrons and linear accelerators to help with their experiments.

What happens when the giant high-energy accelerators are used?

Hooray for Robert Wilson!

Do we really need to accelerate tiny particles? Robert R. Wilson, a Wyoming boy who became the first director of the National Accelerator Laboratory (later to be Fermilab), was asked that by a congressional committee. It was like asking Columbus if he should sail west to find the East. It was like asking Beethoven if there was any reason to compose music as it had never been composed before.

Wilson said, "It has only to do with the respect with which we regard one another, the dignity of men, our love of culture. It has to do with: Are we good painters, good sculptors, great poets? I mean all the things we really venerate in our country and are patriotic about. It has nothing to do directly with defending our country except to make it worth defending."

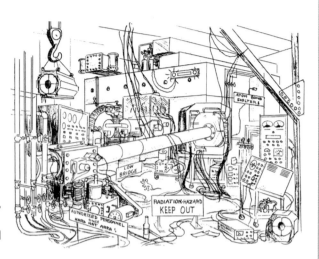

The cyclotron as seen by... the visitor (part of Dave Judd and Ronn MacKenzie's cartoon series "The cyclotron as seen by...").

That's when we discover: *first*, that the protons and neutrons in the nucleus aren't the smallest particles that exist. They are made up of other still-smaller subatomic particles; and *second*, that **high energy can lead to the creation of particles that were not in the nucleus before** you smashed it.

This is amazing. Here it is again: Giant accelerators not only probe the insides of the nuclei; **the energy of incoming particles can be converted into new particles** with a variety of masses. Each particle created, and the energy converted into its mass, is determined by Einstein's famous equation run backward: $m = E/c^2$. Now there is no question about it: Energy and mass are interchangeable. And that helps explain why electrons and photons and other particles appear and disappear in the universe.

Soon the accelerators are producing previously unknown particles. Physicists had thought that with the proton, the neutron, and the electron, they had atoms explained. They learn that protons and neutrons are not the smallest

While Albert Einstein was rewriting physics, James Joyce was trying out a new approach to literary writing, and Sigmund Freud was creating psychoanalysis. Each went inside his head to find insights and intellectual breakthroughs. Einstein came up with his famous *gedanken* (thought) experiments. Joyce wrote in a stream of consciousness (the mind speaking to itself). Freud's realm was the subconscious and unconscious mind, inner thoughts people aren't aware of having. Plato and Aristotle would be cheering all this mind work.

particles in existence. They are made up of still-smaller particles called quarks, which are held together by gluons. **Quarks are basic particles that have mass. Gluons are particles without any mass,** known as zero-mass particles.

Maybe it is the strangeness they are encountering that leads physicists to find playful names for the newly discovered particles. *Gluons* act like glue, so that name is an easy choice. But *quark* comes from a phrase in James Joyce's groundbreaking but difficult book, *Finnegans Wake*. "Three quarks for Muster Mark! Sure he hasn't got much of a bark," writes Joyce. No one has any idea what Joyce means—so it seems a perfect name for a baffling particle—a particle that is actually a good fit in a dynamic century.

As for quarks, they come in six varieties (scientists call them six "flavors"): up, down, top, bottom, strange, and charm quarks. For each quark, there is an antiquark. What happens when a quark meets an antiquark? They annihilate (destroy) each other! (Trying to understand why that happens will lead scientists to consider the earliest seconds of the universe, when there seems to have been a few more quarks than antiquarks. If that imbalance hadn't existed, we wouldn't be here today.)

Cosmic Gall

Neutrinos, they are very small.
They have no charge and have no mass
*And do not interact at all.**
The earth is just a silly ball
  To them through which they simply pass,
Like dustmaids down a drafty hall
  Or photons through a sheet of glass.
They snub the most exquisite gas,
Ignore the most substantial wall,
  Cold-shoulder steel and sounding brass,
Insult the stallion in his stall,
And, scorning barriers of class,
Infiltrate you and me! Like tall
And painless guillotines, they fall

An artist imagined a collision of nuclei that caused quarks and gluons (they make up protons and neutrons) to break free and rearrange themselves into plasma, a hot "soup" of charged particles.

  Down through our heads into the grass
At night, they enter at Nepal
  And pierce the lover and his lass
From underneath the bed—you call
It wonderful; I call it crass.
—John Updike

*Well, hardly at all. J.H.

Friendly Physicists

Rocky Kolb and Dave Finley, who are both physicists at Fermilab, send facts and a "bit of fun" to share with the readers of this book:

The first fact is: the distance a particle travels in one orbit of the Tevatron accelerator at Fermilab: It's "close enough" to the actual figure at 6.28 kilometers per orbit.

The next fact is: The beam of particles goes around at 299,800 kilometers per second, which is very, very, VERY close to the speed of light.

Dividing this speed by the distance (6.28 kilometers per orbit) gives you about 47,739 orbits per second. A slightly more accurate number, which does not round off the speed and distance, is 47,713 orbits per second. So let's use this number to figure out the number of times in all that the particle beam goes around the Tevatron:

First, we convert orbits per second to orbits per hour: 60 seconds x 60 minutes x 47,713 orbits = 171,766,800 orbits per hour.

Now you need another fact: The average particle beam stays in the Tevatron for about 18 hours. So multiply that by the orbits per hour and you get:

18 x 171,766,800 = 3,090,000,000 (rounded off)

So the number of times a particle goes around the Tevatron is typically about 3 billion times. Then, we start over with a new particle beam and do it again.

The main thing to remember here is that subatomic particles exist everywhere in this universe of ours, and each marches to a unique drumbeat. The accelerators lead to the discovery of 47 species of particles, some of which can combine to form composite particles—which means hundreds of other particles.

Some are particles of matter, and some are particles that carry the forces (like gravity, electromagnetism, and two forces that work inside the nucleus—details on page 378). Just how small are these particles? Elementary particles come in different styles and sizes, but if you shrink to one-billionth of one-billionth of a meter, you'll be able to join the quantum dance.

Electrons are big hunks in comparison to neutrinos, which are among the least massive of the crowd. An electron is 10 million times more massive than a neutrino. But neutrinos outnumber all the rest of the known particles by 1 billion to one. Unlike the negatively charged electrons, neutrinos don't carry an electric charge. There isn't much to a neutrino, but there are so many of them in the universe that their mass may be very significant. (Scientists are working on that.)

Neutrinos can go through planets, through a light-

Every charged particle has an antiparticle with an opposite charge and a reverse spin, like quarks and antiquarks. Brought together, matter and antimatter destroy each other in a burst of energy. Could that energy power a space vehicle? *Star Trek* fans say yes; an antimatter reactor fuels the fictional starship *Enterprise* (below). Some scientists say no. Right now, it takes too much energy to produce antimatter.

Chemistry, Charisma, and Peace

The greatest breakthrough in theoretical chemistry in the twentieth century was achieved by one man, Linus Pauling, whose idea about the nature of the chemical bond was as fundamental as the gene and the quantum because it showed how physics governed molecular structure.
—Peter Watson, British business writer and historian, *The Modern Mind: An Intellectual History of the 20th Century*

Many people live full lives. Some live truly active lives. A few people seem to live many lives. Linus Pauling was one such person.
—Pete Moore, British science writer, *E=mc²: The Great Ideas That Shaped the World*

It is 1962, and John F. Kennedy is President of the United States. He and his wife, Jacqueline, have invited America's Nobel Prize laureates to the White House for what everyone knows will be an elegant occasion. President Kennedy begins by saying, "I think this is the most extraordinary collection of talent, of human knowledge, that has ever been gathered at the White House—with the possible exception of when Thomas Jefferson dined alone." Linus Carl Pauling (1901–1994), born in Oregon, is among those being honored. Pauling, who was awarded a Nobel Prize in 1954 for his achievements in chemistry, is often called the greatest chemist since Antoine Lavoisier.

A tall, blue-eyed, cheery man, Pauling was a kid who read a lot and liked to experiment with chemicals. And, since his father was a pharmacist, he had safe ones available. But his carefree childhood ended when he was nine and his father died. Then his mother became seriously ill. Still, Pauling managed to graduate from Oregon Agricultural College

Antoine Lavoisier, a Frenchman who had his head chopped off during the French Revolution, is known as the founder of modern chemistry. For details, see chapters 25 and 26 of *Newton at the Center*, book two in this series.

(later Oregon State University). Money was scarce, and getting through college wasn't easy, but his academic work was outstanding. Oregon hired Pauling as an instructor before he even had his degree; they paid him a much-needed $100 a month. After that, he went off to the California Institute of Technology (Caltech), where he earned a doctorate.

Having been granted a fellowship for postdoctoral work in 1926, Pauling heads for Europe; he studies with Niels Bohr in Copenhagen, Erwin Schrödinger in Zurich, and William Bragg in London. Fascinated by atoms and molecules, Pauling has chosen the right places to learn about them. For the first time in world history, the technology to study the atom is actually available. This young American has taken himself to the discovery centers. He is there with the sleuths as the structure of the atom is uncovered.

Bohr had originally thought of the electron as a hard particle; he has come to realize that a single electron can spread out and seem to surround an atom's nucleus like a cloud. Given that insight into electrons, the quantum physicists are looking at the periodic table of elements (page 123) in a new and profound way. They realize that the number of electrons in its atoms helps determine why one element behaves differently from another. It is *not* chance that makes elements line up in the table as they do.

Scientists are just beginning to understand why some atoms link with other atoms to make molecules. Figuring out that bonding process will lead to a marriage between physics and chemistry. That challenge excites Linus Pauling. He knows (thanks to Bohr and quantum mechanics) that electrons exist in layered shells: 2 in the first shell (except for hydrogen), up to 8 in the second shell, and up to 18 in the third. (After that, it can get complicated.) The molecule story is in those electron shells—especially the outermost shell, where electrons are most free to move from one atom

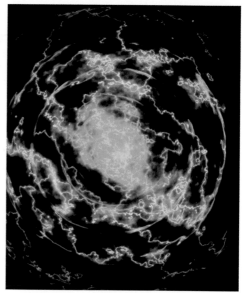

Electrons *seem* to surround the nucleus like a cloud in this modern, color-coded representation of a helium (He) atom. Actually, it is the probability of where an electron could be that is spread out. When you do find the electron—perhaps by using a gamma ray—it is located in only one spot. The colors stand for different densities of local electric charge; an electron is more likely to be in the central blue area (highest density) than the yellow (medium density) and outer red areas (lowest density).

to another. What happens in the outer shell determines if one atom will bond with another atom, making a molecule.

If an atom stands on its own, it's described as **inert**; if it bonds with other atoms to make molecules, it's **reactive**. The number of electrons in the outer shell decides the way an atom will react.

An American chemist, Gilbert Lewis, laid some of the groundwork for all this back in 1916. Lewis figured out that an atom with eight electrons in its outer shell is stable; it is not likely to bond and make a molecule. Then de Broglie came along in 1923 and added a key concept. He made it clear that electrons can be cloud-like waves *without fixed positions*.

Pauling, in 1926, returns to Caltech, builds on the known information, adds experimental data, and transforms chemistry. Here's the atomic picture as he paints it:

Imagine an atom with electrons filling its inner shells and one extra electron just hanging around in its outer shell (that's a picture of sodium). This is a very reactive element, eager to *get rid* of the superfluous electron.

Now imagine an atom with only seven electrons in its outer shell (like chlorine or bromine). It is also reactive, ready to *take on* another electron. **Getting rid of extra electrons, taking in strays, or sharing electrons—that's the way molecules are formed.** It's what makes for the diversity of matter. It's the secret behind bonding.

The process captivates Linus Pauling. For him, understanding and explaining chemical bonding becomes a lifetime passion (one of several). And it is a fascinating process. Atoms are searching for stability, so they either toss out electrons or take them in. When an atom's outermost shell is full, the atom is complete. It's chemically stable and has no urge to bond. Like a hotel with all its rooms filled, no more guests are needed.

But bonding isn't always simple. Benzene and graphite are two examples of an unusual kind of covalent bonding (see page 186). Their atoms share electrons in a way that, at first, baffles chemists. The benzene molecule (C_6H_6) consists

In 1929, Irish-born physical chemist and mathematician Kathleen Lonsdale (1903–1971), a contemporary of Linus Pauling, confirmed through X-ray diffraction the structure and dimensions of the baffling benzene molecule (C_6H_6). As shown here, it's a flat, hexagonal ring with alternating single and double bonds of carbon atoms (C), each attached to a hydrogen atom (H).

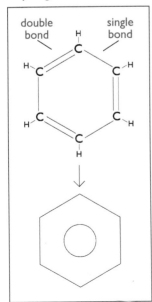

of six carbon atoms locked together in a hexagonal ring; one hydrogen atom is attached to each carbon atom. The ring has single bonds and double bonds alternating; that shouldn't produce a stable molecule. But benzene molecules *are* stable. No one knows why.

Linus Pauling rises to the challenge. He discovers that when atoms are symmetrically arranged in a single plane (as they are in the benzene molecule), then electron waves can spread out over all the carbon atoms. This spreading out of electrons leads to a very stable molecule. Understanding that explains many chemical reactions; it makes them predictable.

Pauling finds the structure of hundreds of molecules. He is especially interested in the repeated three-dimensional lattice patterns found in crystals, like table salt, quartz, and snowflakes. That lattice array is formed when layers of atoms line up in precise, three-dimensional order. He is able to measure the arrangement of atoms in a crystal by using X-ray diffraction, the scattering of X rays.

Pauling knows that the layers of atoms in a crystal act as mirrors, reflecting X rays, which bounce back and can be measured and recorded. Using that reflection—the scattering—it is possible to make a blueprint of the innards of a crystal. For Linus Pauling, science is about structure: Scattering gives him a way to discover the structures of many molecules.

Pauling has other tools in his toolbox. For instance, he can measure the magnetic properties of a substance and that, too, will tell him things about its inner architecture. He uses all the means he has and studies a whole lot of molecules—from metals to crystals to proteins—and publishes their structures. In 1939, Pauling describes what he has learned in a book titled *The Nature of the Chemical Bond* (he dedicates it to Gilbert Lewis).

Then World War II comes along (1941–1945 in the U.S.), and Pauling puts aside his theoretical work to concentrate on the war effort. He designs explosives and missile

In the microgravity of space, crystals grow perfectly, without the tiny flaws that can render X-ray diffraction useless. These "space babies" are proteins.

As you'll recall (from chapter 17), **SCATTERING** is the random reflection of a beam of particles when it hits something, like an atom's nucleus.

Let us examine a crystal…. [T]he equality of the sides pleases us, that of the angles doubles the pleasure. On bringing to view a second face in all respects similar to the first, this pleasure appears to be squared; on bringing into view a third it appears to be cubed, and so on.
—Edgar Allan Poe (1809–1849), "The Rationale of Verse"

propellants for the navy. He invents a device to monitor the oxygen level in planes and submarines (it will later be used in incubators to keep premature infants breathing). His war work is so impressive that President Harry S. Truman presents him with a Medal of Merit.

In 1947—two years after the war's end—Pauling publishes a textbook, *General Chemistry*. Like Lavoisier's eighteenth-century chemistry text, it changes the way chemistry is taught. (It will be followed by a new edition of *The Nature of the Chemical Bond* in 1960.) Pauling's books are enormously influential. Nobel Prize–winner Max Perutz says they transform chemistry into "something to be understood and not just memorized." Thanks to Linus Pauling, modern chemistry will focus on bonding; it will tie itself to physics and the emerging quantum world.

Linus Pauling teaches Caltech students about the structure of molecules. A former student says Pauling's passion for chemistry wasn't just about bonding: "For students, he pushed the 'magic' of the diversity of properties of substances."

Because of his dual interests—in political issues and science—Pauling becomes a public figure and, except for Albert Einstein, perhaps the best-known scientist in America. As a scientist, Pauling argues that radioactive fallout from nuclear weapons tests will increase the incidence of cancer and genetic disorders, including birth defects. As a citizen, Pauling marches in peace parades and speaks out against further nuclear testing.

During the McCarthy era in the United States (roughly 1950 to 1956), Senator Joseph McCarthy goes on a "witch hunt" against those he considers Communists or liberals. Pauling, who is involved in controversial political activities, is out of step with the times. The State Department takes his passport from him. He can't travel abroad; in 1951, he misses an important conference on the "Structure of Proteins" held in England. In 1953, Pauling publishes a book, *No More War!*, which further irritates his critics.

Isolated in California, Pauling studies molecules of living tissue. He is trying to find the secret of heredity. He isn't the only one. In England, two teams have the same goal in mind. They all know if someone can crack that code, we can begin to understand why some of us are healthy and others often sick, and why some of us win races while others stumble along. Understanding how heredity works will help us comprehend the structure of our minds and bodies.

Pauling studies protein molecules because he doesn't think DNA is complex enough to carry the genetic code. He believes proteins may hold the answer to heredity. He bends and folds a sheet of paper (in the way he knows acids bend) and comes up with a model of a kind of coil, or corkscrew spiral; he calls it an alpha helix.

In England, Francis Crick and James Watson know about Pauling's helix. Working at the Cavendish Lab at Cambridge in 1953, they figure out that DNA is a double helix after seeing an X-ray diffraction picture of a molecule taken by physical chemist Rosalind Franklin. A double helix—two spiral strands with connecting rungs—is fully capable of carrying the genetic code. Pauling, unable to travel without a passport, never sees that picture. If he had, would he have been first to discover DNA? No one knows.

Pauling is on to other things. Hearing a doctor talk about sickle-cell anemia, Pauling makes a quick guess that it

Linus Pauling was a theoretician, an experimenter, and a model builder. In other words, he worked both in his head and as a hands-on scientist. That's his alpha helix model of DNA, above.

The Ladder of Life

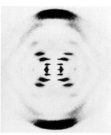

DNA (deoxyribonucleic acid), whose structure is shown in the photo at left, consists of two long chains of nucleotides twisted and joined by bonds like the rungs of a ladder. DNA is present in the nuclei of the cells of all living things. The sequence of nucleotides determines the characteristics of heredity.

In 1953, Watson (left) and Crick pose with their double helix DNA model at the Cavendish Laboratory. It is based in part on an X-ray diffraction of the molecule (right), made by Rosalind Franklin that year.

In 1949, Pauling made the link between sickle-cell anemia and faulty hemoglobin molecules. Due to a genetic flaw, the molecules cause red blood cells to sickle—stretch into a crescent or rod shape. Sickle cells die faster than they can be replaced, leading to anemia.

EMINENCE means fame, high rank, or importance. If you are eminent, you stand above others in some way. Usually the word describes a person, but a mountain peak can be eminent.

might be caused by a defect in the red blood cells. That means the disease is molecular and transmitted by genes. He does the research to confirm this possibility. Pauling's description of this molecular disease opens a whole new field of medical research.

By 1962, Pauling has white hair, which he often tops with a black beret. Along with his eminence in physical chemistry, some call him the founder of molecular biology, and he has also made important discoveries in genetic diseases, biomedicine, biochemistry, and brain function. Pauling has a special interest in nutrition and health. He studies anesthesia and memory. And he continues to be a peace activist.

And so, on the very day of the Kennedy White House gala, Pauling marches outside the gates of the president's house carrying a placard protesting atmospheric nuclear testing. Later, he dons his dinner suit and, with his customary zest, dances, talks, and enjoys the special evening inside the White House.

One year later, on October 10, 1963, the U.S., Great Britain, and the U.S.S.R. (Soviet Russia) sign a limited ban on nuclear testing. That same day, Linus Pauling is awarded a second Nobel Prize, this time the Nobel Peace Prize. It is given to the person thought to have made the greatest contribution to world peace during that year. A member of the Norwegian selection committee comments that the testing ban would not have happened without Pauling's effort. Linus Pauling becomes the only person to ever win two unshared Nobel Prizes.

After his trip to Sweden to receive the honor (his passport has been restored), Pauling hardly seems to take a breath before he is back at work on both scientific research and issues of world peace.

Pauling's continuing peace activism is annoying to many involved in business, security, and military issues. In 1964, he leaves Caltech and becomes a founding member of the Center for the Study of Democratic Institutions (described as a "liberal think tank"). Three years later (in 1967), he is appointed a professor at the University of California at San Diego and after that (in 1969) at Stanford University. But Pauling's political ideas are still too much for some, and he retires under pressure. Undaunted, he establishes the Linus Pauling Institute to research the role of vitamin C in human health (another controversial issue). In 1994, at 93, he dies at his ranch in Big Sur on the California coast.

Pauling has been called a "visionary of science and a prophet of humanity." Not a bad legacy.

Since Albert Einstein seemed to enjoy playing with our perceptions of time, maybe it's appropriate that in reading this book you'll do some traveling back and forth in time. In the next chapter, we'll be in the years just before World War II. In later chapters, we'll go still farther back in time. The timelines will help you stay grounded.

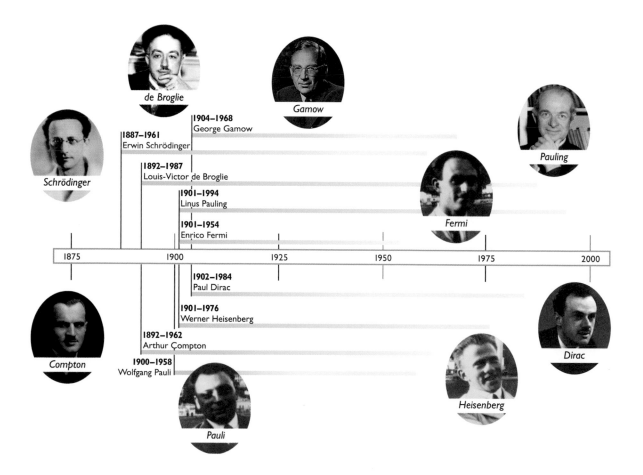

de Broglie

Gamow

1904–1968
George Gamow

Pauling

1887–1961
Erwin Schrödinger

Schrödinger

1892–1987
Louis-Victor de Broglie

1901–1994
Linus Pauling

1901–1954
Enrico Fermi

Fermi

| 1875 | 1900 | 1925 | 1950 | 1975 | 2000 |

1902–1984
Paul Dirac

1901–1976
Werner Heisenberg

Dirac

1892–1962
Arthur Compton

1900–1958
Wolfgang Pauli

Compton

Heisenberg

Pauli

What's in a Bond?

For you, bonding may mean getting married. For an atom, it means finding a way to stick with other atoms.

Here are two basic ways for one atom to bond with another atom:

Ionic bonding involves a giveaway. One atom donates an electron to another atom; the resulting attractive charges can bind the two atoms together or into a crystal.

In solid ionic compounds, ionic bonds take the distinctive shape and structure of a rigid crystal lattice. Sodium chloride's crystals are cubic.

Natural sodium chloride, with its cubic crystals, is called halite or rock salt. Sodium and chloride form an ionic bond (next page).

Metallic bonding (a form of ionic bonding) joins metals in a crystal lattice. Each atom contributes one or more electrons that "swim" more or less freely through the solid, leading to electrical conductivity. Since most elements are metals, this is common.

Covalent bonding is a sharing thing: An electron is shared between atoms. Carbon atoms, which lie at the core of life, bond covalently with up to four other atoms. Atoms of nonmetallic elements are particularly likely to form covalent bonds. Hydrogen and oxygen combine in a covalent bond to form water (see illustration on page 187).

Just to complicate things, some chemical bonds are a mixture of both processes, and some textbooks put hydrogen bonding in a separate category. What's important to remember is that chemistry is a quantitative (measuring) science down to the atomic level.

An **ION** is an atom or molecule that loses or gains one or more electrons, changing it from a neutral state to an electrically charged state. If it loses one or more negative electrons, it has one or more positive charges, making a **CATION**; if it gains one or more electrons, it has one or more negative charges, making an **ANION**. **VALENCE** (from the Latin word for "strength") is the measure of the number of bonds that an atom can form with other atoms. **ELECTRONEGATIVITY** is the tendency of an atom to attract electrons to itself when it is chemically combined with another element. Metals tend to have lower electronegativity (because they often give up electrons freely rather than attract them), and nonmetals, like fluorine, have higher electronegativity.

IONIC BONDING

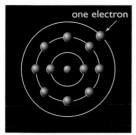
one electron

A sodium atom has 11 electrons: two in its first shell, eight in its second, and just one electron floating around in its outer shell. It's anxious to get rid of that extra electron.

room for one more

Chlorine has 17 electrons: two in the first shell, eight in the second, and seven in the highest energy level. It's on the prowl for one more electron to complete its outer shell.

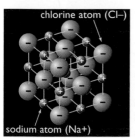
chlorine atom (Cl–)
sodium atom (Na+)

Sodium chloride (table salt) is the product of an ionic bond, a transfer of sodium's lonely electron to the outer shell of a chlorine atom. The sodium atom becomes a sodium ion with the stable electron configuration of neon (atomic number 10), its noble gas predecessor on the periodic table. The chlorine atom takes on the stable electron configuration of argon (atomic number 18), the noble gas that follows it. The sodium and chloride ions stick together by electrostatic attraction, forming a crystal arrayed in a repeating cubic pattern.

COVALENT BONDING

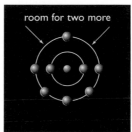
room for two more

An oxygen atom (atomic number 8) has two electrons in its inner shell and six electrons in its outer shell. Two more electrons would complete that outer shell.

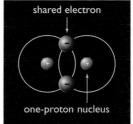
shared electron
one-proton nucleus

Hydrogen normally exists as a diatomic (two-atom) molecule, H_2. Two protons share two electrons equally in what's called a non-polar covalent bond. This setup mirrors the stable electron configuration of helium (atomic number 2), the noble gas right after hydrogen on the periodic table.

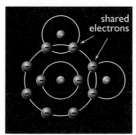
shared electrons

Another covalent bond happens when an H_2 molecule shares its two electrons with an oxygen atom. The result is a water molecule (H_2O). The sharing is unequal and so is called a polar covalent bond. The atom with the stronger electronegativity (oxygen) has a greater ability to attract electrons and acquires a slight negative charge. The other atom (hydrogen) is slightly positively charged, so that the overall charge of the molecule is zero.

Keep in mind that these diagrams look nothing like the atoms themselves. They are simple visual charts to show how the electrons are configured.

To Bond or Not to Bond?
Ask Whether 'Tis a Noble

In 1916, Gilbert Lewis (1875–1946) came up with an explanation for why atoms tend to form certain ions and molecules. Lewis was taking ideas from physics and applying them to chemistry. He said, "Atoms react by changing the number of their electrons so as to acquire the stable electron structure of a noble gas."

The "noble" (inert) gases are helium, neon, argon, krypton, xenon, and radon. Think of them as "snobbish" gases because they rarely combine with other atoms. (There are a few exceptions and, thus, a few compounds of noble gases.)

All the other gases are less than noble, and they do combine with one another. Those commoners seek partners. They are looking for nobility.

Each noble gas has eight electrons in its highest energy level (except for helium, which has two). So Gilbert Lewis's rule is known as the octet rule. Atoms of metallic elements obey the octet rule by losing electrons. Atoms of some nonmetallic elements obey the rule by gaining electrons.

The tendency to either lose or gain electrons when bonding is a measure of electronegativity. In 1932, Linus Pauling proposed an electronegativity scale of elements that's still in use today.

Neon, as in neon lights on a building, is one of the six noble gases, along with helium, argon, krypton, xenon, and radon. All of these atoms have full outer shells and so are stable and chemically inert (unreactive).

Back in 1690, deep-thinking John Locke wrote, "[If we knew] the mechanical affections of the particles of rhubarb, hemlock, opium, and a man, as a watchmaker does those of a watch, whereby it performs its operations; and of a file, which by rubbing on them will alter the figure of any of the wheels; we should be able to tell beforehand that rhubarb will purge, hemlock kill, and opium make a man sleep: as well as a watchmaker can." Ah, Mr. Locke, we've done all that—and much more—by penetrating the world of atoms and molecules. We've used our knowledge to become creative chemists—modern alchemists.

What Is a Metal?

The majority of elements are metals. Metals are usually shiny, malleable (they can be beaten into various shapes), and good conductors of electricity. Those properties come from the chemical bonding that holds the structure of a metal together. And that bonding is determined by electrons.

The outer electrons in metallic atoms in a solid tend to roam more or less freely through the metal. Those free-floating electrons (which carry a negative charge) are attracted to positively charged metal ions. (Pure metals, like copper or iron, consist of closely packed positive ions, also called cations.)

It's those mobile electrons that make metals such good conductors of electric current (which is a flow of electrons) and give metals their shine. Metals are malleable because of their fluid, changeable structure.

Understand bonding, and you'll know a lot about why things are one way or another.

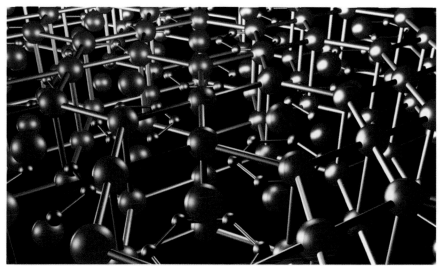

The red balls stand for magnesium (Mg) and the little purple ones for boron (B), which bond into a useful metal called magnesium diboride (MgB_2). In 2001, Japanese physicists discovered that the compound is a rare, high-temperature superconductor, which means it can carry lots of electrical current without resistance.

Energy Equals Mass Times the Square of the Speed of Light or $E=mc^2$

> What impresses our senses as matter is really a great concentration of energy into a comparatively small space.
> —Albert Einstein and Leopold Infeld, *The Evolution of Physics*

Scientific thoughts are written in the language of mathematics. And the most famous sentence in that language is one that Einstein wrote in 1907, two years after his miraculous year (based on one of those 1905 papers). This is it:

$$E = mc^2$$

It is the best-known equation in all of science. E = energy, m = mass, and c = the speed of light in a vacuum. Einstein's formula says that **mass and energy are interchangeable.**

Now, common sense tells you that a ball and the energy it takes to throw a ball are not the same. Yet particle accelerators can take the nucleus of an atom from that ball and, under certain circumstances, change it into energy. Mass, it seems, is very, very concentrated energy.

$E = mc^2$ turned out to be an explosive formula.

What does this formula do to the law of conservation of energy? How about the other conservation law—the law of the conservation of mass—which says that mass can neither be created nor destroyed?

Einstein's formula unites those conservation laws. It shows us that energy can be changed—not only to another form of energy but also to mass and vice versa. But neither can be

Einstein first wrote his famous equation as: $m = E/c^2$. What's the difference? This original equation tells how much mass can be converted from an amount of energy; the famous form explains how much energy is contained in an amount of mass.

Here's an example of the law of the conservation of mass: If you burn a log, the resulting ash and gases will have the same total mass as the unburned log plus the oxygen needed for combustion. Individual particles are destroyed, but the mass of the whole system stays the same.

destroyed. So scientists no longer have to deal with two conservation laws. Einstein's theories turned them into one: **the law of mass-energy conservation**.

Hardly anyone could imagine matter actually being changed into energy or energy into matter. Would you have thought it possible?

Even Einstein has no idea, in 1907, that his simple but astounding formula will energize a generation of scientists. They will make the formula work in the real world, and that will lead to a new kind of power: nuclear power.

Square the enormous speed of light—roughly 300,000 kilometers per second (186,000 miles per second) in a vacuum—and consider the size of the number you get. A little mass goes a long way when multiplied by the square of the speed of light. The power pack in the first nuclear bomb will be the size of a grapefruit. The small change in mass in that bomb will release enough energy to level a city.

Einstein will later ask himself, "If every gram of material contains this tremendous energy, why did it go so long unnoticed?" Then he says, "The answer is simple enough: so long as none of the energy is given off externally, it cannot be observed. It is as though a man who is fabulously rich should never spend or give away a cent; no one could tell how rich he was."

To repeat: $E = mc^2$ says that mass has the potential to be converted into energy and vice versa. "The law permits us to calculate in advance, from precisely determined nuclear weights, just how much energy will be released with any nuclear disintegration we have in mind. The law says nothing, of course, as to whether—or how—the disintegration can be brought about," Einstein writes.

In other words, the formula describes a possibility; it doesn't give how-to directions. Einstein doesn't know if humans will ever actually split a nucleus and convert its mass into energy. That we do it, and during his lifetime, will stun Einstein and everyone else.

Banesh Hoffman, a student of Einstein, writes:

Imagine the audacity of such a step.... Every clod of earth, every feather, every speck of dust becoming a prodigious reservoir of untapped energy. There was no way of verifying this at the time. Yet in presenting his equation in 1907 Einstein spoke of it as the most important consequence of his theory of relativity. His extraordinary ability to see far ahead is shown by the fact that his equation was not verified ... until some twenty-five years later.

In the United States, Edwin Armstrong takes out four patents on FM (*frequency modulation*) broadcasting. He has adjusted the frequency of radio waves to carry the signal instead of changing the amplitude—as with AM (*amplitude modulation*) radio. The result is much-reduced static in radio reception.

Albert Einstein is in Pasadena, California, on his third visit to the California Institute of Technology. At a party hosted by silent-screen film star Charlie Chaplin, the physicist joins three other musicians and plays a Mozart quartet.

In March, the first German concentration camp, Dachau, is constructed on the grounds of an old ammunition factory near Munich. Heinrich Himmler, the new police chief of Munich, says it is for "political prisoners." (By 1945, 10 million humans will have been sent to Nazi camps, many working as slave laborers to aid the German war machine; more than half will be murdered.)

Bruno Walter, the Jewish orchestra maestro, flees Germany.

A mountain of soulless shoes silently testifies to the mass murder of prisoners, mostly Jews, in a World War II German concentration camp.

Einstein and his second wife, Elsa, take a train across the United States before returning to Europe. The train stops in Chicago on March 14 (Albert's birthday), and he is honored at a luncheon. There, he meets Clarence Darrow, the lawyer who defended biology teacher John Scopes in a famous Tennessee trial about evolution. At the next train stop, Albany, New York, Einstein learns that Nazi police have torn apart his Berlin apartment, but his stepdaughter had already sent his scientific papers to a safe place. The German consul in America warns him, "If you go to Germany, Albert, they'll drag you through the streets by the hair." Einstein understands that if he goes home to Germany, he will lose more than his hair.

A 1925 cartoon published during the Scopes "Monkey Trial" imagines the "science" classroom of William Jennings Bryan, the antievolution attorney.

Still, he and Elsa head for Europe; they have family and personal matters to tend to. On board the ship, they learn that in the German city of Ulm, the city administration has ordered the change of a street name; it will no longer be known as Einsteinstrasse (Einstein Street). He hears from Max Planck, who asks Einstein to resign from the Prussian

Academy of Sciences. Planck is no Nazi, but he feels that Einstein's criticism of the new government disqualifies him from holding the prestigious German membership. The day after the ship lands in Antwerp, Belgium, Einstein resigns. The Prussian Academy denounces Einstein as a traitor. Planck, showing his lack of bias, says, "In the history of future centuries the name of Einstein will be celebrated as one of the most brilliant stars that have shone in our Academy." But only Max von Laue opposes the academy's action.

Accepting temporary asylum in Belgium, Einstein is soon playing violin in a string quartet that meets weekly. His good friend, Belgium's Queen Elizabeth, also plays in the quartet. On a visit to England, Einstein and Winston Churchill (the future British prime minister) speak out against Nazi aggression at a meeting chaired by Ernest Rutherford.

Einstein is offered a position at the Hebrew University of Jerusalem, another at Caltech, lecture invitations at several universities, and political asylum in Turkey. He finally accepts an offer from the brand-new Institute for Advanced Study at Princeton, New Jersey, when they agree to also take his assistant. Einstein intends to split his time between the United States and Europe. He doesn't know that he will never see Europe again.

In an Arthur Szyk cartoon, Hitler and his henchmen—Himmler, Göring, and Goebbels—sing "Peace, peace" while secretly arming for war.

The 1933 Nobel Prize for physics is shared by England's Paul Dirac and Austria's Erwin Schrödinger. Schrödinger has succeeded Max Planck at the University of Berlin. Schrödinger is not Jewish, but he hates the Nazis, speaks out against them, and leaves for Ireland.

Germany withdraws from the League of Nations, announcing its intention to rebuild its army. Germany had been secretly arming since 1929, even though it was forbidden to do so under the terms of the Treaty of Versailles signed after the Great War (later called World War I).

In May, thousands of Germans turn out for a torchlight parade that stops at a plaza near the University of Berlin. The marchers watch as books are thrown onto a bonfire. "Albert Einstein, Jack London, Ernest Hemingway, Helen

In 1933, in Berlin's Opernplatz public square, a large crowd burns books that the Nazis have judged "un-German."

Keller, John Dos Passos, Sigmund Freud..." The crowd cheers as each author's name is called and his or her books are burned. German propaganda chief Joseph Goebbels says triumphantly, "Intellectualism is dead."

Adolf Hitler, speaking to Max Planck, says, "If the dismissal of Jewish scientists means the annihilation of contemporary German science, then we shall do without science for a few years!"

Edward Teller, Hungarian-born nuclear physicist, flees Germany for London and then for the United States, where he and George Gamow (see box, page 139) will work together studying radioactive decay. Teller will become an advocate of powerful weaponry, the "father" of the hydrogen bomb, and a controversial figure.

Leo Szilard, another Hungarian-born Jewish physicist, is in England, where he reads H. G. Wells's prophetic novel *The World Set Free*. In it, Wells predicts an atomic bomb. Szilard, standing on a London street corner, has a vision of how one might be possible. He is believed to be the first scientist to figure it out. (See the next chapter for details.)

German physicist Otto Stern heads for the United States (in 1943 he will win a Nobel Prize). In addition to scientists, some 60,000 artists (authors, actors, painters, and musicians) leave Germany and Austria in the next six years. It is the greatest intellectual migration in history.

1934. Hitler signs a non-aggression treaty with Poland, promising not to invade that country. Hitler is stalling for time to build his military forces. He knows the Polish army is more than two and one-half times the size of the German army. Hitler begins courting Italian prime minister Benito Mussolini, who, at first, calls Hitler a "buffoon."

In Mexico, Lázaro Cárdenas becomes president and carries out the liberal reforms promised by the Mexican Constitution of 1917.

Mao Zedong and other Communist leaders begin the Long March across China, retreating from their Nationalist foes. Mao will eventually take over the nation.

Italian dictator Benito Mussolini, *Il Duce* ("The Leader"), invented the fiery fascism that spread through Germany under a Nazi banner.

In a 1973 poster, Chairman Mao presides over "New Year in Yenan." Yenan is the Communist base founded after the Long March.

In the United States, the first modern comic book, *Famous Funnies*, is published. Child star Shirley Temple is a movie box-office sensation.

In England, P. L. Travers, a 28-year-old Australian-born actress, writes a book called *Mary Poppins*. It is a huge success.

In France, Irène and Frédéric Joliot-Curie begin experiments with neutrons, hoping to create new elements. In Italy, Enrico Fermi works on the same idea.

The Nobel Prize in chemistry goes to an American, Harold Urey, for the discovery of heavy hydrogen. He has found isotopes of hydrogen—atoms with extra neutrons. One of these isotopes, deuterium (D), with a nucleus composed of one proton and one neutron, will eventually be used to build hydrogen bombs. Heavy water, D_2O, will also be used to make those fusion bombs.

The first modern comic book featured the futuristic Buck Rogers, who introduced fans to rocket ships, robots, and ray guns.

1935. The Nobel Peace Prize for 1935 is given to German anti-Nazi journalist and author Carl von Ossietzky, who has been thrown into a concentration camp. The Nazi government demands that he turn down the honor. He refuses to do so. The government announces that in the future no German can accept any Nobel Prize. Ossietzky can't leave Germany to accept the prize.

Chemist Wallace Carothers at the DuPont company in Wilmington, Delaware, produces nylon, the first synthetic fiber.

The first radar (*radio detection and ranging*) stations are built in England. The idea goes back to Heinrich Hertz, who, in the 1880s, found that radio waves could be bounced off objects to determine the position of the objects. During the coming war, radar will be a key defense tool in directing fighters to intercept German bombers.

The Nobel Prize in chemistry goes to Frédéric and Irène Joliot-Curie for creating radioactive isotopes in their laboratory. Shooting alpha particles at atoms, they change aluminum atoms into a new radioactive form of phosphorus (radio-phosphorus), boron atoms into radio-nitrogen, and make other transmutations.

In the 1930s, Shirley Temple became Hollywood's first child superstar. Little girls everywhere had their hair curled to look just like Shirley's.

The **EXCLUSION PRINCIPLE** states that no two electrons (or protons or neutrons) in an atom can have the same quantum state—with all the quantum numbers identical.

Nazis forced *Juden* (Jews) to wear a yellow patch identifying them by religion. In concentration camps, identifying numbers were burned into the prisoners' skin.

I hold my breath during the last stretch. I stick with the field, breathing naturally until 30 yards from the finish. Then I take one big breath, tense all my abdominal muscles, and set sail.
—Jesse Owens, Olympic gold medalist

Wolfgang Pauli, who formulated the exclusion principle and also predicted neutrinos, leaves Austria for a professorship at Princeton University. Pauli's godfather was Ernst Mach (the great physicist who never believed in atoms); his mother was Jewish.

Hans Bethe, another physicist with a Jewish mother, was fired from his teaching job in Germany in 1933 and now heads for the United States.

New German laws, called the Nuremberg Laws, take all civil rights from the nation's 600,000 Jews. (Later those laws will be used against the millions of Jews in nations that Germany occupies.) It is the beginning of what the Nazis call "the Final Solution." Jews are not allowed to vote, hold public office, or work at most jobs. Marriages between Jews and non-Jews are considered illegal; those who refuse to divorce are jailed. Few protest when Jewish families disappear from their midst. Gypsies, homosexuals, and those suspected of disloyalty are also swept up in the lethal purge.

1936. German elections give Hitler 99 percent of the vote. German troops occupy the Rhineland, a demilitarized zone on either side of the Rhine River in Germany. Mussolini and Hitler proclaim a Rome-Berlin Axis.

Britain, France, and the U.S. sign the London Naval Treaty; it will lead to the formation of the Allied Powers ("the Allies"), a group of nations opposed to the Nazi government.

The Olympic Games are held in Berlin, Germany. Hitler expects German athletes to win medals and when they do he publicly shakes their hands. Hitler insists that blacks, like Jews, are inferior to the German "race." Ten African-Americans are awarded Olympic medals, including eight golds. One of them, America's Jesse Owens, is the sensation of the Olympics; he wins four gold medals and becomes a world figure. Owens is the son of an Alabama sharecropper. "I wasn't invited to shake hands with Hitler, but I wasn't invited to the White House to shake hands with the President either," he says. Back home, he and his African-American teammates must sit in the back of the bus in many states. (Twenty years later, in 1956, Owens will attend the Olympics as President Dwight Eisenhower's personal representative.)

In England, the BBC (British Broadcasting Corporation) inaugurates television service.

The civil war begins in Spain; it is a preview of the world war to come.

In Russia, Soviet secret police arrest some 5 million citizens. Anyone who questions Joseph Stalin's rule is in danger. Millions will never be seen again in what is known as the Great Purge.

In this ceremonial 1925 painting, the Russian dictator, Joseph Stalin, is reciting an oath of allegiance to the Soviet nation.

In *Modern Times*, Charlie Chaplin's Little Tramp can't cope with machines like the counter-shaft, double-knee-action corn feeder with a synchro-mesh transmission.

Charlie Chaplin stars in the film *Modern Times*.

British test pilot and engineer Frank Whittle and German physicist Hans von Ohain each invent a jet engine; they work separately and know nothing of each other's work. Whittle receives a patent for his turbojet engine in 1930; von Ohain gets his patent in 1936. In 1939, von Ohain's jet will be first to fly.

Carl David Anderson (at Caltech in California) and Victor Franz Hess (at Innsbruck University in Austria) share the Nobel Prize in physics. Anderson wins for his 1932 discovery of the positron. Hess wins for his discovery of cosmic rays. His wife is Jewish, and they will soon flee Austria for the United States; Hess will teach at Fordham, a Jesuit university in New York City.

Eugene Paul Wigner, born in Budapest, Hungary, and educated in Germany, is a professor of physics at the University of Wisconsin. He works out the mathematics explaining what happens when scientists bombard a

In an article in *Physics Today*, Eugene Wigner (shown here with his daughter Erika) is described as a "towering figure" who "decided that physics had a duty to provide a living picture of our world, to uncover relations between natural events, and to offer us the full unity, beauty, and natural grandeur of the physical world."

nucleus with neutrons. This is important information for
nuclear energy theorists. Wigner is one of a brilliant group
of Hungarian physicist/mathematicians who come to the
United States in response to the turmoil in Europe. Later, he
will win a Nobel Prize as well as an Atoms for Peace Award.

1937. Ernest Rutherford (now Baron Rutherford of
Nelson) dies in England at age 66.

Russian Sergei Prokofiev composes *Peter and the Wolf*, an
orchestrated children's story in which each character is
represented by an instrument and a melody.

This famous
painting by Pablo
Picasso shows the
horror of war, in
this case the
bombing of the
village of Guernica
in the Spanish Civil
War.

Pablo Picasso, a Spanish painter living in Paris, paints
Guernica, a soon-to-be-famous mural showing the horror of
the Civil War in Spain.

Max Planck resigns from the presidency of the Kaiser
Wilhelm Institute in Berlin, protesting the Nazis' treatment
of Jewish scientists. Later, his youngest son, Erwin, will be
executed for plotting against Hitler.

Japanese troops seize major Chinese cities. The Chinese
Nationalist leader, Chiang Kai-shek, allies himself with his
former enemies, the Chinese Communist leaders Mao
Zedong and Zhou Enlai, to resist the invaders.

Germany promises not to invade Belgium (a promise that
will be broken).

The French actor and director Jean Renoir makes a film
about the Great War, *Grand Illusion*, which is a plea for peace.

More than one-half million Americans are involved in
sit-down strikes in the Depression-stressed nation.

Only four gaps remain in the periodic table of the
elements between number 1 (hydrogen) and number 92

Grand Illusion was banned
in Germany and Italy and
nearly destroyed by
propagandist Joseph
Goebbels, who called it
"Cinematic Public Enemy
Number One."

(uranium). Element 43 is one of the missing. Italian physicist Emilio Gino Segrè decides to try to manufacture it. He knows that Enrico Fermi has been bombarding elements with neutrons to produce elements with a higher atomic number. He decides to bombard element 42 (molybdenum) with deuterons (each contains one neutron and one proton) in a particle accelerator. The experiment works. Element 43 is named technetium, from the Greek word that means "artificial." It is radioactive and found in certain stars in extremely minute amounts. Technetium turns out to be useful as a superconductor, a hardener in steel alloys, and in nuclear medicine.

Carl David Anderson (see 1932 and 1936) discovers an intermediate subatomic particle; more massive than an electron but less massive than a proton, it is called a **meson** or sometimes a mu-meson. It was predicted by Japanese physicist Hideki Yukawa in 1935.

In December, Japanese troops capture the capital of the Republic of China, Nanking, and for six weeks descend into barbarism, raping and murdering hundreds of thousands of civilians in a massacre condoned by their officers.

Technetium (Tc), atomic number 43, has an isotope widely used for medical diagnosis with a half-life of just six hours.

In September 1937, in Nanking, China, a baby cries amid the ruins of a Japanese bombing raid. The city fell to Japan in December; at least 300,000 civilians were slaughtered.

1938. British prime minister Neville Chamberlain meets with Hitler in Germany, talks of "peace in our time," and agrees to let German troops take the Sudetenland (part of Czechoslovakia). Chamberlain believes he has kept England from war; others, like Winston Churchill, call it appeasement, a way to calm Hitler by agreeing to his demands. German troops march into the Sudetenland.

President Roosevelt urges Hitler and Mussolini to solve their problems peacefully.

British civilians are given gas masks in anticipation of war.

The U.S. Supreme Court rules that the University of Missouri Law

Would you have bought a gas mask in 1938 England? This ad predates the first German air raid on September 6, 1939.

EVERYTHING FOR SAFETY EVERYWHERE

AIR RAID PRECAUTIONS EQUIPMENT

GENERAL CIVILIAN RESPIRATORS
CIVILIAN DUTY RESPIRATORS
GENERAL SERVICE RESPIRATORS
PROTECTIVE CLOTHING
STEEL HELMETS
FIRST AID OUTFITS
OXYGEN ADMINISTRATION APPARATUS
GAS SMELLING SAMPLES
STEEL STRETCHERS
REDHILL CONTAINERS
STIRRUP PUMPS

All materials for Gas-Proofing Rooms, etc., etc.
Manufacturers of breathing apparatus of every description
Established 1819

SIEBE, GORMAN & COMPANY, LIMITED
Gas Mask Contractors to H.M. Government.   The oldest designers and manufacturers of breathing apparatus in the world
WESTMINSTER   BRIDGE   ROAD,   LONDON,   S.E.1

African-American Lloyd L. Gaines sued the University of Missouri, citing the Fourteenth Amendment, to be admitted to law school.

Those egg-shaped, metallic giants on flamingo legs are the 1927 comic book version of H. G. Wells's Martian invaders. His sci-fi classic *The War of the Worlds* was published in 1898.

School must admit "negro students."

War of the Worlds, a radio play produced by Orson Welles (based on a story by H. G. Wells), airs the night before Halloween. An announcement is made that aliens have invaded the United States. It's part of the play, but some people believe it and panic.

Wealthy American industrialist Howard Hughes flies around the world in 3 days, 19 hours, and 17 minutes.

Abraham Pais, 20, who will become a particle physicist and biographer of Albert Einstein and Niels Bohr, is studying for his Ph.D. in Amsterdam. Because he is Jewish, he must go into hiding. Unlike Anne Frank, Pais is lucky and survives.

President Roosevelt recalls the American ambassador from Germany; Germany recalls its ambassador in Washington—ending normal relations.

Alan Turing, a graduate student at the University of Cambridge, publishes a paper that turns out to be the birth announcement of the digital computer.

Ukrainian-born physicist George Gamow suggests that nuclear fusion is the source of the Sun's energy and that the

This is a photo as I would wish myself to look all the time. Then I would maybe have a chance to come to Hollywood.
—Anne Frank, 10 Oct. 1942

nuclei of hydrogen (the lightest element) fuse into helium nuclei (the next heaviest element) in that fiery orb. In the process there is a loss of combined mass that turns, explosively, into energy. Gamow doesn't know exactly how it happens, but Hans Bethe (now an American; see 1935) figures it out, explaining why the Sun shines.

German laws now prohibit Jewish children from attending

public schools, Jewish teachers are dismissed, and Jews have their passports taken from them. On November 9, mobs attack Jewish businesses, synagogues, and homes in what is called *Kristallnacht*, the Night of Broken Glass. Hitler blames the Jews for the damage and sends them a bill for the cleanup costs. Germany and Italy are now refusing to let Jews leave.

In Italy, racial laws are passed copying those in Germany.

Breaking a tradition of secrecy, Niels Bohr tells Italian Enrico Fermi that Fermi will win the Nobel Prize in physics. Laura Fermi, who studied engineering and science before marrying Enrico, is Jewish. Fermi takes his whole family with him to Stockholm; from there, as the story goes, "they get lost and can't find their way back to Italy." Fermi joins the physics department at Columbia University.

Austrian-born Lise Meitner—a professor in Berlin since 1917, co-discoverer of the radioactive element protactinium (number 91) and known for her work in nuclear physics—flees Germany. Einstein compares her to Marie Curie. She's a world-class physicist at a time when women are mostly shut out of the field. Niels Bohr helps make Meitner's rescue possible by finding university work for her in Sweden.

On *Kristallnacht*, German citizens took sledgehammers to Jewish homes and businesses, leaving streets littered with broken glass. Jews were beaten to death and synagogues set afire, as documented here, in Munich.

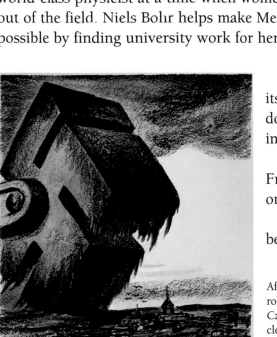

1939. Breaking its promise not to do so, Germany invades Poland.

Britain and France declare war on Germany.

World War II begins.

After the Nazi war machine rolled into Austria and Czechoslovakia, it was clear that Poland would be "NEXT!", the title of this powerful 1939 cartoon.

The Fission Vision

The new era really began in 1933, when Leo Szilard had the inspired idea that a chain reaction could be initiated by the bombardment of an element with neutrons.
—Michael White and John Gribbin, *Einstein: A Life in Science*

[In London] while living on his savings, Szilard had no other scientific or academic burdens and deadlines. No family. No close friends. No household chores. When he wanted to think about the chain reaction, he could. And did. For days and nights at a time.
—William Lanouette with Bela Silard, *Genius in the Shadows: A Biography of Leo Szilard, the Man Behind the Bomb*

It is September 12, 1933, and London's morning paper, the *Times*, carries an article about a speech by Britain's celebrity scientist, Ernest Rutherford. The great physicist (now *Lord* Rutherford) describes "the discoveries of the last quarter century." He talks of "bombarding atoms" and about the "transformation of elements." This is news to most readers of the *Times*; elements are supposed to be unchangeable. When it comes to the atom as a source of power (as imagined in H. G. Wells's popular science-fiction novel *The World Set Free*), Rutherford dismisses the idea, saying, "The energy produced by the breaking down of the atom is a very poor kind of thing. Anyone who expects a source of power from the transformation of these atoms is talking moonshine."

Leo Szilard (SIL-ard), newly arrived in England, reads the morning news. A 35-year-old Hungarian who served in the Austro-Hungarian army during

Leo Szilard, shown here near Oxford, England, in 1936, had broad abilities. He submitted 29 applications to the German patent office, many in partnership with Einstein.

Hungary's capital, Budapest, is a combination of the hilly Obuda ("Old Buda") and Buda on the west bank of the Danube River and the flat, modern city of Pest on the east bank. The cities were unified in 1873, about the time this panorama was painted.

the Great War (that's World War I), Szilard was saved from sure death when the flu kept him from a battle in which his fellow soldiers all perished.

After the war, in 1919, Szilard left his native Budapest to study at the University of Berlin. There, he intended to pursue engineering, but Max Planck, Max von Laue, and Albert Einstein were among the physics professors. Szilard was smart enough to change his major to physics.

When the university didn't offer a course he wanted to take, Szilard persuaded Einstein to teach it in a seminar. When Max von Laue gave him a problem to pursue, young Szilard decided it wouldn't lead anywhere. So, after just a year of university physics, he discovered an "unsolvable" physics problem and solved it in an original paper. Then he was afraid to show it to von Laue (after all, he was supposed to do what his teacher had told him to do). Instead, he showed Einstein what he had found.

"That's impossible," said Einstein. "That cannot be done."

"Well, yes, but I did it," said Szilard, who explained his ideas to Einstein, and "[Einstein] liked this very much."

The next day, Szilard had the courage to take his paper to von Laue, who accepted it as a Ph.D. thesis. Not long after that, Szilard joined the faculty at the University of Berlin.

Szilard has a mind that dazzles. When he isn't experimenting or theorizing in physics, he invents things.

The "unsolvable" problem was posed by James Clerk Maxwell. It involved an imaginary demon that seems to defy the second law of thermodynamics. Curious? Read chapters 37 and 38 of *Newton at the Center*, book two in this series.

Besides that, he keeps trying to save humanity (mostly by being involved with do-good organizations). Szilard cares deeply about moral issues. Many of Europe's Jewish physicists believe that the Nazis are a momentary madness. They assume they can wait out Hitler. Szilard thinks otherwise. He stuffs all his savings into his shoes and leaves Germany. So that's why he is in London in 1933. He is living on those savings, hoping to find productive work. Years later, the French biologist Jacques Monod will describe Szilard as a "short fat man...his eyes shining with intelligence and wit...generous with his ideas." But in 1933 he is still slim and boyish with thick, curly dark hair and a soulful, Bohemian air.

By nature, Szilard is a contrarian. He loves to take the opposite view of whatever he hears. And, that morning in 1933 when he reads the *Times*, Szilard finds Rutherford's thoughts on nuclear energy "rather irritating because how can anyone know what someone else might invent?" Rutherford must be wrong, he says to himself. But why? Szilard's mind focuses on atoms and nuclei and their potential power; he thinks of little else.

Hitler youth parade in Germany in the 1930s. To many boys, the drills, uniforms, and hype made military life seem heroic and war patriotic. Hardly anyone considered the morality or purpose of it all.

Many...people took a very optimistic view of the situation. They all thought that civilized Germans would not stand for anything really rough happening. The reason that I took the opposite position was... [because] I noticed that Germans always took a utilitarian point of view. They asked, "Well, suppose I would oppose this thinking, what good would I do?...I would just lose my influence."...You see, the moral point of view was completely absent, or very weak....And on that basis I reached in 1931 the conclusion that Hitler would get into power, not because the forces of the Nazi revolution were so strong, but rather because I thought that there would be no resistance whatsoever.
—Leo Szilard, quoted in *Leo Szilard: His Version of the Facts*

The original "Eureka!" moment struck Greek philosopher/mathematician/engineer Archimedes while he was taking a bath, as depicted here in a sixteenth-century woodcut. Read about it in chapter 17 of book one, *Aristotle Leads the Way*.

"In the days that followed, [Szilard] pondered Rutherford's declaration in a routine favored for serious thought: long soaks in the bathtub and long walks in the park," says his biographer William Lanouette. "Szilard walked and wondered that chilly September, seeking with each quick step a way to disprove the 'expert' Rutherford."

Standing on a London street corner, Szilard has a "Eureka!" moment. Here are his words describing it:

As the light changed to green and I crossed the street, it suddenly occurred to me that if we could find an element which is split by neutrons and which would emit two neutrons when it absorbs one neutron, such an element, if assembled in sufficiently large mass, could sustain a nuclear chain reaction.

(Read that paragraph a few times to be sure you have it.) Szilard has imagined the two steps needed to free the energy in an atom's nucleus: one, a nuclear chain reaction, and two, a critical mass of the right element to set off and sustain it. Does he envision the process we now call "fission," the splitting of nuclei? Yes, he seems to, but Szilard's ideas are not yet clearly expressed. And he doesn't know which

When it comes to chain reactions, getting the mass just right is critical. Some neutrons leak from the surface of whatever you're testing. Those leaked neutrons are out of the game; they can't cause further fission events. So, if you start with a small chunk of an element, you may not have enough neutrons to maintain fission. A big chunk will go off prematurely. A just-right amount is called a "critical mass."

Fermi and Szilard: Neat Must Work with Messy

Fourteen-year-old Enrico Fermi was browsing near a statue of Giordano Bruno (the philosopher who was burned at the stake in 1600 for his scientific beliefs). It was market day in Rome, and outdoor stalls were filled with paintings, books, food, clothing—a mishmash of things. Enrico was grieving. His older brother had just died, and they had been inseparable. The boy needed something to think about other than his brother. So when he found two old books on physics (written in Latin), he bought them. Then he read them straight through. From then on, there was no question about it: He would be a physicist.

A few years later, when Fermi applied for a fellowship at the University of Pisa, the examiner who read his competitive essay was astounded. He said the work would do credit to a doctoral

Student Emilio Segrè (left) met Professor Enrico Fermi (right), just four years older, and quickly switched from engineering to physics. In 1927, the newly acquainted young men spent this day at the beach with a friend.

In this courtyard at the University of Pisa, politics and scientific ideas have been and are actively debated.

candidate. Enrico was 17. Two years after that, he was teaching his professors. In a century filled with outstanding physicists, Enrico Fermi was among the greatest. He became a leading force among the physicists who built the first nuclear bomb. Here are a few comments from his peers:

J. Robert Oppenheimer said that Fermi had "a passion for clarity. He was simply unable to let things be foggy. Since they always are, this kept him pretty active."

"My greatest impression of Fermi's method in theoretical physics was its...simplicity....He stripped it of mathematical complications and of unnecessary formalism. In this way, often in half an hour or less, he could solve the essential physical problem involved," said Hans Bethe.

"His teaching was exemplary, minutely prepared, clear, with emphasis on simplicity and understanding of the basic ideas, rather than generalities and complications....We would knock at his office door, and if free, he would take us in, and then he would be ours until the question was resolved," said Jack Steinberger.

"Fermi was a rigorous academic whose life centered on a brilliant physics career; he had little interest in politics....A homebody,...he awoke at 5:30 each morning and spent the two hours before breakfast polishing his theories and planning the day's experiments," writes William Lanouette in *Scientific American*.

As for Szilard, Lanouette says, "The bachelor Szilard rarely taught, published infrequently and dabbled in economics and biology....A late sleeper, he often appeared at Columbia only in time for lunch, after which he would drop in on colleagues, posing insightful questions and suggesting experiments *they* should try." This "odd couple" had to work together to build a bomb. It wasn't easy for either of them.

Like the ancient alchemists, each is trying to transmute (change) elements in the laboratory. By inserting extra particles into the nucleus, they hope to get new elements. The winner of the race will gain international acclaim and probably a Nobel Prize.

In Paris, in 1933, Irène and Frédéric Joliot-Curie shoot alpha particles at aluminum and produce a radioactive isotope in their laboratory. It's a big achievement. Marie and Pierre Curie won the 1903 Nobel Prize in physics for their work in natural radioactivity; their daughter and son-in-law receive the 1935 prize in chemistry for artificially creating radioactive elements.

At about the same time, the talented Enrico Fermi, a professor at the University of Rome, begins a series of experiments by propelling electrons at nuclei. Nothing much happens. Then he tries sending protons; they are repelled by the positive electric charge of the protons in the nucleus. After James Chadwick discovers neutrons, Fermi starts tossing neutrons at nuclei—and something does happen.

Since the neutron has no charge, it is *not* repelled by nuclei (just as Szilard hypothesized). Neutrons don't need to overcome an electrical barrier, as protons do. The strong force attracts and welcomes them. Rutherford thinks the neutron is too lethargic—slow-moving—to start any action. Lise Meitner has already discovered that neutrons are more likely to be absorbed by a nucleus when they move slowly.

Fermi discovers the same thing accidentally when he does his experiments on a wooden table—and gets better results than on a marble table. He figures out that the nuclei of elements in wood (especially hydrogen) must slow the neutrons by colliding with them. He guesses that paraffin (wax) nuclei will do the same thing, so he puts paraffin filters between the neutron beam and the target. After Fermi describes his encouraging results, Lise Meitner writes to him (on October 26, 1934): "Enclosed is a small notice currently

Enrico Fermi's goal in 1934 was to produce new, heavier elements by adding neutrons. The process that he didn't understand then, and wasn't looking for, is nuclear fission. Elements 93 and 94 were discovered in 1940 by Emilio Segrè, Edwin McMillan, and Philip Abelson. Lise Meitner predicted these new elements, but she didn't have a laboratory to do the experimenting.

Being electrically neutral, [the neutron] encounters no electrostatic barrier to penetrating the nucleus. Indeed, slow neutrons often find their way into nuclei more efficiently than fast ones, much as a slow cricket ball is easier to catch.
—Philip Ball, *The Ingredients: A Guided Tour of the Elements*

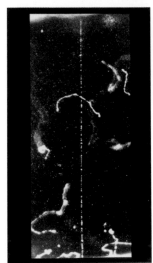

In a 1923 experiment, high-speed beta particles (electrons) left faint, intermittent tracks in a cloud chamber. Slower electrons made the thicker squiggles. Electrons can be knocked out of atoms by X rays or gamma rays or emitted during radioactive beta decay.

The elements above number 92 (uranium) are called *transuranic* ("beyond uranium") elements. They are all radioactive and synthesized (created in laboratories) except for trace amounts of natural neptunium (number 93) found in uranium ore.

in press... from which you can see that by quite different means I have arrived at similar conclusions... to yours."

Fermi begins whamming all the elements he can put his hands on (except hydrogen and helium). His colleague Emilio Segrè says, "[We] discovered about forty new radioactive substances." They are in unexplored territory, using neutrons to bring on radioactivity. When a nucleus is bombarded, beta particles (electrons) are sometimes emitted. This is weird. There are no electrons in the nucleus. How could this happen? It seems that electrons (and neutrinos) are created as a neutron converts to a proton during radioactive decay. Fermi has found an interaction that explains radioactivity. Called the "weak force," it is one of nature's four forces. The weak force (which is a whole lot stronger than gravitation) only operates inside the nucleus. (See page 378 for more on the four forces, or interactions.)

When Fermi decides to bombard nuclei of the metal uranium with neutrons, it is a fateful decision. He is hoping to create a still heavier element, *beyond* uranium, a "transuranic" element. He sends neutrons (slowed just a bit) off toward uranium atoms. Some are absorbed.

Has Fermi done what he set out to do: produce a new element in his laboratory? "We thought that we had produced transuranic elements," Segrè will write later. "In this we were in error, at least in part; while it was true that transuranic elements were formed... what we had observed was something quite different."

I. I. Rabi, a physicist who grew up in New York City, later describes the process: "When a neutron enters a nucleus, the effects are about as catastrophic as if the moon struck the earth. The nucleus is violently shaken up by the blow, especially if the collision results in the capture of the neutron. A large increase in energy occurs and must be dissipated, and this may happen in a variety of ways, all of them interesting." But Rabi is writing later, in 1970, when what happens is understood.

In 1934, Fermi doesn't know it, but **he has split the**

In this cloud chamber, a uranium nucleus split into two fragments, which shot off to the right and left. A single neutron caused the fission. Having no charge, it didn't leave any glowing "bread crumbs," but its presence is evidenced by the protons (short lines, every which way) that it disturbed.

uranium nucleus! He is playing with uranium **fission**. Since he isn't looking for fission, he doesn't discover the small "daughter" nuclei that recoil with high energy when he experiments. His detector has a window of aluminum foil that stops any secondary nuclei produced by fission. Fermi isn't aware that Szilard has been searching for an element that will sustain a chain reaction. Uranium is that element. So far, no one is making the right connections. (Actually, Ida Noddack, a German chemist, makes a good guess, but no one pays attention to the paper she writes. And Lise Meitner is making interesting conjectures, but she doesn't put the clues together.)

Still, what Fermi does is big news in the tight scientific world involved in nuclear research. (In 1938, Fermi will win a Nobel Prize in physics for his work on nuclear reactions produced by slow neutrons.)

The seeds of atomic power are now blowing in the scientific wind. In his 1935 Nobel acceptance speech, Frédéric Joliot-Curie calls for a next step in nuclear research. He talks about splitting the nucleus and says it could lead to explosive nuclear chain reactions and also to "the enormous liberation of usable energy." No one realizes that Fermi has already split a nucleus.

As for "usable energy" from the nucleus? Except for Leo

If it weren't for a piece of aluminum foil, Enrico Fermi might have discovered fission in 1934 and Germany (Italy's Axis ally) might have developed nuclear weapons in time to use them during the Second World War (and there might have been many, many more deaths).

Presidential Power

Although Einstein did not foresee that nuclear energy would be released in his time, he thought it was scientifically possible. He did not know about the discovery of fission, but he learned... [in] fifteen minutes. I was very impressed that he realized in such a short time the problem of the atomic bomb as a possibility. And he was very much aware of the political problem.

—Eugene Wigner (1902–1995), Hungarian-American physicist

[Einstein] was willing to do anything that needed to be done. He was willing to assume the responsibility for sounding the alarm even though it was quite possible that the alarm might prove to be a false alarm. The one thing most scientists are really afraid of is to make fools of themselves. Einstein was free from such a fear.

—Leo Szilard (1898–1964), Hungarian-American physicist, *Bulletin of the Atomic Scientists*

In 1938, Otto Hahn and Fritz Strassmann produce fission by bombarding uranium nuclei in a German laboratory. Niels Bohr isn't the only European who learns about it. Paul Harteck, a patriotic German scientist, lets Nazi officials in on the information. Harteck, who trained with Rutherford in England, makes it clear to the Germans that this "newest development in nuclear physics" has the potential to produce an "explosive many orders of magnitude more powerful than the conventional ones" and that the "country which first makes use of it has an unsurpassable advantage over the others."

Hitler's propaganda chief, Joseph Goebbels, is delighted to hear that an experiment done in Germany may lead to a bigger-than-ever-before explosive.

In a 1945 cartoon by World War II cartoonist Daniel R. Fitzpatrick, a menacing atomic bomb threatens the world.

As for Bohr, he does more than bring news of fission to the United States; he also reports that Germany has stopped Czechoslovakia from exporting uranium (in 1938, Czechoslovakia falls under Nazi control). "And this alarmed us," says physicist Eugene Wigner later. The nuclear physicists are quite sure that the scientists still in Germany, especially Bohr's protégé Werner Heisenberg, are capable of making a bomb.

Fritz (Friedrich Georg) Houtermans, a German physicist working in Switzerland (who has been imprisoned by both Stalin's police and Hitler's Gestapo), is frightened by Hitler's goal of world conquest. (In 1942, Houtermans will send a telegram to Enrico Fermi in Chicago, saying, "Hurry up! We [he means Germans] are on the track.")

Because he understands the bomb's potential, Leo Szilard is frantic. By July 1939, there are more and more rumors that German scientists are working on uranium fission. What would you do if you had knowledge that might give the Allies the bomb before Hitler can make one? Szilard decides to call on Albert Einstein; he knows Einstein has a voice that commands attention.

Einstein is now living in Princeton, New Jersey. At first he imagined he would stay there only a few months, perhaps settling at Oxford in England. But his wife, Elsa, loves the town and says, "The whole of Princeton is one great park with wonderful trees." Einstein seems to agree and searches out musical neighbors with whom he can play his fiddle. (He plays classical music with enthusiasm, but he isn't really much of a violinist.) A perennial student, he takes a course in topology (a kind of geometry) at Princeton University. He has an office, a few colleagues, an assistant, his work— and that is enough.

But in the summer Einstein misses his lake house in Caputh, Germany. (The Nazis confiscated the house and his beloved sailboat—in other words, they stole them.) So Einstein and his wife rent a summer cottage at Peconic, way out on Long Island, New York. And that's

Albert Einstein rehearses with a chamber music orchestra in Princeton, New Jersey. He began playing the violin at age six.

Summers with Professor Einstein

Thomas Lee Bucky was at the Einsteins' cottage on Long Island when Leo Szilard made his famous visit on July 12, 1939. A housekeeper told the 20-year-old that Professor Einstein was having important guests and to stay out of the way. (Only much later did he learn the importance of that day.) Bucky spent eight summer vacations with the Einsteins. His father was Einstein's doctor and a close personal friend.

Tom was 13 the first time he met Einstein. It was 1932, and the Bucky family had been invited to dinner at Einstein's home in Caputh, a suburb of Berlin. Tom wrote later:

It was a great event, and I was terribly excited at the prospect of visiting such a celebrity at home. When we shook hands, he must have felt my fright and awe....

Then he said, "I have something to show you." He went to his desk and returned with a Yo-Yo, at that time the schoolboy rage of Berlin. He tried to show me how it worked, but he couldn't make it roll back up the string. When my turn came I displayed my few tricks and pointed out to him that the improperly looped string had thrown the toy off balance. Einstein nodded, properly impressed by my skill and knowledge.

The next day, Tom went to a toy store, bought a new yo-yo for the professor, and sent it to him as a Christmas present. By return mail he received a page-long poem of thanks that began:

Santa Claus doesn't like to visit
Rickety ladies and old gentlemen. [It rhymes in German.]

"Although he spent more time with my father than with any other friend, it was always 'Professor Einstein' and 'Doctor Bucky,'" said Bucky. "The formality seemed to make both of them comfortable." That formality didn't prevent the professor from having a good sense of humor, as Bucky recalls in an article in *Harper's* magazine:

During the early part of the war when Einstein was a consultant to the Navy (he was trying to figure out the laws that governed waves of detonation) I asked him if the admirals had offered to put him in uniform. The vision of himself in Navy garb so amused him that he broke into his loud, staccato laugh...

His humor could even bend his ordinarily inflexible attitude that a fact was a fact, and no amount of human wishing could alter it. Einstein's sister Maja lived at his Princeton house for several years. Like Einstein, she was a gentle person. Her tender regard for all living creatures made her a vegetarian. But she had one painful conflict: she loved hot dogs. After listening to Maja bemoan her problem, Einstein resolved the dilemma by decreeing

Eugene Wigner said that if the Manhattan Project could have been run on ideas alone, no one but Leo Szilard would have been needed.

where Szilard goes to see Einstein on July 12, 1939.

Einstein respects Szilard. He once described him as "intelligent and many-sided . . . especially rich in ideas." The two are fellow inventors.

Since Szilard doesn't know how to drive, his friend Eugene Wigner does the driving. The two Hungarian physicists get lost when they reach Peconic, and no one they ask seems able to help them. They are about to return to New York City when they see a six-year-old boy and ask him if he knows where Professor Einstein lives. He does.

that, in Maja's case, a hot dog was a vegetable....

When I bought my first car, an old Model A Ford coupe, Einstein was visiting us in New York, and I asked him to go for a ride. He climbed into the rumble seat and off we went down Fifth Avenue with motorists and pedestrians doing double-takes as they caught sight of him, a smile on his face and his hair flying in the breeze....

As he kept his mind free of destructive emotions, so did he avoid cluttering his life with material things. He believed in simplicity, so much so that he used only a safety razor and water to shave. When I suggested that he try shaving cream, he said, "The razor and water do the job."

"But Professor, why don't you try the cream just once?" I argued. "It makes shaving smoother and less painful."

He shrugged. Finally, I presented him with a tube of shaving cream. The next morning, when he came down to breakfast, he was beaming

Composer Philip Glass honored his childhood hero with a 1976 opera titled *Einstein on the Beach.*

with the pleasure of a new, great discovery. "You know, that cream really works," he announced. "It doesn't pull the beard. It feels wonderful." Thereafter, he used the shaving cream every morning until the tube was empty. Then he reverted to scraping his face with water.

Once, when Tom Bucky was sick and in a hospital, Einstein visited and created a sensation. Tom recalls:

Within a few minutes the corridor outside my room was crowded with people. The only person who did more than slowly pass and peer was the hospital rabbi. He couldn't resist the chance to meet Einstein. He began apologies, confessing he had no right to impose, but Einstein stopped him. "Oh no, you have rights," he assured the chaplain. "After all, you work for a very important boss."

Thomas Bucky became a physician, practicing in New York City and in Weston, Connecticut.

Einstein, in an undershirt and scruffy, rolled-up pants, serves them tea and cookies on the big screened-in porch of the cottage. He doesn't know about Szilard's patent for a nuclear chain reaction or of the discovery of fission, but when Szilard explains, Einstein quickly understands that a nuclear bomb is a possibility. Szilard tells him there is evidence that the Nazis are stockpiling uranium.

Einstein is a pacifist, but the monstrosities occurring in his homeland are making him rethink his hostility to war.

What does Szilard want from Einstein? He asks Einstein

A grapefruit-sized "button" of uranium-235 is the material for both nuclear reactor fuel and a nuclear weapon explosive.

War? Peace? and the Bomb

Scientists on both sides of the war were struggling with the moral and ethical implications of a nuclear bomb. In Germany, Otto Hahn opposed scientific secrecy. He said, "It would be worse for the entire world, even for Germany, if Hitler were to be the only one to have [the bomb]."

His colleague, Carl Friedrich von Weizsäcker, wrote later, "At that time [1939] we were faced with a very simple logic. Wars waged with atom bombs...do not seem reconcilable with the survival of the participating nations. But the atom bomb exists. It exists in the minds of some men. According to the historically known logic of armaments and power systems, it will soon make its physical appearance. If that is so, then the participating nations, and ultimately mankind itself, can only survive if war as an institution is abolished."

to warn his friend, Belgium's Queen Elizabeth, to keep her nation's large source of uranium ore from the Germans. Einstein agrees but thinks a letter to the Belgian ambassador might be a better idea. They work on the letter together.

Back in New York, Szilard talks about all this with economist Alexander Sachs, who happens to be an unofficial advisor to President Franklin Delano Roosevelt. Sachs says the letter should go to the American president, not the Belgian ambassador, and that he will take it to the president himself.

All at once, the importance of the letter is transformed. Szilard writes a four-page draft of a new letter and mails it to Einstein (it is now July 19). Einstein asks him to come back to Peconic.

This time Edward Teller (another Hungarian) acts as chauffeur. Again, they sit on the porch, and Einstein, writing in German, pares down the letter to basic points. It's that letter (translated by Szilard) that Alexander Sachs delivers to Roosevelt. It will change the world's history.

Here are some words from Einstein's letter of August 2, 1939, to President Roosevelt (the complete letter is on page 225):

> *Sir:*
> *Some recent work by E. Fermi and L. Szilard, which has been communicated to me in manuscript, leads me to expect that the element uranium may be turned into a new and important source of energy in the immediate future.*

Einstein urges Roosevelt to initiate further research. To make the president understand the need for haste, he closes with this warning:

Germany has actually stopped the sale of uranium from the Czechoslovakian mines . . . [that] might perhaps be understood on the ground that the son of the German Under-Secretary of State . . . is attached to the Kaiser-Wilhelm Institute in Berlin where some of the American work on uranium is now being repeated.

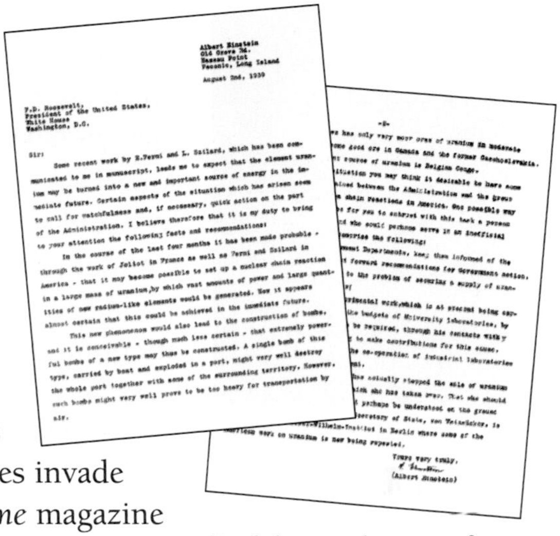

Albert Einstein
Old Grove Rd.
Nassau Point
Peconic, Long Island

August 2nd, 1939

F.D. Roosevelt,
President of the United States,
White House
Washington, D.C.

Sir:

Some recent work by E.Fermi and L. Szilard, which has been communicated to me in manuscript, leads me to expect that the element uranium may be turned into a new and important source of energy in the immediate future. Certain aspects of the situation which has arisen seem to call for watchfulness and, if necessary, quick action on the part of the Administration. I believe therefore that it is my duty to bring to your attention the following facts and recommendations:

In the course of the last four months it has been made probable - through the work of Joliot in France as well as Fermi and Szilard in America - that it may become possible to set up a nuclear chain reaction in a large mass of uranium, by which vast amounts of power and large quantities of new radium-like elements would be generated. Now it appears almost certain that this could be achieved in the immediate future.

This new phenomenon would also lead to the construction of bombs, and it is conceivable - though much less certain - that extremely powerful bombs of a new type may thus be constructed. A single bomb of this type, carried by boat and exploded in a port, might very well destroy the whole port together with some of the surrounding territory. However, such bombs might very well prove to be too heavy for transportation by air.

-2-

Yours very truly,
(Albert Einstein)

Read the complete text of Einstein's famous letters to President Roosevelt online. Here's one source: http://www.hypertextbook.com/eworld/einstein.shtml.

Alexander Sachs asks for an appointment with the president. FDR has other things on his mind. On September 1, German tanks and planes invade Poland in a massive, coordinated attack that *Time* magazine calls a "*Blitzkrieg*." This is the word the Germans use for "lightning war." The name sticks. Two days later, Great Britain and France declare war on Germany. Belgium and the United States declare their neutrality. The president has no time to see his economist friend Alexander Sachs.

Meanwhile, in a secret September 16 meeting in Berlin, Otto Hahn and some other German physicists are given an up-to-the-minute report on uranium research in the United States and Britain. They discuss a just-published paper by Princeton's John Wheeler and Copenhagen's Niels Bohr that suggests that U-235 is the isotope of uranium that will get fission going. The Nazis ask Werner Heisenberg to investigate the potential for nuclear weaponry.

Before E-mail

In the 1920s, Max Born (1882–1970) made the University of Göttingen in Germany a center of physics second only to Copenhagen. But, as a Jew, he had to flee Germany when Hitler came to power. He headed for the University of Cambridge in England in 1933. Then he became a professor at the University of Edinburgh in Scotland.

He and Einstein were good friends, who, when they were together, talked physics and played violin concertos. Apart, they wrote letters through 40 years of friendship. Born, who was also a Nobel Prize–winning physicist, collected those letters in *The Born-Einstein Letters, 1916–1955.* That wonderful little book was first published in 1971 and reissued in 2005.

Otto Hahn won the Nobel Prize in chemistry in 1944 for the artificial fission of heavy nuclei, but Lise Meitner had to tell him what he had found.

Secret Science

At its best, the scientific community is democratic and open. New ideas are described and shared in scientific journals where, if those ideas have merit, they quickly find their way around the globe. Being first to publish means getting credit for a discovery; so, like newspaper reporters fighting for a scoop, scientists vie to publish first.

As for secrecy, most scientists understand that keeping breakthrough ideas hidden just doesn't work for long. If something is known to be possible, others will figure it out.

Top-secret research on a nuclear bomb has begun, but except for a few high-level scientists and a handful of government officials, almost no one is in on the secret. To most of those who are involved, a weapon based on the formula $E = mc^2$ seems like a long shot. It's not a high priority and the funding is modest.

The theoretical concept—the idea behind the bomb—is well known to all scientists who understand nuclear fission. That means German, Italian, and Japanese as well as Allied physicists now understand that neutrons can indeed break apart a uranium nucleus. But no one knows if theory and actuality will mesh.

Niels Bohr, who probably understands the atom better than anyone in the world, believes the technological problems are too difficult to solve in time for this war. Despite Max Born's concerns, Bohr hasn't been arrested, but he is isolated in Denmark and doesn't realize that in the United States attention is being paid to the paper he and John Wheeler wrote suggesting that U-235 is the key

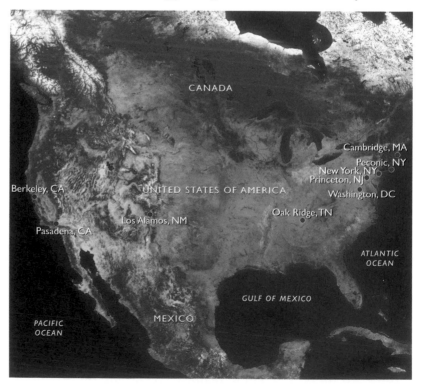

uranium isotope (the element Szilard had searched for). U-235 makes up 1 part in 140 of natural uranium; the other 139 parts are U-238 (which doesn't support a chain reaction).

At the University of Minnesota, Alfred Nier is attempting to separate U-235 from U-238. (This isn't easy at all—the details are in the next chapter.) In England, Otto Frisch and Rudolph Peierls are also working on a separation system and defining other crucial areas that need to be explored. And at Columbia University in New York, after some months of stewing while waiting for government money, Szilard and Fermi are finally at work.

Fission has been proved, but no experiment has confirmed Szilard's idea of a nuclear chain reaction. If it can be shown to be possible, the bomb should be a go, although a mountain of engineering and technological challenges will need to be overcome. Fermi knows that neutrons must move at just the right speed to hit and split a nucleus. If they go too fast, they speed through the atom. Heavy water, with one proton and one neutron in the hydrogen nucleus, will slow them down. But heavy water is expensive and difficult to produce.

> Regular water has just one proton in the hydrogen nucleus, so it absorbs neutrons and stops a chain reaction.

A Norwegian firm has been making most of the world's small supply of heavy water. British agents tell the Norwegians that heavy water has important war uses, without explaining why. In January of 1940, Germany offers to buy all the heavy water the Norwegians have on hand and sign a contract for increased future production. The Norwegian firm holds back. In March, it gives most of its stock of heavy water to the French, refusing payment. On April 9, Germany invades Norway. After that, courageous Norwegians sabotage production as best they can.

On May 10, 1940, 77 German divisions and 3,500 aircraft attack neutral Belgium, the Netherlands, and Luxembourg.

In June, construction begins on a new building at the Kaiser Wilhelm Institute in Berlin. It contains a laboratory and a special 6-feet-deep, brick-lined pit. Water is placed in the pit and then a large aluminum canister with alternating

> In February 1944, 39 drums of heavy water were loaded onto a ferry, the *Hydro*, heading for Germany. Norwegian commandos blew up the ferry. Later, Germans said, "The main factor in our failure to achieve a self-sustaining atomic reactor before the war ended" was the lack of heavy water.

Heavy Hydrogen Hype

Talk about stripped down and basic: The hydrogen atom has a single proton as its nucleus, and that's all. That nucleus is circled by a single electron. There are no neutrons—none—in a normal hydrogen nucleus.

As element number 1, hydrogen (H) is the simplest, lightest, and most abundant of the elements. About 93 percent of all the atoms in the universe are hydrogen, which makes up 76 percent of the universe's mass. Most of the stars, including our Sun, are primarily hydrogen. And most of the hydrogen atoms in the universe are in diatomic (twosome) molecules as the gas H_2.

Since there is so much of it, it's a surprise that hydrogen comprises only about $\frac{1}{20}$ of the air in Earth's atmosphere. That's mostly because hydrogen gas is lighter than air, and it usually rises to the top of the atmosphere and escapes. And the hydrogen on our planet? It is usually found combined in molecules, as with oxygen in H_2O.

Put a lot of pressure on hydrogen—maybe one-half million or more times the pressure of the Earth's atmosphere—and it will turn solid. You can find it on Jupiter as a liquid with metallic properties. On Earth, hydrogen gas is

This heavy water is 11 percent heavier than regular water, thanks to a single neutron in each deuterium (D) molecule that's not present in common hydrogen.

commercially produced by breaking apart water. It is used in the chemical and food industries (check a good chemistry text for details).

Hydrogen has a sibling, an isotope: D, or deuterium. (Most isotopes don't have names; this one is unusual.) Deuterium's nucleus has *one neutron* along with the one proton. (Remember, regular hydrogen nuclei have no neutrons.) Heavy water is D_2O, or deuterium oxide. About 1 in every 10,000 water molecules in Earth's lakes and oceans is deuterium oxide. *So, heavy water is in regular water, but it is very rare.* Water that is more than 40 percent D_2O is toxic. Extracting deuterium from water is expensive and difficult; making it artificially is expensive and difficult.

Why is it needed for nuclear fission? Why not regular water? If you send neutrons through regular water, they are likely to be absorbed—joining with protons to make deuterium nuclei. That doesn't happen with heavy water because the nuclei are already deuterium. That's why D_2O is useful in nuclear science. It slows down, but doesn't absorb, neutrons.

layers of uranium oxide and paraffin. To fool the curious, the building is nicknamed the Virus House.

Heisenberg visits Bohr in Copenhagen and draws a picture of a heavy water reactor. (Was Heisenberg trying to impress, scare, or warn Bohr? That is still being argued.) Bohr gives the drawing to the British. On the night of February 28, 1943, nine British and Norwegian commandos break into the *Vemork* heavy water plant, capture a guard, and set off explosives, temporarily destroying the Nazi supply of heavy water.

Szilard finds something besides heavy water to slow the neutrons. It is pure graphite. Szilard, who studied chemical engineering in Germany, realizes that most graphite includes small amounts of boron and that boron "poisons" the graphite by absorbing neutrons. Fermi runs new tests that show that, without boron, graphite is an ideal moderator. (The German scientists never figure out that it is the boron in graphite that makes it unworkable.)

Fermi and Szilard order 4 *tons* of boron-free graphite. It arrives at the Pupin Lab at Columbia in carefully wrapped bricks. ("The physicists... started looking like coal miners," Fermi writes later.) They need to set up a test track and see how far and how fast the neutrons will diffuse through the graphite.

Graphite (below) is made of carbon atoms arranged in interlocking sheets (above). Pencil "lead" is actually graphite.

Moderating Mass

To a nuclear physicist, a moderator is any substance that slows down a neutron, thus making it more likely that it will enter an atom's nucleus.

How do you slow down a neutron? Let it hit something. In an elastic collision, some of the kinetic energy of the neutron will transfer to whatever it collides with.

The best moderators have a mass close to that of a neutron. A billiard ball transfers a lot of its energy when it hits another billiard ball, but not when it bounces off a bowling ball. In the same way, neutrons transfer energy when bouncing off protons in hydrogen or deuterons in deuterium.

But (this is a big "but") the moderator needs to do more than slow down a neutron; it has to resist absorbing it. Heavy water and pure graphite work on both counts. But paraffin has two flaws as a moderator: It has hydrogen atoms that absorb neutrons, and it melts at a relatively low temperature. Both are bad for nuclear piles!

Theirs is a strange collaboration. These two scientists can't stand each other. Fermi is a family man, elegant, methodical, hardworking, and painstaking. Szilard, a bachelor, sleeps late, is messy, has little patience with details, and has a mind that leaps from idea to idea but has a hard time settling down. Both are brilliant. And both know it is important for them to collaborate on this project. Szilard seems to see the large concepts; Fermi makes them happen. Later, Szilard says, "If the nation owes us gratitude—and it may not—it does so for having stuck it out together as long as was necessary."

In England, Frisch and Peierls are convinced that a bomb can be built and quickly; they say that in a report written in the spring of 1940. Among other things, they say that a uranium sphere should be made in two parts, "which are brought together first when the explosion is wanted. Once assembled the bomb would explode within a second or less." Radiation, a byproduct of the bomb, would be "fatal to living beings even a long time after the explosion." And protection from such a weapon would be "hardly possible." They figure out the size of the necessary critical mass. So does James Chadwick. Years later, Chadwick will write, "I realized then that a nuclear bomb was not only possible—it was inevitable. Sooner or later these ideas could not be peculiar to us."

On October 1, 1940, Albert Einstein became an American citizen.

British intelligence intercepted a telegram from Niels Bohr to a colleague asking about Maud Ray. "Maud" must be code, said the sleuths. It could stand for *Military Applications of Uranium Disintegration.* A MAUD commission was empowered to study a nuclear bomb's military potential. Its 1941 report included the finding of Frisch and Peierls that *the bomb could be small enough to drop from an airplane.* That convinced FDR to fully fund the bomb. But Maud Ray was no code; she was a former housekeeper for the Bohrs!

I have often been asked why I didn't abandon the project there and then, saying nothing to anybody. Why start on a project which, if it was successful, would end with the production of a weapon of unparalleled violence, a weapon of mass destruction such as the world has never seen? The answer was very simple. We were at war, and the idea was reasonably obvious; very probably some German scientists had the same idea and were working on it.
—Otto Frisch, *What Little I Remember*

The Frisch-Peierls report brings focus and urgency to those working on what has become a combined American-British-Canadian effort. The atomic bomb now moves into top-priority status.

Meanwhile, Glenn Seaborg in Berkeley, California, identifies element 94 and names it plutonium after the planet Pluto (which was named for the Greek god of the underworld). Seaborg finds that plutonium undergoes fission when hit by slow neutrons. Like uranium, it can be used to make a nuclear explosive.

Now the hurdle is to get enough fissile material—either uranium-235 or a plutonium isotope—to bring about an explosion. In each case it's a drip by drip, or milligram by milligram, process.

It will take never-built-before nuclear reactors and tremendous power to produce enough of those fissile isotopes for a critical mass to make a bomb. The Allied scientists have an incentive—the fear that the Nazis may build a bomb first. **On December 6, 1941**, a new government committee, the Top Policy Group, meets to consider major funding of the bomb project.

Just before dawn the next morning, **December 7, 1941**, Japanese commander Mitsuo Fuchida lifts his plane skyward from the deck of an aircraft carrier some 440 kilometers (about 273 miles) north of Hawaii. He is part of an attack squadron of fighter planes, torpedo bombers, dive bombers, and high-level bombers. Cheered by sailors and officers on the

FISSILE MATERIAL is any material whose atomic nuclei can be fissioned (split) by slow-moving neutrons.

On December 7, 1941, American warships were lined up in Pearl Harbor, Hawaii—an easy target for Japanese bombers.

aircraft carriers, Fuchida leads the first group of bombers. "We could see that the sky over Pearl Harbor was clear.... I peered intently through my binoculars at the ships riding peacefully at anchor. One by one I counted them. Yes, the battleships were there all right, eight of them!...It was 7:49 a.m. when I ordered my radio man to send the command, 'Attack!'" By 7:53 a.m., he writes, "The effectiveness of our attack was certain, and a message, 'Surprise attack successful!' was accordingly sent to the *Akagi* [the flagship of the Japanese attack fleet]. The message was received by the carrier and duly relayed to the homeland."

The story will be told again and again. Twenty-one ships are destroyed or damaged in the surprise raid. Nearly 2,400 Americans are killed. That same day, Japanese forces attack the U.S.-held islands of Guam, Wake Island, and the Philippines. On December 8, the U.S. Congress declares war on Japan. Three days later, December 11, Germany and Italy declare war on the United States.

On the day of the Chicago squash court test, news arrived that, by then, the Nazis had murdered at least 2 million European Jews.

Szilard and Fermi move to the University of Chicago. There, Fermi builds a "nuclear pile"—a tall, layered sandwich of graphite and uranium—in an unused squash court under a grandstand at a university playing field.

On December 8, 1941, newspapers around the world all blared the news: "The U.S. Declares War."

Fermi's "pile" contains 6 tons of uranium, 50 tons of uranium oxide, and 400 tons of graphite blocks. Fifty-seven layers thick, it is arranged (according to Szilard's plan) in a nearly spherical shape about 7 meters (23 feet) wide and just about as high; it fills the squash court.

It is bitterly cold in Chicago on December 2, 1942, when the pile is readied for the vital test. Fermi, Szilard, and their colleagues watch from a gallery where measuring devices monitor the neutron emission and safety rods can be inserted into holes in the pile if the reaction gets out of hand. They begin in the morning, and by two in the afternoon everyone is watching the neutron counter. Except for its clickety-clacks, there isn't a sound in the room. Then Fermi raises his arm. "The pile has gone critical," he says. The neutron intensity will now double every two minutes. Fermi waits a minute, and another, and another—4.5 minutes in all—and then he says, "Zip in," the signal to insert the safety rod (called "Zip") into the pile. Fermi has done what he set out to do.

The scientists break open a bottle of Chianti wine and toast the first planned nuclear chain reaction. It is the breakthrough that makes everything else possible.

This painting depicts the first self-sustaining nuclear chain reaction, on a University of Chicago squash court, December 2, 1942, at 3:36 p.m. George Weil (bottom center) operates the control rods while Enrico Fermi and his team look on.

Yoshio Nishina (1890–1951), Japan's leading physicist, had studied with Niels Bohr. In 1943, he and his colleagues announced that an atomic bomb was possible but that no one could make it in time to use during the current war.

Manhattan on a Mesa

The very fact that knowledge is itself the basis for civilization points directly to openness as the way to overcome the present crisis.... [R]eal cooperation between nations on problems of common concern presupposes free access to all information of importance for their relations. Any argument for upholding barriers for information... based on concern for national ideals or interests, must be weighed against the beneficial effects of common enlightenment and the relieved tension resulting from openness.
—Niels Bohr, Danish physicist, "Open Letter to the United Nations," June 9, 1950

The war made it obvious by the most cruel of all arguments that science is of the most immediate and direct importance to everybody. This had changed the character of physics.
—Victor Weisskopf (1908–2002), Austrian-American physicist, J. Robert Oppenheimer Papers

In 1943 (at the height of World War II), a set of keys is delivered to Niels Bohr by someone in the Danish underground resistance movement. The keys have come from the British intelligence service. One of them has a hollow center and contains a half-millimeter piece of microfilm with a coded message on it. Bohr reads the message under a microscope. It is a letter from James Chadwick, the English physicist who discovered the neutron. Chadwick urges him to come to England where he is needed for work on "scientific matters." Bohr knows "scientific matters" has to do with nuclear weaponry, but he isn't ready to leave his homeland. He is working hard trying to save endangered Europeans.

Bohr is still skeptical that a bomb can be made quickly. He doesn't know details of what is happening in the United States, where many of his former students are hard at work on a nuclear bomb. He does know, after Werner Heisenberg comes for a visit, that the Germans are gathering materials for high-level nuclear research, and he passes that information on to the British.

In return for food from Danish farms, Denmark's Jews have been left alone by the Germans. But many Danes are actively sabotaging Nazi efforts; when that begins to hurt, the Nazis blame the Danish Jews. Bohr gets word that the Gestapo (German police force) is about to arrest him. He climbs into a fishing boat and is rowed to Sweden. Once there, he works on the docks, aiding a boatlift bringing other Danes to still-safe Sweden.

The Germans are now eager to kidnap Bohr to make sure his scientific knowledge doesn't reach the Allies. The English want him in England. In the United States, his former students and colleagues long for his advice. Bohr wants to stay in Stockholm and work with refugees. England's feisty prime minister, Winston Churchill, takes action. He sends a Mosquito bomber to Stockholm, and Bohr is persuaded to climb into the tiny plane. Bohr sits on top of the bomb door in the rear of the plane where, unable to figure out how his oxygen mask works, he soon faints. Fortunately, someone is able to wake him when the plane lands in England.

The U.S. effort to build a nuclear bomb is given the code name Manhattan Project. Some of the world's greatest physicists and mathematicians are in on the secret. Just to make sure it stays secret, the U.S. Army sends them to a remote location, a newly built compound at Los Alamos, 2,100 meters (almost 7,000 feet) above sea level in New Mexico's Sangre de Cristo Mountains. Before, only a boys'

Why did Heisenberg visit Bohr in the midst of war? Did he want to warn Bohr that the Germans were developing a nuclear bomb? Did he want information from Bohr as to what the Allies were doing? Bohr was half-Jewish; was Heisenberg worried about Bohr's safety, or about his own safety? We are uncertain about all that.

Prime Minister Winston Churchill was a skilled orator who, even in the war's darkest moments, was able to convince the British people that they would be victorious.

That big "A" on the Manhattan Project seal stands for atomic, but the correct scientific term for the atomic bomb is *nuclear fission bomb*. It's the nuclei of the atoms that split.

What better place to hide a secret laboratory than on a high desert mesa surrounded by natural beauty (at right and opposite page)? Los Alamos borders one of the world's biggest volcanic calderas, the Valles Caldera in the Jemez (HAY-mez) Mountains of New Mexico. To the south is Bandelier National Monument, a wilderness peppered with thousands of ancestral Pueblo homes, including an apartment complex carved into a cliff. When the Manhattan Project kicked into gear, the park closed, and the brand-new Bandelier lodge housed scientists and soldiers.

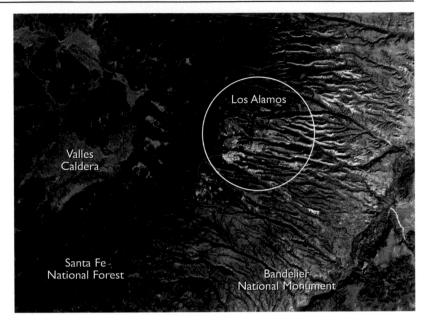

school had been there, hidden away on a sandstone mesa: a bit of flat tableland speckled with pine and aspen and scrub.

Some 55 kilometers (34 miles) northwest of Santa Fe, it is, geologically, a fascinating site bordering a spectacular hole in the ground. A volcano blew its top here aeons earlier, forming the verdant Valles Caldera. Nearby, cliffs tell the natural history of the land in chronological layers of colored rock. Streams splash with fish, while mountain goats and sheep help sustain the few Native Americans and Mexican Americans who live in the neighboring hills. But geology and nature are not on the minds of most of those who come to Los Alamos: speed, secrecy, and physics are.

The U.S. Army quickly builds wooden structures at Los Alamos, painting most of them dull green. They, and the old school buildings, become apartments and houses and

In Los Alamos we were working on something which is perhaps the most questionable, the most problematic thing a scientist can be faced with. At that time physics, our beloved science, was pushed into the most cruel part of reality and we had to live it through. We were, most of us at least, young and somewhat inexperienced in human affairs.... But suddenly in the midst of it, Bohr appeared at Los Alamos. It was the first time we became aware of the sense in all these terrible things.... Every great and deep difficulty bears in itself its own solution.... This we learned from him.

—Victor Weisskopf, Austrian-American physicist

laboratories. Nothing is fancy. The unpaved streets are either muddy or sandy. Coal provides the only heat in houses and stoves. Barbed wire and security fences ring the settlement. The people who come—the scientists, mathematicians, engineers, technicians, and helpers—receive code names along with instructions not to discuss their work with anyone, not even family members. Military censors read their mail.

Living in primitive huts and trailers (below) under a continuous cloud of secrecy was the downside of being a founding scientist at Los Alamos National Laboratory in the early 1940s. The perks were the spectacular mountain locale (left) and the opportunity to change the course of scientific and world history.

Most of the scientists have been city dwellers; hardly any of them has ever before encountered the American West. Some love it. Others, like urbanite Leo Szilard, are unsettled by how remote it is. Szilard says, "Everyone who goes there will go crazy." But they don't go crazy; they work hard—and sometimes, for relief, they hike and ski and play hard, too. What they share is horror at this war begun by totalitarian nations intent on ruling the world. They believe they have a mission: to build a weapon—a nuclear bomb—that will destroy Hitler's venomous regime and end the war as quickly as possible. Early in 1943, moving vans begin bringing scientists, staff, and their families to Los Alamos. This will be the think tank: Most of the actual bomb and its components will be built elsewhere.

To those in the field, many of the physicists and

POLYGLOT means "many tongues," but for many of the physicists at Los Alamos, German was the language of science. For all of them, math was a universal tongue.

J. Robert Oppenheimer married (in 1940), became a father (in 1941), and launched Los Alamos laboratory (in 1942) while in his mid-30s. His second child was born there in 1944.

The one thing that runs through all the scientists' narratives [of Los Alamos] is enthusiasm. As hard as they worked building the bomb, they also had a great deal of fun.

—Jeremy Bernstein, *Portrait of an Enigma*

mathematicians who come to Los Alamos are legendary: Anderson, Bethe, Bohr, Fermi, Feynman, Frisch, Peierls, Rabi, Segrè, Szilard, Teller, Ulam, Weisskopf, Wheeler, Wigner, von Neumann, and more—a core of about 100 physicist/mathematicians. It is a polyglot group. Many are or will be Nobel laureates. Some scientists represent America's allies Britain and Canada. (One of the physicists is a Communist spy, Klaus Fuchs, who is secretly sending reports of their work to Moscow.) Except for the superstars, it is a very young group; most are in their 20s.

New York–born J. Robert Oppenheimer is the scientist in charge. He has founded two great schools of theoretical physics: one at the University of California at Berkeley, the other at Caltech in Pasadena. Oppenheimer has an awesome mind along with elegant manners, a skinny frame, brilliant blue eyes, and a sense of fairness and humor that make others trust him. "Oppie," as he is called, began his physics training at Harvard, where he was admitted to graduate study during his first year as an undergraduate. Then he studied with J. J. Thomson at Cambridge, with Bohr in Copenhagen, and with Max Born at Göttingen. By the time he returned to Harvard, he had written more than a dozen physics papers, including one with Born.

For pleasure, Oppenheimer reads Sanskrit (a language of India). Laura Fermi, the wife of physicist Enrico Fermi, says Oppie "turned out to be a marvelous director, the real soul of the project." This is the job of a lifetime, and Oppenheimer does it superbly. Later, he will be accused of arrogance, left-wing leanings, insensitivity, and more—but that is elsewhere; here and now he shines.

At first the enterprise is under the Office of Scientific Research and Development, headed by Vannevar Bush (the initial funding is $6,000). When the United States enters the war (in December 1941), the U.S. Army is given joint responsibility, and the project becomes top priority under the command of Brigadier General Leslie R. Groves. (It is Groves who picks Oppenheimer to lead the scientific endeavor.) Eventually, $2 billion will be spent. At its peak,

Like Los Alamos, the laboratory at Oak Ridge, Tennessee, was a secret city that sprang up quickly and quietly on a massive, never-been-done-before scale. At left is the K-25 plant, one of three huge processing facilities on the site, photographed near the end of World War II. Inside, some 12,000 workers and thousands of pumps and converters are coaxing precious teaspoonfuls of uranium-235 out of uranium-238. An even larger plant labeled Y-12, operated by 22,000 workers, used electromagnets to process U-235.

130,000 people are employed on the project, and except for Fuchs and one other spy, they keep their activities secret. Never, in the world's history, has so much money and talent been brought to such a concentrated scientific effort.

While Los Alamos is the brain center, at a huge facility built in Oak Ridge, Tennessee, fissionable uranium-235 is separated from the much more abundant but non-fissionable isotope uranium-238. (U-235 splits readily when entered by a neutron; U-238 actually derails a reaction.)

Separating those isotopes is a big challenge. It can't be done chemically because different isotopes have essentially identical chemical properties. One way to do it physically is to cycle a beam of uranium atoms in a magnetic field, where the tiny difference in mass separates the atoms into slightly different circular orbits. That sounds easy, but it requires facilities on the scale of giant oil refineries as well as huge amounts of electric power. None of this has been done before. Wrong turns are inevitable. One costly effort to separate uranium isotopes at the University of California at Berkeley produces one gram of pure uranium. A week's output from Oak Ridge can be carried in a briefcase.

In the town of Hanford in the state of Washington (another built-from-scratch mini-city), uranium-238 is turned into plutonium-239 (another fissionable isotope).

Remember, an **ISOTOPE** is an atom with a different number of neutrons in its nucleus but the same number of protons and electrons as its sibling atoms.

A plutonium bomb was detonated at the Trinity test site near Alamogordo because of the uncertainty that it would work. The simpler bomb with U-235 was more certain to work and was not tested before use at Hiroshima, Japan.

If you are a scientist you believe that it is good to find out how the world works; that it is good to find out what the realities are; that it is good to turn over to mankind at large the greatest possible power to control the world and to deal with it according to its lights and values....It is not possible to be a scientist unless you believe that it is good to learn. It is not good to be a scientist, and it is not possible, unless you think that it is of the highest value to share your knowledge, to share it with anyone who is interested. It is not possible to be a scientist unless you believe that the knowledge of the world, and the power which this gives, is a thing which is of intrinsic value to humanity, and that you are using it to help in the spread of knowledge, and are willing to take the consequences.

—J. Robert Oppenheimer, speech to the Association of Los Alamos Scientists (November 2, 1945)

This is done in gigantic nuclear reactors using a process Fermi pioneered with his nuclear pile. There will be two bombs: one using uranium, the other plutonium.

Since everything depends on human effort, the drama of the participants' lives is a big part of the story. Brooklyn-born Richard Feynman spends each weekend hitchhiking from Los Alamos to Albuquerque, where his adored young wife, Arline, is dying of tuberculosis. (New drugs, not yet available, will soon change the response to that disease.) Feynman, at 25, is the youngest group leader on the project. He and Arline write letters to each other in codes that change with each letter, delighting in the frustration they cause the army's censors.

"What do you care what other people think?" Arline asks him. And Dick Feynman never does care. He is an original and one of the most brilliant physicists that America has produced.

In England, James Chadwick asks his Austrian colleague Otto Frisch, "Would you like to work in America?"

"I would like that very much," says Frisch. A week later, he has packed one suitcase and is on the night train to London. There, he goes through a ceremony that makes him a British

Otto Frisch, a key actor in the nuclear bomb drama, was also a concert-level musician. He had a weekly slot playing piano on KRS, the radio station at Los Alamos, where he was identified as "our musician" to hide his Austrian identity.

citizen. Then he gets a passport and is sent to Liverpool, where he climbs on a ship. After a train trip from New York, Frisch finally arrives in New Mexico where he is greeted by pipe-smoking Robert Oppenheimer. Oppie, wearing his signature flat-top hat and a smile, says, "Welcome to Los Alamos, and who the devil are you?" (Remember, it was Frisch, with his aunt, Lise Meitner, who figured out fission.)

When Fermi arrives, he brings with him the nickname his Italian colleagues use. To them he is "the Pope"; his judgments seem infallible.

Edward Teller, another of the brilliant Hungarians, comes to Los Alamos with his wife, Mici, their two-month-old son, Paul, and a concert grand piano. Teller plays the piano with the same intensity he brings to physics. Mici leads a mini-rebellion, convincing a group of women to sit under trees the army plans to bulldoze. They win: The trees stay.

Teller is soon obsessed with the idea of nuclear fusion (instead of fission). He knows that fusion can lead to a "super" bomb, one even more powerful than the fission bomb everyone else is working on. Teller believes the super bomb should be built. That will lead to conflict with Oppenheimer and others and much controversy in the future.

A Polish mathematician, Stanislaw Ulam, does the math to make the super bomb possible. His wife, Françoise, will later write:

> *Engraved on my memory is the day when I found him at noon staring intensely out of a window in our living room with a very strange expression on his face. Peering unseeing into the garden, he said: "I found a way to make it work."*

Political Science

After the war, J. Robert Oppenheimer became head of the Institute for Advanced Study at Princeton, but, in 1954, he lost his security clearance and position at the Atomic Energy Commission because of his liberal views. Edward Teller, the chief

The hawkish Edward Teller flashed this ID photo at Los Alamos.

witness against him, came to dominate the political direction of science. He urged more powerful weaponry—including the hydrogen bomb and the missile defense system nicknamed Star Wars.

Meanwhile, Einstein worried about his friend Oppenheimer. He said that Oppenheimer knew more about atomic energy than anyone, and it bothered Einstein that Oppenheimer was taking the attacks against him so seriously.

"Oppenheimer is not a gypsy like me," he told his friend Johanna Fantova. "I was born with the skin of an elephant; there is no one who can hurt me." Fantova was curator of maps at the Firestone Library at Princeton. Her diary, which includes many stories about Einstein, is at that library.

Math whiz Stanislaw Ulam (left) chats with theoretical physicist Richard Feynman (right) during an outing to Bandelier National Monument.

Ulam's "different scheme" was to use a fission bomb to compress thermonuclear material (deuterium) to the point of ignition of a fusion reaction. The challenge was to keep the reaction from fizzling out or, at the other extreme, igniting the whole atmosphere with a chain reaction. "It is still an unending source of surprise for me to see how a few scribbles on a blackboard or on a sheet of paper could change the course of human affairs," Ulam wrote in *Adventures of a Mathematician.*

"What work?" I asked.
"The Super," he replied. "It is a totally different scheme, and it will change the course of history."

In England, Bohr and his 22-year-old son, Aage (OH-wuh), learn details about "Tube Alloys"—the code name for the bomb. After seven weeks of briefing, they leave for the United States. "From first to last, the deeply creative and subtle and critical spirit of Niels Bohr guided, restrained, deepened, and finally transmuted the enterprise," writes Oppenheimer.

Bohr sees things differently. "They didn't need my help in making the atom bomb," he says of his colleagues at Los Alamos. He now knows that the bomb is going to happen. But how will the world react after it explodes? That's the thought that obsesses him. "That is why I came to America," he says. He realizes that this bomb can lead to the end of humanity as we know it; or, he believes, it can be the impetus that forces humans to solve their differences in new ways.

Many of those at Los Alamos are beginning to question their participation in this venture. Bohr brings them the possibility of a better world—where nations and scientists

The Little Fellow in the Back

One day at Los Alamos, Dick Feynman got a call from James Baker. Baker said he and his father wanted to meet with him. Feynman knew that Baker was a code name and that James was really Aage and that Baker's father, Nicholas, was Niels Bohr. Why would the renowned Bohr want to get together with Feynman, who was very much a junior scientist on the project?

"They were very famous physicists, as you know. Even to the big shot guys, Bohr was a great god," Feynman wrote later of the father and son. At the time, Feynman didn't have a clue as to why Bohr wanted to see him.

"So at eight o'clock in the morning, before anybody's awake, I go down to the place. We go into an office in the technical area and he says, 'We

have been thinking how we could make the bomb more efficient and we think the following idea.'

"I say, 'No, it's not going to work.... Blah, blah, blah.'"

So, for about two hours, they argued and agreed and disagreed—about scientific ideas and technological details. Why did they choose to have a brain session with Richard Feynman? It seems that, when Bohr had been at Los Alamos before, he had said to his son, "Remember the name of that little fellow in the back...[because] he's the only guy who's not afraid of me, and will say when I've got a crazy idea. So next time when we want to discuss ideas, we're not going to be able to do it with these guys who say everything is yes, yes, Dr. Bohr. Get that guy and we'll talk to him first."

are forced to cooperate because of the unthinkable horror of nuclear war.

Bohr sees the bomb in terms of his philosophy of complementarity: "Every great and deep difficulty bears in itself its own solution," he tells the physicists. To Austrian theoretician Victor Weisskopf he says, "This bomb may be a terrible thing, but it might also be the 'Great Hope.'" He believes the fearsome weapon, with its Earth-destroying possibility, also has the potential to end war.

But Bohr fears an arms race when the war ends. He doesn't want any nation to stockpile nuclear weaponry. He believes Soviet Russia should be told about the project. He sees the openness of the scientific world as a model for the world in general.

Neither Bohr nor the Allied political leaders realize that the Russians already have the secrets of the bomb. But Bohr understands that, even without the help of spies, Russian scientists will soon be able to produce a bomb. He knows that knowledge doesn't stay hidden for long in the world of science. Once something is known to be possible, others soon figure it out. The seventeenth-century English poet John Milton explained that centuries before in *Paradise Lost*:

Given Joseph Stalin's paranoid personality, the repressive Russian system, American fear of Communism, and the knowledge gap between politicians and scientists— an arms race between the U.S. and the Soviet Union became inevitable. This Cold War cartoon, captioned "Handle with care!", warns of the danger as both nations began to build stockpiles of nuclear bombs.

> *The invention all admired and each, how he*
> *To be the inventor missed; so easy it seemed*
> *Once found, which yet unfound most would have thought*
> *Impossible.*

Bohr goes to see Winston Churchill, hoping to convince the prime minister to lead the way into an open world society without scientific secrets but with international controls. Churchill, who has been a great wartime leader, is no scientist. He is busy planning for D-day, the invasion of German-occupied France. Share secrets? Have an open world society? Churchill thinks Bohr is some kind of nut. He writes a memo to his aide: "Bohr ought to be confined or at any rate made to see that he is very near the edge of mortal crimes."

In July 1944, while Bohr is in Washington staying at the Danish embassy, he works on a memorandum to President Roosevelt. "The whole enterprise," Bohr writes, "constitutes . . . a far deeper interference with the natural course of events than anything ever before attempted and . . . it will bring about a whole new situation as regards human resources. . . . [W]e are being presented with one of the greatest triumphs of science and engineering, destined deeply to influence the future of mankind."

Both Oppenheimer and Bohr believe that if the bomb is used it will make the case that war—as it has been known—must end. Bohr's vision is of "openness" and an end to national secrecy. Oppenheimer has faith that Roosevelt will use the United Nations to help establish a new kind of sovereignty over these weapons of mass destruction. Neither knows that FDR has only months to live.

Leo Szilard (below) urged Congress to shift control over the nuclear program from the military to civilians. In a September 1945 meeting, he predicted the Soviet Union would make bombs. ("Reds Have Atom Bomb" shouts the headline below, four years later.) He also suggested joint controls with Moscow, atomic-energy control under a world government, and relocating large city populations as a defense. The Atomic Energy Commission was formed as a result of that meeting.

Niels Bohr has the instincts of a diplomat. Leo Szilard does not. "[He is] the kind of man that any employer would have fired as a troublemaker," Leslie Groves says of the Hungarian thinker. That there is conflict between the two of them is not surprising. Mercurial Szilard seems to enjoy baiting the military. He is especially annoyed that General Groves won't let physicists working on one phase of the project talk to those working on another phase. Szilard feels open discussion breeds ideas. Later, he says that by limiting talk and interaction between physicists, the bomb was delayed by a year.

Groves writes a letter to the U.S. attorney general, calling Szilard an "enemy alien" and suggesting he "be interned for the duration of the war." Then he puts Szilard under surveillance. The Hungarian physicist is tailed each time he leaves Los Alamos. An army counterintelligence report from that surveillance states: "The

Subject . . . has a fondness for delicacies and frequently makes purchases in delicatessen stores, usually eats his breakfast in drug stores and other meals in restaurants, walks a great deal when he cannot secure a taxi. . . . He is inclined to be rather absent-minded and eccentric." When Szilard meets Wigner in Washington, the agent reports that the two Hungarians tour the Supreme Court and later sit by a hotel tennis court where they "pulled off their coats, rolled up their sleeves and talked in a foreign language."

Unlike Bohr and Oppenheimer, who believe in complementarity and that good may come from the bomb (and that it can't remain an untried secret), Szilard now wants to stop it. When he thought Hitler might build one first, he believed it necessary. But the Nazis haven't built a bomb, and he has changed his mind.

On March 25, 1945, Szilard visits Albert Einstein at his house on Mercer Street in Princeton, New Jersey. He brings the draft of another letter for Einstein to send to President Roosevelt; this one alerts the president to the need for dialogue between "scientists who are doing the work" and "members of your Cabinet who are responsible for formulating policy." Einstein signs the letter, Szilard sends it, and Eleanor Roosevelt replies to Szilard early in April, agreeing to meet on May 8. That meeting is doomed not to happen.

On April 12, President Franklin D. Roosevelt, age 63, is in Warm Springs, Georgia, taking a few days to relax and work on a speech. It is the eighty-third day of Roosevelt's fourth term as president. He tells an artist who is sketching him, "I have a terrific headache." They are the last words he will say. The president has died of a massive stroke. The whole world mourns.

FDR is succeeded by the vice president, Harry S. Truman. A former senator from Missouri, Truman has never been told about the Manhattan Project.

In Germany, the war is now hopelessly lost, but Hitler insists that the German people fight to their deaths. He

President Franklin and Mrs. Eleanor Roosevelt ride together at Hyde Park, their (now open to the public) estate near the Hudson River in upstate New York.

In this colorized photo, a soldier raises the Red Flag over Berlin, which fell to the Soviets on May 2, 1945.

denounces Nazi leaders who wish to negotiate with the Allies. Then Russian troops enter Berlin. On April 30, Hitler kills himself in an underground Berlin bunker.

May 8 is VE Day, named for Victory in Europe. The European phase of World War II is over.

In June, Szilard drafts a petition urging President Truman "not to resort to the use of atomic bombs in this war unless the terms which will be imposed upon Japan have been made public in detail and Japan knowing these terms has refused to surrender." Copies of Szilard's petitions are signed by more than 150 Manhattan Project scientists. A counterpetition garners two signatures.

After the war, Szilard left physics and became a molecular biologist. He was appointed Professor of Biophysics at the University of Chicago in 1946. Ten years later, he helped found the Salk Institute for Biological Studies in La Jolla, California. In 1959, Szilard received the Atoms for Peace Award. He died in La Jolla on May 30, 1964.

Arline Feynman dies on June 16, and Oppie insists that Dick Feynman take some time off.

Feynman is at home when he gets a call that says, "The baby is expected." He knows what that code phrase means and flies back at once, arriving just in time to climb into a bus heading 340 kilometers (211 miles) south to an edge of the Alamogordo Bombing Range, a rattlesnake- and tarantula-infested flat desert trail known in Spanish days as *Jornada del Muerto*, the Journey of Death.

There, engineers have built a tower of steel 30 meters (98 feet) high, adding a $20,000 winch that lifts what is called "the baby" or "the test device" to a high wooden platform.

At 5:30 a.m. on July 16, 1945, the moment they have planned for comes.

We were lying there, very tense, in the early dawn, and there were just a few streaks of gold in the east; you could see your neighbor very dimly. Those ten seconds were the longest ten seconds that I ever experienced. Suddenly, there was an enormous flash of light, the brightest light I have ever seen or that I think anyone has ever seen,

says I. I. Rabi, one of the Nobel laureates.

Richard Rhodes, in *The Making of the Atomic Bomb*, describes what happens after the chain reaction begins when

fission multiplying its prodigious energy release through eighty generations in millionths of a second, tens of millions of degrees, millions of pounds of pressure... The radiant energy loosed by the chain reaction is hot enough to take the form of soft X rays; these leave the physical bomb and its physical casing first, at the speed of

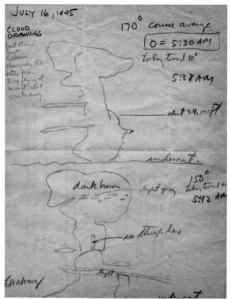

Physicist Luis Alvarez witnessed the world's first nuclear bomb explosion from a B-29 airplane, where he drew these two sketches of the shifting mushroom cloud.

Melba Phillips Didn't Laugh

What would happen to nuclear energy after the war? Most physicists were concerned. Melba Phillips (below), who had been one of Oppenheimer's graduate students at the University of California at Berkeley, helped organize the Federation of American Scientists in 1945 to consider that opportunity. Women physicists were a rarity in the 1940s. Phillips (1907–2004) had done groundbreaking work in early nuclear physics and was a professor at Brooklyn College and a research scientist at Columbia University. In 1952, she lost both jobs for refusing to testify before the U.S. Senate's Internal Security subcommittee over alleged Communist activities (this was during the Senator Joseph McCarthy era).

In 1987, Brooklyn College publicly apologized for the firing and in 1997 established a student scholarship in her name. By then, Phillips had written, edited, and co-authored several physics textbooks, had become the first woman president of the American Association of Physics Teachers, and was a legendary science professor at the University of Chicago.

Phillips was one of a number of physicists called "un-American" by zealots who feared Communism after World War II. The FBI had a 1,500-page file on Albert Einstein labeling him "an extreme radical." Of Joseph McCarthy, Albert Einstein said, "America has a sense of humor. In time, we will laugh at this man."

This child that the scientists have produced has the potential to destroy life on Earth as we know it. After the war, many Los Alamos scientists worked to limit nuclear energy to peaceful uses. Einstein, who took no part in the Manhattan Project, says that if he had known where his first letter to Roosevelt would lead, he would never have signed it.

light.... *What the world sees is the shock front and it cools into visibility.... Further cooling renders the front transparent; the world if it still has eyes to see looks through the shock wave into the hotter interior of the fireball....*

Feynman remembers:

Time comes, and this tremendous flash out there is so bright that I duck.... Then finally, a big ball of orange, the center that was so bright, becomes a ball of orange that starts to rise and billow... and get a little black around the edges, and then you see it's a big ball of smoke with flashes on the inside of the fire going out.... [A]fter about a minute and a half, there's suddenly a tremendous noise—BANG, and then a rumble, like thunder—and that's what convinced me [that the bomb had really exploded].

At the Trinity test site two months after the nuclear blast, Oppenheimer (center, in the wide hat) examines what's left of the 100-foot (30-meter) steel tower. Most of it vaporized instantly. The intense heat melted the surrounding sand, turning it into green glass cinders dubbed "Trinitite."

Oppenheimer says:

We waited until the blast had passed, walked out of the shelter and then it was extremely solemn. We knew the world would not be the same.... I remembered the line from the Hindu scripture, the Bhagavad-Gita: *Vishnu is trying to persuade the Prince that he should do his duty and to impress him he takes on his multi-armed form and says, "Now I am become Death, the destroyer of worlds."*

The steel tower has vaporized, only a few melted bits of metal are left, and the asphalt pad has turned into glass cinders as green as jade. The scientists expected a blast that would be equal to 5,000 tons of TNT. What they get is the equivalent of 20,000 tons.

Feynman will later say:

After the thing went off, there was tremendous excitement at Los Alamos.... I sat on the end of a jeep and beat

On the eve of battle, the warrior prince Arjuna and Lord Krishna speak of life and death in the *Bhagavad-Gita*, a poem in the great Hindu epic *Mahabharata*. The god Vishnu (above) was believed to have taken human form as Krishna. Most Hindus see this poem, written in Sanskrit, as a supreme expression of their religion.

Using Radio Waves

Raymond Gram Swing was a crusading journalist in the 1930s and 1940s and an early radio voice warning Americans about Hitler. Swing interviewed Albert Einstein just after the atomic bomb was dropped. In the broadcast, Einstein compared nuclear energy to sunlight. "In developing atomic or nuclear energy, science did not draw upon supernatural strength, but merely imitated the action of the sun's rays." He said he was not the father of the bomb, that he had only suggested that it was possible in theory. "It became practical through the accidental discovery of chain reactions, and this was not something I could have predicted."

Einstein, like his friend Niels Bohr, believed the secrets of the bomb should be shared, but not with Russia or the United Nations—rather, with a world government founded by the three great military powers: the United States, Great Britain, and the Soviet Union. Did he fear the tyranny of a world government?

"Of course I do. But I fear still more the coming of another war or wars. Any government is certain to be evil to some extent. But a world government is preferable to the far greater evil of wars." Einstein had little faith in the United Nations. He felt it lacked real power.

He said the world was not yet ready for the nuclear age, but he added, "It may well intimidate the human race into bringing order into its international affairs, which, without the pressure of fear, it would not do." And that, so far, has proved to be true.

An article based on the interview was published in the November 1945 issue of the *Atlantic Monthly*.

On August 7, 1945, the day after the Hiroshima bombing, the world at large learned what an "atomic bomb" was—and could do—from news accounts (above). The second bomb created this radioactive mushroom cloud (right), rising more than 11 miles (18 kilometers) high over Nagasaki, Japan. Nearly 75,000 people died instantly.

drums. . . . But one man, I remember, Bob Wilson, was just sitting there moping.
I said, "What are you moping about?"
He said, "It's a terrible thing that we made."
You see, what happened to me [Feynman], what happened to the rest of us—is we started for a good reason, then you're working very hard to accomplish something and it's a pleasure, it's excitement. And you stop thinking, you know; you just stop.

The Manhattan Project might have been the most ambitious scientific endeavor humans have ever undertaken. Was it a success? You can decide that for yourself. As a work of science and technology, it was spectacular. The German effort, with Heisenberg and Hahn participating, never got very far. The German scientists believed it was impossible to make a bomb in time for this war, so they never convinced Hitler to give them enough money to make it happen. But the Allies didn't know that.

Within two weeks, bombs are dropped on the Japanese cities of Hiroshima and Nagasaki. A few days after that, the war is over.

At first there is euphoria and celebration at Los Alamos. All the hard work has paid off; the Manhattan Project has accomplished what it set out to do. But it doesn't take long for scientists, politicians, and others to begin thinking about the awfulness of the bomb. In Hiroshima, some 70,000 Japanese were killed instantly and another 100,000 will die later from burns and radiation sickness.

Einstein is at a small cabin on Saranac Lake in New York's Adirondack Mountains on August 6, 1945. Helen Dukakis, his longtime secretary (as the position was then called), is listening to the news on the radio and hears that a new kind of bomb has been dropped on Japan. ("I knew what it was because I knew about the Szilard thing in a vague way," she will later tell a writer.) When Einstein comes for his morning tea, she tells him.

"*Oh Weh!*" he gasps. It is a German wail of anguish that goes beyond words, but is often translated as "Woe is me!" or, "Oh my God!" Then he says, "*Ach!* The world is not ready for it."

A month later, Philip Morrison, one of the physicists at Los Alamos, flies over Hiroshima. "We circled finally low... and stared in disbelief. There below was the flat level ground of what had been a city, scorched red.... One bomber and one bomb, had, in the time it takes a rifle bullet to cross the city, turned a city of three hundred thousand into a burning pyre. This was a new thing."

As a weapon, it worked; it ended a murderous war.

As a milestone in the human drama, it finished one era and began a new one.

As for science, which before had always seemed to focus on making the world better, well, science had lost its innocence.

An injured child and woman, each clutching a morsel of bread, stand on a bleak Nagasaki street a month after the nuclear explosion.

At ground zero in Hiroshima, those few buildings not flattened were concrete structures reinforced for earthquakes. One is now a peace memorial.

Quantum Electrodynamics? Surely You're Joking

We can imagine that this complicated array of moving things which constitutes "the world" is something like a great chess game being played by the gods, and we are observers of the game. We do not know what the rules of the game are; all we are allowed to do is to *watch* the playing. Of course, if we watch long enough, we may eventually catch on to a few of the rules. *The rules of the game* are what we mean by *fundamental physics*.

—Richard P. Feynman (1918–1988), American physicist, "Mainly Mechanics, Radiation, and Heat"
(*Lectures on Physics*)

Every scientist knows that the real fascination of research lies in the totally unexpected development, the revelation that tears the fabric of supposedly sacred theory.

—Julian Schwinger (1918–1994), American physicist, speech at Columbia University (1978)

The theoretical physicists know there is a problem that needs solving. They have this great theory—quantum mechanics—and it works astonishingly well, but not always. Sometimes it gives nonsense answers. It just isn't all there. Quantum mechanics doesn't quite work in the relativistic world, which means at or near the speed of light.

So right after the war, in the spring of 1948, 28 world-famous physicists get together at the Pocono Manor Inn in the Pocono Mountains in northeastern Pennsylvania. Oppie (that's J. Robert Oppenheimer) helps raise the money to pay for their rooms and train fare. They come because they are passionate about physics, and they know there are important questions to be answered. No one keeps a transcript (a written record) of the meeting. It is the kind of informality among physicists that made Niels Bohr's institute in Copenhagen so productive. That informality won't be around much longer as science and government grow and become intertwined.

In 1948, Oppie opted for the luxurious Pocono Manor Inn (above), rather than a lecture hall, to shoot the breeze informally about quantum physics.

Paul Dirac is one of the stars. He is known for his intense shyness, and he rarely says anything, but his brilliant mathematical equations are words enough. Dirac has posed the problem and set the goal for this gathering. It is (to simplify things a bit) to explain the action and interaction of light and matter so that it makes sense (mathematically and otherwise) in quantum terms; or, to put it another way, to explain how electromagnetism works on the photon-electron level. To do this, Dirac has seen the need for a new field, which he names "quantum electrodynamics." His younger colleagues shorten the name to QED.

As you'll recall, photons are light, and so travel at the speed of light (c); electrons, while very fast, do not.

That's what brings the scientists to the Pocono Mountains. They expect to hear new theories of QED. Many of the physicists believe QED will dramatically extend our understanding of the way the universe works. There is a sense of anticipation at the meeting; there is excitement and expectation.

It happens that three young physicists have each worked out solutions to Dirac's problem (they will share a Nobel Prize in 1965). Two of them—Julian Schwinger of Harvard and Richard Feynman of Cornell University— are at this meeting. The third, Sin-Itiro Tomonaga of Japan, isn't here. Schwinger and Tomonaga have based their work on

Shy Paul Dirac (left) leans back and listens to the outgoing and sometimes outlandish Dick Feynman at a 1962 conference. These opposite personalities first found common ground in 1948, when Feynman came up with an elegant answer to Dirac's question about quantum electrodynamics.

Those Relatives: Galileo and Albert

> One must divide one's time between politics and equations. But our equations are much more important to me, because politics is for the present, while our equations are for eternity.
>
> —Albert Einstein speaking to mathematician Ernst Straus (1922–1983)

A Newtonian physicist, by presuming space and time to be absolute, is forced to conclude that the speed of light is relative—it depends on one's state of motion....Einstein, by presuming the speed of light to be absolute, was forced to conclude that space and time are relative—they depend on one's state of motion.

—Kip S. Thorne, American physicist, *Black Holes and Time Warps: Einstein's Outrageous Legacy*

The possibility of time-travel...was among Einstein's early discoveries. It was one of the principal conclusions of Special Relativity...and his first strong clue that physicists would have to abandon the old notion of absolute time, ticking away everywhere at the same rate.

—Nigel Calder, British science writer, *Einstein's Universe*

I t was Galileo, not Einstein, who first stated the principle of relativity. Galileo was obsessed with motion, and understanding motion is the key to understanding relativity.

Galileo described a thought experiment—belowdecks on a smooth-sailing ship without portholes—where you can walk and jump or play a game of darts, and nothing will seem different from when you do those same activities on firm land. A goldfish in a glass tank doesn't have to change its way of swimming to adjust to the speed of the ship. Nor will you need to adjust your strokes when you play Ping-

Read more about Galileo's thought experiment in chapter 7 of *Newton at the Center*, book two in this series.

Leaping on Earth

Why, if the Earth is turning, don't you land a few feet away from your starting point when you jump straight up into the air? (Remember, that was one reason the ancient philosophers decided the Earth didn't turn.)

Galileo figured that out. He said that while the Earth is turning, you and I are passengers. Imagine yourself a passenger in a train or a plane moving uniformly with respect to Earth. Jumping up and down brings you back to the same place. You keep up with the vehicle that carries you—and Earth is a vehicle. You may not think of yourself as moving while standing on the ground, but someone in a spaceship away from Earth can watch you going along for the ride as our planet moves. Earth is turning, and it also is speeding around the Sun while our solar system is in motion around the center of our galaxy.

Do you feel any of that? Of course not: **Uniform velocity is seen with respect to something else, not felt directly.** It is comparative: one object to another. With velocity, the observer is the key. That's what makes it relative. That whole concept came to us from Galileo, who was a young scientist in the year 1600. More than 300 years later, Einstein said the same thing. So what's new?

Relative motion was much faster for Albert (riding trains) than it was for Galileo (mostly on foot or horseback or boat). It's even faster for us (using rockets). And Einstein knew something Galileo didn't know: He knew that light's speed has the same value (in a vacuum) for every (inertial) observer.

These "star trails" show how night-sky objects appear to move, due to Earth's rotation, over seven hours. The North Star is near the bull's-eye.

Pong in a moving ship's lounge. The laws of physics are the same in the hold or lounge of that uniformly moving ship as they are on land.

Galileo used that observation to explain why we don't feel Earth's motion as it turns on its axis and moves around the Sun. In a uniformly moving vehicle (a car, a plane, or on Earth), you can't detect motion (unless you look out a window). (Note: *Uniformly moving* is the key. If your car lurches, speeds up, hits a bump, or changes direction, you'll know you are moving.)

See if you can come up with an experiment that will detect motion in an enclosed, smoothly moving vehicle.

Uniform motion is steady motion in a straight line. We follow a curved path around the surface of the Earth as it rotates, and the Earth follows a curved path around the Sun, so our travel on Earth isn't uniform motion. However, during the time it takes for an observation, your direction of motion *changes very little* along those curves, so it's almost the same as traveling uniformly in a straight line.

(Enclosed means you can't see out the window.)

I'll save you some work: **It can't be done. There is no possible experiment that will tell you that you are in a moving vehicle if the motion is uniform.** You don't believe me? Find that experiment, and you'll be the next Einstein. Then think about this: If you *cannot* detect motion, then motion has no meaning, unless it is relative to something else. That's Galileo's concept of relativity, and that's where Einstein starts in formulating his own theory.

Galileo's concept of relativity explains uniform motion—for Newton, for Einstein, and for us. But how about measuring motion? How do you do that if motion is relative?

That's easy, say Galileo and Newton. You find something that you choose to be "at rest," and you measure your motion (your speed) in comparison to that thing. Do you want to clock the speed at which you run in kilometers or miles per hour? Just choose any stick or person or building as an "at rest" for a comparison.

How about measuring the speed of Earth or of the stars or a rocket in space? With respect to what? What do you use as an object at rest for a comparison? In Galileo's time, the big scientific argument was this: Is it the Earth or the Sun that is stationary at the center of the world? (Galileo gets house arrest for his answer to that question.)

By Isaac Newton's day, the natural philosophers agree: The Earth and planets move around a central Sun. But Newton knows the Sun also moves. So what is uniquely stationary?

Newton comes up with an unmoving frame of reference called "absolute space" against which any form of motion can be measured. **"Absolute space, in its own nature, without regard to anything external, remains always similar and immovable,"** says Isaac. As for James Clerk Maxwell, he agrees with Newton *on this point*. This is what he says: "Absolute space is conceived as remaining always...immovable. The arrangement of parts of space can

Curious about Galileo's arrest? Read chapters 8 through 10 of *Newton at the Center*.

Remember, **NATURAL PHILOSOPHER** was the term used before "scientist" took hold.

no more be altered than the order of the portions of time. To conceive them to move from their places is to conceive a place to move away from itself." That explains the ether: It is absolute space.

Ah, that all seems simple and sure until Albert Einstein comes along and tries to ride a light beam. He wants light to be his "at rest," his reference point.

Whoops, he can't do it. He's chosen the one thing that never slows down or speeds up (in a vacuum). That light beam won't slow for him no matter where he places himself. Light (in a vacuum) always moves at the same speed as measured by every (inertial) observer. Its speed is no different for a man sitting in his living room chair in Iowa or an astronaut in a spaceship circling the Moon. And nothing else can go as fast. James Clerk Maxwell got that right.

But what about unmoving space—the ether? The ether was the uniquely stationary "at rest" that made Newton's laws of motion work in the greater universe. Einstein realizes something is wrong with that concept. He spends years worrying about the ether and Newton's absolutes of time and space.

In 1905, Einstein blasts through the mental barrier that has thrown most scientists off course. He does it by ditching the ether (absolute space). A universe without ether turns out to be no big deal. Trains and ships and other moving things have always been measured with respect to Earth (Earth being the frame of reference). Einstein shows that in a universe where only relative motion can be observed, anything (the Sun, the

The happy travelers on this scenic train ride (ca. 1910) are moving relative to the ground, which appears to be stationary to them. Yet the ground is moving (the Earth is spinning and also traveling around the Sun) to an observer in space. All motion is relative; it depends on the observer's frame of reference.

the motion of everything else (including the motion of time) in relation to it.

For Newton, Earth was "at rest." He never considered space travel, but you can. If you want to travel very fast and into space, then a stationary object on Earth may not be the best point of reference. The idea here is that we are all in motion on our spaceship Earth, so all motion is measured relative to something else moving. In order to make sense of this, you have to single out something as an "at rest." Usually we choose Earth and treat it as a stationary point of reference, but we don't have to pick Earth.

You're moving faster than, farther than . . . than what? Than your chosen point of reference. Maybe it is your spaceship. **Relativity is about comparisons between things that are all moving.** Understanding this can stretch your head. It isn't the way we normally think. But work at it, because that is the way the universe is: a place where nothing is really at rest, or central, so **everything is in motion relative to everything else.** You get to pick the center. You get to pick the "at rest."

We didn't realize we needed a new theory of relativity as long as we kept our sights Earth-bound or crawled along at the fastest speeds of Einstein's youth—the speed of trains. It was when we started looking beyond our small planet, when we began to consider our place in the greater universe, when we thought in terms of space travel, that we came to realize that the science of the past—Newtonian science—was leading us astray. Here in our global village called Earth, classical science works well (if you don't move too fast). But that science doesn't correctly explain cosmic forces—like gravitation—and it doesn't correctly explain time and space when there is relative high-speed motion.

The big idea in special relativity (the subject of one of Einstein's pioneering 1905 papers) is that there **is no single, central, universal spot from where all measurements are made.** You are the center of your universe, and time and space measurements are relative to where you are and how fast you're moving. If "aliens" exist in a far-distant galaxy, they are the center of their universe. Nature's laws (call them laws of physics, if you prefer) are the same for all observers everywhere in the universe.

If you're sitting at a desk on Earth, you see the world from the solid Earth; that's your frame of reference. If you're sitting inside a rocket speeding through space, your frame of reference (the rocket) is moving in respect to Earth. In both situations, think of yourself as a stationary observer, while everyone elsewhere is moving. As an observer, you can perform experiments and make deductions—the rules of physics stay the same for you sitting at your desk or sitting in a rocket. (If you throw a ball to the ceiling from your desk chair or from your seat in the accelerating rocket, it will fall back in the same way.)

This is a huge idea, but just one more step in the process that Copernicus began when he moved Earth from the center of the solar system. After that, we just assumed the center was the Sun. Then Einstein came along. He said: **There is no center.** Or, put another way, the center is wherever we choose it to be.

Albert Einstein is still a teenager when he first realizes something is not right with the classical theories of motion. Those theories and what is known about light are in conflict.

To a physicist, a **FRAME OF REFERENCE** is a set of lines or planes used to describe the position of an event plus a clock to fix the time of that event. Think of the frame of reference as a place (a room on Earth, the inside of a rocket) with its own clock. Locate an event with respect to this frame of reference: 1 foot from the floor, 2 feet from the east wall, 3 feet from the north wall, and at 9:23 a.m.

You can date the modern scientific view of the universe to Galileo (with thanks to Copernicus, who hatched the idea a century earlier). When Galileo pointed his telescope at Jupiter and discovered it had moons, he realized that the universe might not be Earth-centered. It took 300 years to go from that start to Einstein's theories.

This famous photo, taken December 25, 1968, gave earthlings a new, breathtaking frame of reference. Three *Apollo 8* astronauts witnessed the first view of space from lunar orbit, including what Frank Borman called an "Earthrise" over the Moon's bleak horizon. Our blue marble of a planet suddenly looked beautiful but small and fragile. It was no longer the center of everything but rather a tiny oasis of life in the vast, black void of space.

Einstein was concerned with "a paradox upon which I had already hit at the age of 16: if I pursue a beam of light with the velocity 'c' (speed of light in a vacuum), I should observe such a beam as a spatially-oscillatory electromagnetic field at rest. However, there seems to be no such thing." (Oscillatory means moving back and forth, like a pendulum.)

Einstein spends 10 years (from the time he is 16 to 26) wrestling with the problem. That's when he tries to put himself (mentally) on a light wave—traveling at light's speed—and that's when he finds, to his surprise, that time is not the same for everyone (nor is matter). Time is not a law of physics; time is not a constant. Time does not flow as Newton said, "without relation to anything external." It depends on where you are and how fast you're moving. **Newton had it wrong.** In a world where Newton is revered, that is hard to believe.

Einstein pays attention to Maxwell's equations, which tell him that **the speed of light has the same value for all observers everywhere.** The speed of light (in a vacuum) flows "without relation to anything external." **The speed of light (in a vacuum) is invariant.**

Go with Einstein on this, and it will create situations that may make you shake your head and say, "It can't be."

For instance: We tend to think that if two people have perfectly functioning, synchronized watches and each goes off in a different direction, then if they meet up again in a few years and compare their watches, the watches will still agree (unless one is malfunctioning). But Einstein realizes this is

You can see this famous surrealist painting, *The Persistence of Memory* (1931), at New York's Museum of Modern Art. Just what is the artist, Salvador Dalí, saying about time?

not the case if one of them is traveling at extremely high speeds outward and back with respect to the other one.

If you move close to the speed of light (as observed by your Earth-bound friends), time and other things will seem the same to you as they do on Earth; but to your slow-moving friends back on Earth, your time will have slowed down. It won't just seem to have slowed down. It actually will have slowed *relative to their time*. Time isn't a constant; it isn't invariant; it's always relative to a frame of reference.

> ### A Thank-You from Dr. Einstein
>
> On January 15, 1931, Albert Einstein gave a speech to an audience of physicists in Pasadena, California. Albert Michelson (who would die four months later) was in the audience. This is part of what Einstein said:
>
> *You, my honored Dr. Michelson, began this work when I was only a little youngster, hardly three feet high. It was you who led the physicists into new paths, and through your marvelous experimental work paved the way for the development of the Theory of Relativity. You uncovered an insidious defect in the ether theory of light, as it then existed, and stimulated the ideas of H. A. Lorentz and FitzGerald, out of which the Special Theory of Relativity developed. Without your work this theory would today be scarcely more than an interesting speculation; it was your verification which first set the theory on a real basis.*

Yes, this seems bizarre. You experience one thing; people watching you from another frame of reference see something else. And this really happens. It's a phenomenon that isn't easy for our Earth-bound minds to accept. We don't naturally think that way because we don't normally travel fast enough so that our stay-at-home friends can observe time's flexibility and its effect on our watches.

Einstein doesn't let his normal experiences limit him. He uses his mind to the fullest. He becomes aware that Newton's laws of motion and Maxwell's theory of electromagnetism are only part of the story. Something is wacky when you try to fit them together. Einstein sets out to find an explanation. That's when he comes up with what comes to be called special relativity.

That theory falls into place when Einstein assumes that the speed of light (in a vacuum) is always measured to be the same and therefore other things must change relative to light. Our idea of fixed space (an ether) is wrong. And so is the idea of a fixed time, the same for every observer. Einstein's new ideas are astonishing.

Light Does Its Own Thing

Is Einstein thinking about light? No wonder he can't catch the pitched ball.

AS SMART AS HE WAS, ALBERT EINSTEIN COULD NOT FIGURE OUT HOW TO HANDLE THOSE TRICKY BOUNCES AT THIRD BASE.

Imagine a baseball pitcher winding up and throwing a ball at 90 miles per hour. You can clock that pitch exactly. Now put that same ballplayer on the front platform of a train going at 60 miles per hour. Have him pitch his fastball in the direction that the train is moving. Forget air resistance. How fast will the ball travel? Yes, you add the speed of the ball (90 mph) and the speed of the train (60 mph) to get the answer: 90 + 60 = 150 miles per hour. That's how fast the ball goes if you're watching it from a nearby window.

But that kind of addition doesn't work with light. You can't add its speed to anything else. A light pulse *always* moves at the same speed for all observers. It's invariant. To a physicist, *invariant* is a code word for "has the same numerical value for all observers in uniform relative motion."

Observe a jet from your window. If you measure the speed of a flash of light shining forward from its nose, you will get the standard speed of light (about 300,000 kilometers or 186,000 miles per second in a vacuum)—*not* the speed of light plus the speed of the jet. Shine the beam out the tail of the jet, and you will still measure the same standard speed of light—not the speed of light minus the speed of the plane. That will be true for you, the observer on the ground, and also for the pilot of the plane. She will measure the same speed—the standard speed—for light flashes shot out of the front and back of her plane.

Light doesn't follow our everyday rules for low-speed things. Einstein didn't try to explain why; he just said it is so. No matter which observer measures the speed of light in a vacuum, the value is always the same. When Einstein understood that fact, he realized that other things must change from one observer to another or the universe would be stuck, static, unchanging. And he knows it isn't. So what changes?

Einstein has a breakthrough idea when he thinks about a well-known ratio for speed. He knows that the speed of anything is the distance it travels divided by the time to move this distance ($s = d/t$). Consider that ratio. If the speed of light (in a vacuum) is fixed, then distance and time must change; they must be flexible, otherwise nothing would change. Einstein realizes that **instead of time and distance as sure things, the sure thing is the speed of light.**

In the past, everyone believed that time and distance were absolutes. We humans were quite certain that an hour was an hour and a meter was a meter

The Speed of Sound: A Clue to Light

Sound travels at 343 meters (1,125 feet) per second through air (at room temperature). Visible light, like all electromagnetic radiation, travels at about 300 *million* meters (186,000 miles) per second in a vacuum. **Sound must have a transmitting medium**; it can't travel in a vacuum. Light can.

The speed of sound varies with the carrying medium. Sound goes through air at one speed. It goes through wire or a piece of wood at different speeds. (Air? Wire? Wood? When they carry sound, they are the transmitting mediums.)

Light can also travel in some mediums, and then its speed does vary. It travels more slowly in water than in a vacuum. Scientists have been able to slow light way down in the laboratory in some controlled mediums. No one has been able to make it go faster than its standard speed in a vacuum. Maxwell says that can't be done.

everywhere in the universe. But we were wrong. Even the great Isaac Newton was wrong about this one. Why? Our senses fool us. For slow-moving objects on this planet of ours, changes in time and distance are so slight we don't notice them. We might never have noticed if we hadn't begun to think about the super-fast particles in our accelerators, about the total universe, and about space travel. When we began to take our minds beyond our planet, our perspective changed.

Einstein's realization transforms our worldview. It sends Isaac Newton from the pitcher's mound to the physics outfield. (Note: Newton is still very much in the game in our everyday, slow-moving world.) Ideas that have been accepted from the first moments of recorded human thought will need rethinking. Time, which seems to be the same for everyone, isn't always the same. My hour might *not* be your hour. It depends on our relative motion.

This information—which most nineteenth-century scientists overlooked—helps us create what we think of as the modern world. So the key idea is worth repeating. Here it is in science talk:

> It is impossible to travel faster than the speed of light, and certainly not desirable, as one's hat keeps blowing off.
> —Woody Allen

The speed of light in a vacuum always has the same value as measured by different observers in relative motion.

Or, in everyday language, when different moving observers measure the speed of light, they always get the same number.

That is a huge concept to digest, but Einstein isn't finished. When he realizes that it is time and space that must change for an observer, an astounding idea leaps into his head: **Time and space can't be separate from each other,** as all the scientific thinkers before him believed.

Newton, Euclid, and everyone else had considered time a fourth dimension, completely distinct from space. When measuring distance, they often also measured the time it takes to get from one point to another. That's what we do when we figure out the speed of a pitched ball. Before Einstein had his breakthrough idea, time seemed totally separate from the three dimensions of space. Einstein's formulas showed something else. They showed that time, the fourth dimension, isn't separate from those three spatial dimensions. We exist in a four-dimensional fabric woven of *spacetime*.

That word **CONTINUUM** is often used in today's science (and science fiction). A continuum is a sequence or progression that has no gaps.

Is Einstein being sucked into a black hole? You'll have to ask David Grossman, the software engineer and electric bass guitar player who created this picture.

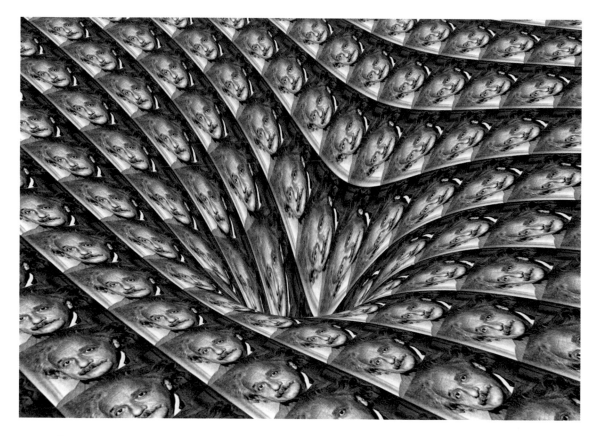

"The non-mathematician is seized by a mysterious shuddering when he hears of 'four-dimensional' things, by a feeling not unlike that awakened by thoughts of the occult. And yet…" said Einstein, "the world in which we live is a four-dimensional space-time continuum."

Euclid's geometry and Newton's physics deal with space *and* time (a division that seems natural on Earth);

Einstein's physics deals with spacetime. It isn't easy to visualize spacetime. We only become aware of the unification of space and time at very high speeds. But you can know how spacetime behaves, you can learn to describe it mathematically, and you can picture it graphically with spacetime diagrams. So welcome to spacetime! It's the environment you inhabit.

Whenever we talk of motion, we are comparing frames of reference. I can look out the window of an airplane (the plane is my frame of reference) and see that I'm moving fast *relative* to the cars inching along on Earth (the drivers' frame of reference). But I'm not moving at all relative to the person in the next seat.

To be clear: Some properties of light (like frequency and direction) do vary as observed in different frames. It is only light's *speed* that is invariant (in a vacuum): The speed is the same when observed in every inertial frame.

she observes the same light pulse, and she also measures the distance between the same two events (emission and the later event of absorption). But she is in a high-speed vehicle, and the distance she measures is not the same as your measurement; it is smaller. What's going on here? You and another observer see and measure the same events but get different measurements.

Maxwell tells us that the speed of that light pulse in a vacuum has the same value for *all* observers, no matter what their relative motion. Wait a minute! If the separation between the events is different for different observers in relative motion, but the light pulse travels between events at the same speed for both observers, then *the time it takes will not be the same.* That makes sense: Light going at a set speed will take

---

Ha-ha! Scientific Humor!

A physicist I know tells the following joke:

The training of flight attendants is amazing. On a plane moving 1,100 kilometers per hour, an attendant can pour coffee into a cup held a vertical distance 10 centimeters below the spout, and every drop will hit the center of the cup, which has moved 45 meters horizontally in the 0.14 seconds during which the drop *was in the air. Imagine being able to pour every drop into a cup 45 meters away! Can the greatest basketball players do anything like that?*

Of course, all this is from the reference frame of an observer on the ground, the physicist adds. The passenger in the seat sees the attendant simply pouring coffee into a cup, as usual.

different amounts of time to travel a different distance!

Why is that hard for us to grasp? Because we don't travel at very high speeds, so we don't encounter this reality.

Galileo understood that the space (the *separation*) between events occurring at different times is different for observers in relative motion. And **all motion is relative motion.** But until Einstein came along, nobody thought about the possibility that the *time* between events would be different for different observers.

This now-proved idea—that **the *time* between the same events can be different for different observers in relative motion**—is central to special relativity. The diagram on the next page—of *one* light-bounce clock observed in two frames—should help you see why the time is different.

The equation $d = vt$ will help you see what's going on. The speed (v) stays the same. The distance (d) is a different value for the two observers. That means time (t) has to be different for them, too, in order for the equation to balance.

Faster than a Speeding Bullet

Put yourself on a streetcar moving at 90 percent of the speed of light. Inside the streetcar, everything seems normal. When you look outside the window, you can tell you are going fast,

but the ride is smooth, and you can't feel it. Inside the streetcar, your watch ticks normally. But, wonder of wonders, to those on the sidewalk, your watch has slowed way down.

Looking quickly at the streetcar from the side as it zooms by, an observer sees the car and the passengers compressed—squeezed flat to a few inches in width. But when the streetcar slows down, you all appear normal to observers.

Why haven't you seen this bizarre effect? Because you've never seen any object pass by at anywhere near the speed of light. But if you did, it would be clear that, in relative terms, size has no absolute meaning. Nor does time.

According to Einstein's Theory of Relativity, a stationary person looking at an object that is moving close to the speed of light observes the following: Clocks on the object appear to slow down, lengths along the direction of motion of the object appear shorter (as with the squished streetcar at left), and masses increase.

Look first at the left-hand diagram. It shows a light-bounce clock made of two parallel mirrors a vertical distance *L* apart. A light-bounce clock uses a flash of light that bounces up and down between these mirrors; the clock "ticks" every time the light flash returns to the *bottom* mirror.

A friend, let's call him Casey, carries this light-bounce clock inside a rocket. His frame of reference, his "at rest," is his seat in the fast-moving rocket. From that frame, he watches the light bounce vertically up and down. The time between ticks for him is the time that it takes light to travel an up-down distance of *2L*.

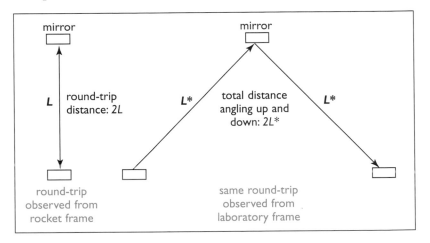

Now Casey's rocket is whizzing past his pal Eli, who is watching the rocket through a super-telescope from a laboratory on Earth. Eli's frame of reference, his "at rest," is Earth. Eli can see right through the rocket's glass walls and observe the light-bounce clock. The path of light that he sees is shown in the right-hand drawing. Eli sees the light flash begin at the bottom mirror on the left, angle up to the right a distance L^*, bounce off the upper mirror, and angle back down an equal distance L^* to the right and hit the bottom mirror.

What's going on? In Eli's frame (observing the rocket from an Earth lab) the mirror moves before the light flash bounces back. In Casey's frame (inside the rocket) the mirror doesn't move. So Casey sees one thing, and Eli sees something else.

Did you get lost trying to follow the path of light as seen from inside a rocket and by an observer watching from Earth? Almost everyone does the first time through. Relativity describes very fast action that we don't normally deal with in our daily lives. To understand it—unless you are a genius—will take several readings. Draw the action for yourself; that will help.

But look! The distance L^* is greater than the distance L. Light travels a longer distance between ticks for Eli than it does for Casey. The faster the rocket's speed, the greater the distance the light travels—as Eli observes it (on the right) but not as Casey observes it (on the left).

Here's the payoff idea: *The speed of light is the same* for Casey in his rocket frame as it is for Eli in his laboratory frame. *But the distance that light travels between clock ticks is not the same.* That means the time between ticks on the clock in the rocket is greater for Eli (watching from Earth) than for Casey (a passenger in the rocket).

Yes, it may seem crazy, but analyze the pictures again and you'll see it makes sense. **Because the speed of light in a vacuum is invariant, the separation between two events (measured between clock ticks) can be different for different observers.**

Is that time difference between ticks only true for light-bounce clocks? Those clocks are mostly in scientific laboratories. What about ordinary clocks?

Let's have Casey carry every possible kind of clock with him. In addition to the light-bounce clock, he has a spring-mechanism clock, a grandfather clock, a quartz clock, and

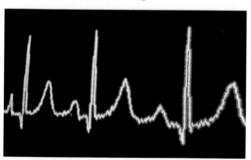

an atomic clock. These clocks are all "at rest" in Casey's frame inside the rocket. Casey verifies that they all tell the same time in his frame.

Now Eli, with his super-telescope in the Earth lab, observes all the clocks inside the rocket. What does he see? All of Casey's clocks (even his body clock, the ticks of his heart) take a longer time between ticks. This is not conjecture. Experiments have proved that this actually happens. So say good-bye to Newton's concept of absolute time and hello to relativity.

It's impossible to illustrate relativity scientifically on a two-dimensional page, since spacetime is four-dimensional, but here's an illustrator's take on a time warp at light speed.

Heartbeats (left) are just another kind of ticking clock. If you travel through space at very fast speeds, an observer on Earth will perceive your heart beating more slowly than if you had stayed on Earth. (To you, your heart will beat normally.) This means that your friends on Earth will observe you to age more slowly than they are aging. This difference in aging is called a paradox (see chapter 33), but it's not really a paradox. It is just the way things are in our universe!

Math Matters:
Euclidean and Non

Do all parallel lines go on forever, never crossing? M. C. Escher's *Other World* (1947) poses the geometrical question every which Euclidean way. The non-Euclidean answer is: no.

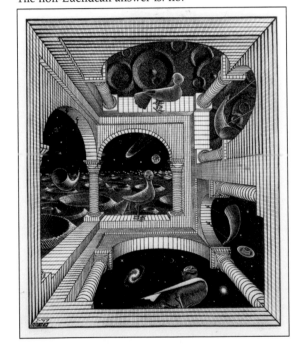

Where there is matter, there is geometry.
—Johannes Kepler (1571–1630), German astronomer

As you'll recall, Einstein the schoolboy was a fan of Euclid (see chapter 1). Euclid based his monumental study of geometry on five postulates, which are propositions or axioms presumed to be true. They seem to require no proof and could be considered self-evident. But one of the postulates wasn't as self-evident as the others. It was the fifth postulate about parallel lines, lines that Euclid claimed never meet. That postulate says: Whenever L is a line and P is a point not on L, there is a unique line through P that is parallel to L. *The idea of parallelism implies lines that go on forever.* But no one was able to prove that concept. To thinking mathematicians, it was bothersome.

In the nineteenth century, some mathematicians (look up Lobachevsky, Bolyai, and Gauss) found that Euclid's fifth axiom wasn't needed. Instead, they devised new "curved" geometries in which any number of lines can be drawn through a point, all parallel to a given line that does not include that point.

Then Georg Friedrich Bernhard

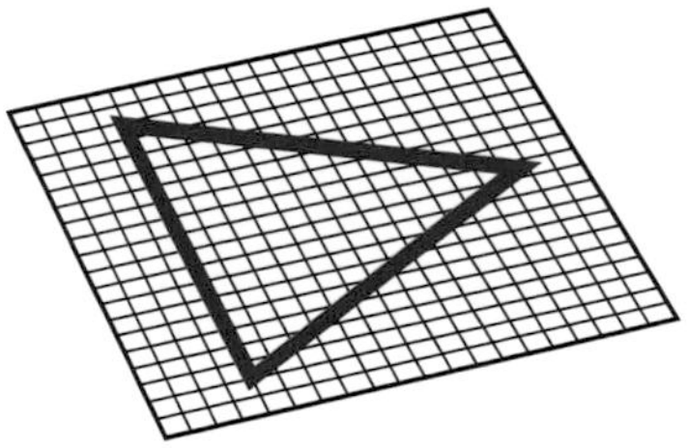

On a flat surface, the angles of a triangle always add up to 180 degrees. (That's Euclidean or plane geometry.)

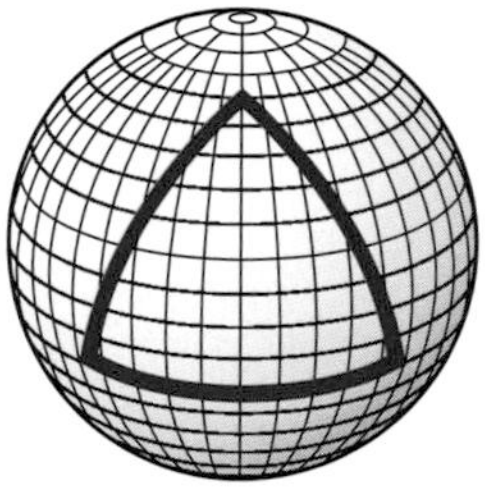

On a positive curve, like the surface of a sphere, the angles add up to more than 180 degrees.

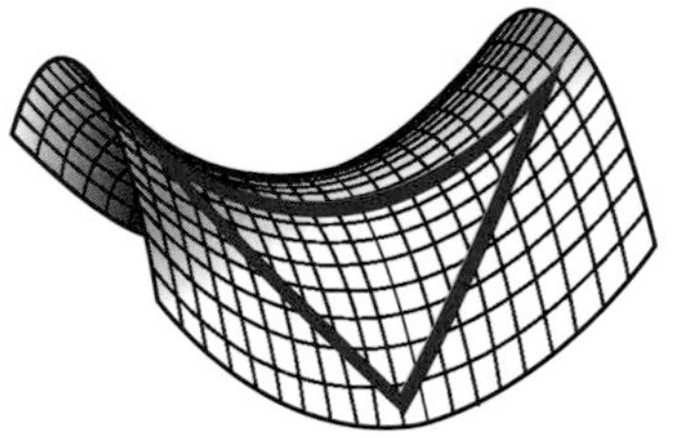

On a negative curve, like a saddle shape, the angles add up to less than 180 degrees.

Riemann (1826–1866) appeared with another kind of non-Euclidean geometry. In his math, it is *impossible* for any two lines to be parallel: All lines intersect. That isn't all. In Euclid's geometry, if you add the three angles of a triangle, you always get 180 degrees. In Riemann's geometry, those angles add up to a different total (see diagrams). On a sphere, they add up to more than 180 degrees. On a saddle shape, they add up to less than 180 degrees. This was new mathematics.

Riemann's geometry led to a universe with many dimensions—not just three or four. When he devised this universe, his ideas seemed to be pure, abstract math. Then Albert Einstein came along and applied them to the real world of four dimensions (length, width, height, and time). Today, many scientists are working with multiple dimensions of space in an attempt to further understand the universe. To do that, they are inventing new forms of mathematics.

Lorentz the Transformer

How much is $\frac{2}{3}$ plus $\frac{5}{6}$? Don't be too sure you know the answer.

The regular laws of addition don't work at or near the speed of light. Suppose I'm in a spaceship traveling at $\frac{2}{3}$ the speed of light and I throw a ball. I have a strong arm: The ball goes $\frac{5}{6}$ the speed of light with respect to my spaceship. If an outside observer looking at my ball adds the speed of the throw ($\frac{5}{6}$) and the speed of the spaceship ($\frac{2}{3}$ or $\frac{4}{6}$), she'll get $1\frac{1}{2}$ times ($\frac{9}{6}$) the speed of light for the speed of the ball.

Since nothing can move faster than light, that answer is unacceptable. So what do we do? We need a new law of addition. It happens that we have one. Dutch physicist Hendrik Lorentz (1853–1928) devised laws of arithmetic, called **Lorentz transformations**, that are non-Euclidean and work at or near the speed of light. Einstein found them very helpful.

Because the Special Theory of Relativity describes motion where gravity can be ignored—in what is called *flat* spacetime—Einstein could use Euclid's approach. But when Einstein later tackled the problem of motion near gravitating bodies, he needed to go beyond Euclid, beyond Lorentz, and beyond special relativity. He improved on Riemann to describe not just curved space but curved space*time*.

Timely Dimensions

No matter how hard you chase after a light beam, it still retreats from you at light speed. You can't make the apparent speed with which light departs one iota less than 670 million miles per hour, let alone slow it down to the point of appearing stationary. Case closed. But this triumph over paradox was no small victory. Einstein realized that the constancy of light's speed spelled the downfall of Newtonian physics.
—Brian Greene, American physicist, *The Elegant Universe*

Does something about a clock really change when it moves, resulting in the observed change in tick rate? Absolutely not! Here is why: Whether a free-float[ing] clock is at rest or in motion in the frame of the observer is controlled by the observer. You want the clock to be at rest? Move along with it! Now do you want the clock to move? Simply change your own velocity! This is true even when you and the clock are separated by the diameter of the solar system.
—Edwin F. Taylor and John Archibald Wheeler, *Spacetime Physics*

Einstein's 1905 paper on special relativity, which fills 31 handwritten pages, carries the very uninspiring title, "On the Electrodynamics of Moving Bodies." (In German, as he wrote it, it is: "*Zur Elektrodynamik bewegter Körper.*") Except for the key word *moving*, that title doesn't begin to indicate Einstein's intent. **He is redefining the nature of time and space.** The mental effort involved in writing this paper leaves him exhausted. Albert spends the next two weeks in bed while his wife, Mileva, checks the paper for mathematical errors.

Everyone thinks of the visible world as having three dimensions: length, width, and height (or up, along, and across). But why must we stop with those three spatial dimensions? Thinking deeply about space and motion, Einstein discovers that space merges with time (and vice versa). Before, everyone had thought of time as a separate dimension.

Einstein says **time is an interwoven fourth dimension**. Energy and matter, as well as electricity and magnetism, are all unified in nature in the four dimensions of space and time.

After the paper is published, Einstein expects some kind of reaction from the physics community; after all, if what he says is true, it is an idea that will overthrow much of what the professors are teaching. Perhaps that is why most physicists ignore his paper. But those who do notice are at the top of the profession. Max Planck is the first to comment. (He is acting editor of the journal that printed Einstein's manuscript.) Planck was skeptical when Einstein suggested that light comes in quantum units, but this space-time idea seems to appeal to him. He sends a letter asking a few questions. He is taking special relativity seriously and will soon teach a seminar on the subject.

Wilhelm Roentgen (the X-ray man) writes the patent clerk, asking him for a reprint of the paper. So does Max Born, at that time a graduate student in Breslau, who will become an important physicist and a good friend. Then someone else notices. It is Hermann Minkowski (1864–1909), who was Einstein's professor at college, the ETH in Zurich, Switzerland. He has a hard time believing his "lazy dog" student Albert Einstein has actually written this paper. Minkowski takes Einstein's theory, applies some new mathematics, and extends Einstein's idea.

It is Minkowski who merges space and time mathematically in a marriage that produces spacetime. Minkowski says, "From now on, space by itself, and time by itself, are doomed to fade away into mere shadows, and only a kind of union of the two will preserve an independent reality."

As the painting *Einstein Juggling with Time* (2000) suggests, time in this universe of ours does not tick to one giant clock. Einstein is clear about that: "Time cannot be absolutely defined, and there is an inseparable relation between time and ... velocity." Inseparable relation? Linking time and velocity links time and space. After he figures this out, Einstein writes that he feels he has tapped into "God's thoughts."

Speaking to an audience in Cologne, Germany, Minkowski tells them not to think of time and the three spatial dimensions as separate variables—like *x*, *y*, and *z*, plus *t* for time. Instead, they should **think of spacetime as a woven entity with four *inter*dependent dimensions**. Minkowski, who dies soon after doing this important work, rarely gets the credit he deserves for contributing to relativity.

Einstein, the audacious 26-year-old patent clerk, has rejected Isaac Newton's idea that time and space are separate and unchanging. If he is correct, a great many hallowed scientific ideas are going to have to go. Albert Einstein is sure he is right.

Proud parents Albert and Mileva pose with their son, Hans Albert.

In 1906, the patent office promotes Einstein and gives him a 30-percent raise in salary. (It's not because of his 1905 papers but because he is a terrific patent clerk.) He and Mileva and their two-year-old son, Hans Albert, now have some financial security. A year later, Einstein applies for a position as a *privatdozent* at the University of Bern. (A *privatdozent* is a private teacher who has the right to teach at a university. It's a European tradition and the lowest rung on the academic ladder.) At first his application is rejected, but in 1908, Einstein gets that appointment and begins a series of lectures in Bern (mostly to a small group of his friends). He keeps the patent office job; he needs the salary.

Meanwhile, Max Planck is paying attention to special relativity. He sends his assistant Max von Laue (who will win a Nobel Prize) to meet the unknown thinker. Von Laue (according to the story) enters a hall and asks a young man there to introduce him to the author of relativity theory. The young man, of course, is Albert Einstein. Later, describing that occasion to a friend, Einstein says that von Laue was the first real physicist he had ever met. His friend thinks, Didn't you ever look in a mirror?

Einstein's ideas are beginning to attract some attention among the elite in the physics community. Finally, in 1909, he is offered a full-time job as an associate professor of theoretical physics at the University of Zurich (a different institution than the ETH, which is also in Zurich). He can leave the patent office and concentrate on physics. Einstein's boss is amazed; he had no idea that his clerk had explained Brownian motion, discovered photons, and developed relativity—all in his spare time at the patent office.

When Einstein arrives for lectures in Zurich, his pants are too short, and his dark, curly hair is unruly. None of that—now or ever—seems important to him; he is obsessed with science, and he tries to share his enthusiasm with his students. He encourages them to stop him in the middle of a lecture if they have a question. Einstein's easy informality is not the norm in German-speaking universities. One student writes, "After the first few sentences, he captured our hearts."

Zurich is a friendly place for Albert and Mileva. It's the city where they first met and where they went to college. In Zurich, in 1910, the Einsteins have a second son, Eduard.

When Albert and Mileva lived in Prague (left), the city was part of the Austro-Hungarian Empire. Czechoslovakia became a nation in 1918 after World War I, with Prague as its capital. That nation split into the Czech Republic (which now has Prague) and Slovakia in 1993.

The Metamorphosis (1915), by Franz Kafka (above), begins, "One morning, as Gregor Samsa was waking up from anxious dreams, he discovered that in bed he had been changed into a giant cockroach." Below is an illustration from a 1946 edition of the horrifying novella.

Then the University of Prague (Czechoslovakia) has an opening for a full professor; it's a prestigious job with a good salary. University officials ask Max Planck if they should consider hiring Albert Einstein. Planck writes to them, "If Einstein's theory [of special relativity] should prove to be correct, which I expect, he will stand as the Copernicus of the twentieth century."

The Einsteins are soon on their way to Prague. It's a gorgeous old city, but ancient hatreds have split the university faculty into Czech and German divisions. Formal, correct behavior is expected of professors; Einstein must purchase an imperial sword and uniform to wear at his installation ceremony. He describes Prague in a letter to his friend Michele Besso as a city of "ostentatious luxury side by side with creeping misery on the streets." In that setting, Einstein's casual attitude seems provocative. He doesn't realize he is supposed to formally call on the older faculty members; some think he is insulting them by not doing so. In addition, Albert's Jewish roots and Mileva's Serbian background make them outsiders. Mileva has given up her career, she has two little boys to tend, and she has no friends in Prague; their marriage begins to unravel.

Meanwhile, Albert needs stimulating conversations. He finds them in Prague's coffeehouses. In 1911, 32-year-old Albert Einstein and 28-year-old Franz Kafka are both participants in a Jewish discussion group where the talk mostly focuses on art, literature, and philosophy. (If we could only be there and hear what they have to say!) Kafka works during the day at an insurance company; at night, he writes stories. His friend Max Brod joins them. Brod plays piano and is soon accompanying Einstein and his violin. Fascinated by Einstein, Brod uses Albert as a model for the character of Johannes Kepler in the novel he is writing, *The Redemption of Tycho Brahe*.

In the fall of 1911, Einstein is invited to Brussels, Belgium, to attend the first Solvay Conference (see page 160). The fact that Einstein is asked to attend means he is now accepted among the first rank of physicists. In Brussels, he sees his friends Max Planck and Hendrik Lorentz and meets, for the first time, Poland's Marie Curie, England's Ernest Rutherford, France's Louis-Victor de Broglie, and Holland's Heike Kamerlingh Onnes (who has discovered superconductivity). Of the 24 who attend, Curie is the only woman; she and Einstein quickly become friends. Albert, the youngest person there, is asked to deliver the concluding lecture. For him, Solvay is a dream experience. He has found a community of talent where nationality has no meaning and goodwill prevails.

Mileva, who had hoped for a career in physics, has been left behind. She writes to Einstein at Solvay, "I would have loved only too much to have listened a little and to have seen all those fine people. . . . It has been an eternity since we have seen each other. Will you still recognize me?"

After 16 months in Prague, the ETH lures Albert back to Zurich, the city that he and Mileva love. Twelve years earlier, his ETH professors wouldn't recommend him for any job in the field; now they have asked Einstein to be a full professor. What must he be thinking?

Newton thought of motion as following curved lines in flat space. Einstein thought of motion as following straight lines in curved space.

A Man in a Red Hat

There was a young lady named Bright,
Who traveled much faster than light,
She started one day
In the relative way,
And returned on the previous night.
—Arthur Henry Bullen (1857–1920), British editor, poet,
and founder of Shakespeare Head Press

Albert Einstein is trying to understand motion, so he takes his mind where it can experience speed. You can go with him on his thought experiment. He begins by climbing aboard the fastest vehicle of his day: a train. (In 1908, Wilbur Wright flies a plane in France as a crowd watches and cheers, but that is three years after Einstein published his paper on relativity.)

Take yourself to a train station where **a man with a *red* hat** is sitting on a bench watching the trains go by. He is sitting still. He seems to be at rest.

Now, a train whizzes down the track. It happens to have glass walls, so you can see right inside where **a man with a *green* hat** is sitting on a train seat. Is he at rest, or is he in motion? If you get on the train he seems to be at rest (he has actually fallen asleep). All of Newton's laws say the man in the green hat is at rest. But, from the platform, you can see him speeding past you. So, is the man in the green hat on the train sitting still, or is he moving? (Think "frames of reference" before you answer that.)

Riding on a train, are you at rest or in motion?

In the twentieth century, Einstein's stunning theories, which allow us to glimpse a master plan that seems to guide the universe, were mostly studied by scientists and the intellectual elite. Now, in the twenty-first century, elites are out of fashion. Mind stretching is for all of us.

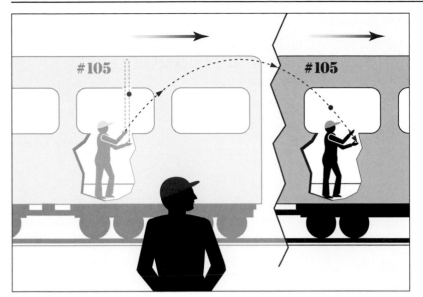

Fellow passengers on a moving train watch a man in a green hat toss a ball straight up and down (inset below). To an observer standing outside the train, looking in through the window, the ball appears to travel in an arc (left). So which is it: Does the ball go up and down? Or does it travel in an arc? Answer: The motion observed depends on the frame of reference of the observer. In other words, it's relative.

Now, back to the man in the red hat on the station platform. He stands up and watches as the glass-walled train comes by for a second time. We're standing there with him.

The man in the green hat, inside the train, is now awake, and he has stepped into the aisle. He throws a ball up toward the ceiling, and it falls back down into his hand. Everyone inside the train is watching; they can see the ball go straight up and come straight down.

You and I and the man in the red hat are standing on the station platform, and we *don't* see the ball moving straight up and down. As the train comes into view, we watch the man in the green hat throw the ball up. By the time he catches the ball, the train has moved past us to the end of the platform. We see the ball go up in one place, make an arc, and come down somewhere else.

So does the ball go straight up and down—as it appears to the people in the train? Or does it make an arc—as we on the platform clearly see it do?

What's going on here? Can the same event be seen differently by different observers? Yes, says Einstein, it depends on where you are *and how fast you're moving* when you do the observing. (Galileo and Newton would agree

Newton said: Objects at rest remain at rest unless acted upon by a force. That makes sense. The chair you're sitting in isn't going to move unless you push or pull it. Newton's laws work in our everyday world. But, says Einstein, nothing is at rest in the universe. Our planet, our solar system, and our atoms are all moving relative to something else. Today, a physicist would put it this way: *Motion has no meaning* except in relation to something else.

with Einstein so far.) Our view of things is relative. Imagine that! Einstein does. He refuses to be trapped in a conventional mind. You don't have to be trapped there, either. Relativity tells us that our ideas on space, time, and motion need expanding, especially if we are to consider the universe beyond our planet.

Here's more to think about: Einstein imagines a clock on a train and a girl standing outside watching the train go by. To the passenger on the train, the time that passes between two successive clock ticks is one second, but to the girl watching from the embankment, it is slightly more than one second. (The time difference is much too small to be easily measured at Earth speeds. You have to rely on scientists who have made the measurements with very high-speed particles and super-sensitive clocks.)

A person on the train and an observer on the embankment disagree about the time for one tick on the train rider's watch. Why? "Every motion must be considered as a relative motion," says Einstein, who further explains:

   a. The carriage is in motion relative to the embankment.

   b. The embankment is in motion relative to the carriage.

You will see that more clearly if you climb into a super-rocket. As you accelerate and reach a very high speed, an Earth-bound observer who is measuring and recording what he observes will be aware of things happening to you and to a pencil you hold in your hand—things he would not easily observe if you were riding a much slower vehicle (like a motorcycle). Your friend, the observer on Earth (who has fantastic sight and can see through the walls of the rocket), will observe you and your pencil get narrower and narrower and time on your watch go slower and slower. Your body's clock, recorded in your heartbeats, will slow down, too. This won't be a trick of his mind. He will actually observe it.

Narrow may be the best word to describe what an observer will record happen to you as you zoom by on your super-fast rocket and approach the speed of light. He will observe you shaped like a paper doll—wide as ever but flat.

EMBANKMENT? It's an old-fashioned word that can mean a mound of earth that supports a road or train tracks. I used the image because it is Einstein's.

Keep in mind, you are moving relative to an observer, and he is moving relative to you. Neither of you is stationary, and there is no stationary measuring rod.

Hold a pencil in your hand and look at its shadow on a table or on a wall. Is the shadow the same size as the pencil? Can you make that shadow change size? Imagine that the pencil is four-dimensional spacetime and our three-dimensional world is the shadow world, and you'll be thinking as Einstein did.

STAPLE TOP EDGE

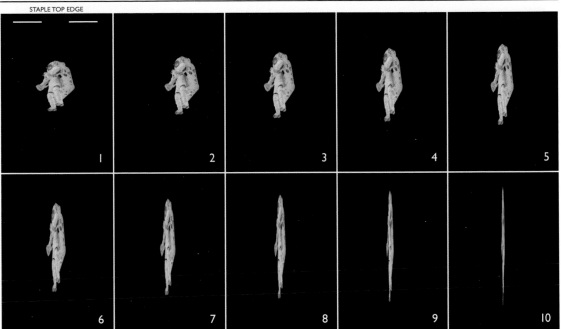

Your pencil will be shorter than it was. If, however, you are sitting sideways in the rocket, then he will observe you flattened in that direction: the direction of the motion.

To make this even more odd and mind-reeling, you, sitting in the rocket, won't be aware of these observations taking place on Earth. Everything will seem normal to you. It's *the observer* watching you from outside the rocket who will note your watch and body functions slowing down and your body getting narrower. And you will observe the observer shrink and his watch slow down (because, from inside your spaceship, you seem to be at rest, and *he* seems to be speeding).

"It is [that balancing of appearances]—that each party should think the other has contracted—that is so difficult to realize. Here is a paradox . . . that is too absurd for fiction, and is an idea only to be found in the sober pages of science," English astronomer A. S. (Arthur Stanley) Eddington will write in 1920, after studying Einstein's work.

Why have you never seen this bizarre effect? Because you've never seen anything (except light) pass by you at anywhere near the speed of light. But it actually happens.

Talk about bizarre effects (above)—this one at extreme gravity rather than extreme speed. It's called *spaghettification*, a fitting word for what would happen to an object—or person!—within the grasp of a black hole. Because of a huge difference in gravitation at the feet (closer to the hole) and the head (a little farther away), the astronaut's body would stretch out thin, like a strand of spaghetti. At the same time, forces squeeze him from the sides as he is funneled toward the center of the black hole. To turn this into a flip book: 1. Make two copies. 2. Cut out the 20 squares. 3. Stack them in order (1, 1, 2, 2, etc.). 4. Staple the top edge. 5. Flip the pages front to back.

To repeat: *Nothing that has mass can go as fast as light.* Today's space travel is nowhere near the speed of light.

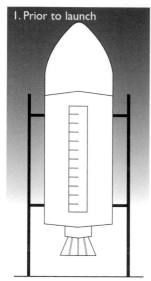

1. Prior to launch

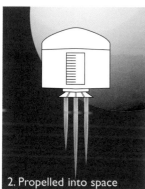

2. Propelled into space

To an observer on Earth, a spaceship traveling at near light speed would appear shorter along the direction of motion (nose to tail) than it was while sitting on the launchpad. Likewise, all objects inside the ship, including a ruler, would also appear shorter to the outside observer. Astronauts inside the ship observe no change in length from liftoff to light speed.

In relative terms, length has no absolute meaning (nor does time).

In Newton's physics, the distance between two measured points—as on a ruler—will always be the same, even if we view the ruler (or rod) as it moves. But that's not true in Einstein's physics. "The rigid rod is...shorter when in motion than when at rest, and the more quickly it is moving, the shorter is the rod," says Einstein. This is not an optical illusion. Each observer's instruments measure a different length for the rod. But the rod won't change for someone "at rest" with it in the same frame of reference—for instance an astronaut inside the rocket. An observer who can see through a window of the soaring spaceship measures the rod to be shorter than it was when the rocket was parked on Earth. (The rocket itself is shorter, too.) Are you baffled? Of course, to an Earth-bound mind, this is startling. It just happens to be true.

Once Einstein understands that time and distance (or length) vary with motion, he asks himself: why not mass? He finds that it also varies. **Mass increases with motion.** That seems like a wild idea, even to Albert Einstein. He wonders if "the good Lord does not laugh at it and has led me up the garden path."

Actually, "the good Lord" seems to have given this man great gifts of mind and spirit; Einstein is using them to tread a path filled with wonders. This idea, that mass increases with motion, leads him to his awesome formula, $E = mc^2$. Before Einstein, mass and energy were thought of as separate, and each had its own conservation law. We now know that mass and energy are just expressions of the same thing. $E = mc^2$ is sometimes called an equivalence formula. Said another way, mass and energy have the potential to be converted: One can turn into the other.

Like dollars and pesos, mass and energy are two forms of the same thing. You need an exchange rate to convert dollars into pesos. In $E = mc^2$ (that's Einstein's handwriting, below), the square of the speed of light is the "exchange rate" for mass and energy.

$$\left(E = mc^2\right)$$

Global Positioning Systems (GPS) in cars keep track of where you are and find the best route to where you are going. Some, like this one, even give voice directions. How do they work? With satellite signals, gyroscopic sensors, digital maps, and Einstein's general relativity equations. Without those equations, GPS would be lost.

If all of this seems strange, remember the round-Earth theory and how long those who believed in a flat Earth struggled with it. Relativity, like the round-Earth theory, has passed many scientific tests. But it does take mind work to understand it. Should you bother? You don't need to if you're not going anywhere. Today, the Global Positioning System (GPS), a satellite navigation tool, guides airplanes, ships, and some automobiles; GPS couldn't work without Einstein's formulas (read why on page 311).

If you want to be a thinking citizen of the twenty-first century, it will help if you understand that time and space are woven together and that matter and energy are ultimately the same thing.

As for special relativity, it isn't a complete theory. Special relativity only deals with situations where the gravity of nearby objects can be ignored. Where gravity deflects objects a bit, special relativity is only an approximation. Where gravity is intense, special relativity gives answers that are wrong. Our friend Albert is thinking about that. He's thinking harder than ever. He's working on a new theory that includes gravity; it will be his greatest achievement and the most remarkable theory in modern physics.

What Einstein's Ideas Tell Us

- There is no absolute reference point in the universe.
- Every place is a correct place from which to observe the universe.
- The center of the universe is wherever we choose it to be.
- There is only one thing whose measured speed is the same for all observers: light.
- There are no instantaneous interactions between objects that are far apart. In other words, nothing can go faster than the speed of light.

The Paradox of the Twins

Time present and time past
Are both perhaps present in time future,
And time future contained in time past.
—T. S. Eliot (1888–1965), American-British poet, from *Four Quartets*,
"Burnt Norton"

Einstein's general theory of relativity transformed space and time from a passive background in which events take place to active participants in the dynamics of the universe.
—Stephen Hawking, British physicist, *The Universe in a Nutshell*

Your body is a clock like any other—a kind of internal clock—and as it ticks away, it ages. Suppose one lifetime is exactly 1 billion billion ticks of a light-bounce clock. After 1 billion billion ticks of your clock, you will die. But, move at very high speeds and (when observed from a laboratory on Earth) there is a longer time between ticks of your internal clock than there would be if you had stayed at home. The time of 1 billion billion ticks of your clock will be stretched out and will compare with *more than* 1 billion billion ticks on the clocks of your Earth-bound friends.

From Earth, **they will observe that you have a longer life than they do**. Do you want to live for three centuries as measured on Earth? Easy: Leave Earth at high speed and come back three centuries from now (centuries as observed and measured on Earth). Your space-voyaging clock, when measured from Earth, will stretch between ticks, which

As you'll recall from chapter 30, a light-bounce clock counts one "tick" of time as the round trip of a laser beam to and from a mirror.

"IF EINSTEIN IS CORRECT, WHEN WE GET BACK, MY CAR WILL HAVE BEEN DOUBLE-PARKED FOR 320 YEARS."

means it will register a lot fewer than 300 years. If your relative motion is fast enough, the elapsed time on your clock could be 10 years. When you come back to Earth, you will still be young, but all your friends will be long dead. Sorry!

A strange-but-true occurrence explained by Einstein's Special Theory of Relativity is called **the paradox of the twins**. It depends on the clock within you.

Imagine that you have a twin. You go on a space journey, but your twin stays home. You spend 10 years (in rocket time) traveling at very high speeds first away from Earth and then back to Earth. While you are in flight, it will seem to you that time is ticking normally. But, as observed from

> When you are courting a nice girl, an hour seems like a second. When you sit on a red hot cinder, a second seems like an hour. That's relativity.
> —Albert Einstein

Earth (your rocket has glass walls), motions inside your spaceship are proceeding at snail speed: Your heart has slowed way down, your aging process is dawdling, and your watch has turned slowpoke. When you come back to Earth, you are as astonished as Rip Van Winkle was when he woke from his nap. You are 10 years older than when you left, but your twin, who stayed on Earth, is a senior citizen (or

Here's the Paradox

Try telling a friend that a stay-at-home twin will be older than his rocket-riding brother when the traveling twin returns to Earth. "No way!" says your friend if she knows some physics. "Each observer in relative motion records the clock of the other observer to be running slow. *So each should find the other younger when they get together again.*" But you both know this is impossible, and that's the paradox.

Can the paradox be solved? Yes. Algebra can do it (using the Lorentz transformations, explained on page 283).

In the meantime, consider this: The motions of the twins are not the same. The stay-at-home twin is at rest in an inertial frame where special relativity rules. In contrast, the traveling twin changes direction: First, he moves away from home, and then he reverses his path to head back home. If the turnaround is sudden, the traveling twin will wham his head. His bruise will be clear evidence of a difference between the motions of the twins. This difference complicates the analysis. Relativity tells us to trust observations made in a single inertial frame. An analysis *from the inertial frame of the stay-at-home twin* correctly tells us that the traveling twin will arrive home younger.

Einstein's theories have passed every test. Highly accurate atomic clocks placed in jets lose time (a few billionths of a second) compared with clocks on Earth. This hydrogen maser clock, with an accuracy equal to a gain or loss of 1 second in 30 million years, flew aboard Gravity Probe A, a rocket experiment to test relativity.

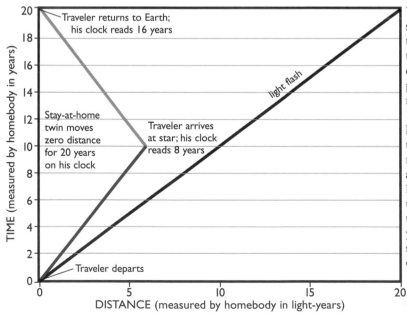

This graph plots the spacetime journey of the twin who leaves Earth and travels close to the speed of light. At the turn-around point, the traveler's clock reads eight years. It reads 16 years when he returns home. The grounded twin travels zero distance. (He moves up the vertical axis—time—on the graph.) His clock reads 20 years at the return of his traveling twin, who is four years younger than he is. The straight purple line traces out a light flash's journey.

maybe dead of old age). Time has passed differently for each of you.

Now imagine that you are the stay-at-home twin, and you are able to observe your astronaut sister speeding away from the surface of the Earth. (She's in that glass-walled rocket, and your telescope is terrific.) You will see the hands on her watch

What do an imagined astronaut traveling at near-light speed (conceived on a computer) and the cyclist have in common? Each is at rest in his own frame.

move more slowly than those on your Earth-bound wrist (but in her spaceship she won't know that by looking at her watch). When your twin returns to Earth, the time lapse on her round-trip clock will be less than the lapse on your stay-at-home clock. This time, it is she who will look younger than you. Yes, this does seem unreal, but it actually happens. It even happens when you ride your bicycle, but all Earth travel, whether on your bike or at supersonic speeds, is relatively slow compared to the invariant speed of light. The frame of reference differences are so very, very tiny that we don't notice them on our clocks or in our heart rates. However, they happen, and they can be measured (it takes many, many digits after a decimal point).

How can we know if all this is true and *real*?

A physicist friend tells me that scientists don't worry about defining reality (that's for philosophers to consider). Science deals with the creation of models (or theories), which are then checked by observation.

In other words, scientists start with hypotheses and then experiment to prove or

How to Be a Genius

In a Berlin interview in 1929, just after his most productive years, Albert Einstein famously said of scientific achievement, "Imagination is more important that knowledge." Although he didn't like to recognize it, Einstein started with a knowledge base. His German schooling and his remarkable parents gave him a solid foundation in math and science that fed his astonishing imagination. Begin with information, ask the right questions, apply some imaginative leaps, and you are on the path to creativity.

Einstein also had a character trait—call it audacity, nerve, or maybe impudence—that made him thumb his nose at accepted ideas. That trait made him enormously aware of the importance of freedom in all fields of endeavor, but especially in science.

Don't think ivory-tower philosopher when you think of Einstein, think clock tower. Working in the patent office, he dealt with a practical problem: How do you coordinate clocks in different towns, where trains and telegraphs are making it important to do so for the first time in history? Einstein reviewed patents for clocks and other devices to coordinate time as part of his everyday work.

disprove them. **Science looks for experimental verification.** Einstein's hypothesis is this: When we observe clocks moving past us, we observe them ticking more slowly than clocks sitting next to us on Earth.

How do we test the hypothesis? If we want to know about time in different frames of reference, we observe your clock "at rest" in a see-through spaceship as it tears past a series of clocks synchronized in my "at rest" Earth frame. Is your clock in the spaceship running slowly compared with my clock in an Earth lab?

I check your clock as it whizzes by on the way to outer space and then as it whizzes by on the way back to Earth. The lapse on your round-trip clock is less than the lapse on my stay-at-home clock. And you look younger than I do as you zip by on your return.

But can we actually prove that? I haven't seen any glass-walled spaceships. And if you try a high-speed reentry back into Earth's frame, you won't live to tell about it. (This is not a good way to keep your youthful looks.) Scientists have found an easier proof. They send *particles* zooming at close to the speed of light and then observe what happens to them. And what happens to them confirms Einstein's Special Theory of Relativity.

Decaying particles are observed to have half-lives that are

The rate of decay in subatomic particles obeys the Theory of Relativity—time (and thus decay) slows for extremely fast-moving particles. In 2007, physicists at the CERN lab in Switzerland gained a new particle observation tool named ALICE—*A L*arge *I*on *C*ollider *E*xperiment. In this cross-section simulation, accelerated ion beams collide in the center, creating new particles that decay into other particles.

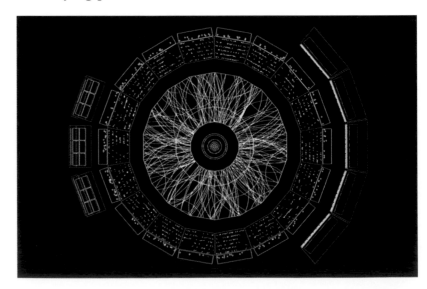

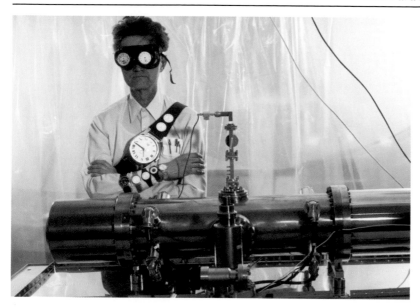

That's Jim Gray, keeper of the NBS-4 atomic clock at the National Institute of Standards and Technology (NIST), having a little fun with timepieces. One second used to be figured as a fraction of Earth's daily period of rotation. In 1955, when it was realized that our planet's rotation is slowing, the second was redefined using a particular frequency of light in the spectrum of the cesium-133 atom.

longer in our everyday frame than their high-speed frame. (That high-speed frame can be called a "fountain of youth" frame when observed from Earth's frame.) Atomic clocks taken on very fast rides in space keep proving special relativity's hypothesis. Today, no one seriously doubts Einstein's hypothesis. To a scientist, proofs like these offer the only acceptable meaning of "real."

Why didn't we notice relativity long ago?

Newton's laws—the laws of gravity and motion of classical physics—are fine in most instances that we face. Even when flying in an airplane, there's no need to worry about time dilation or length contraction. At the speeds we travel, they aren't concerns. But Newtonian science fails in the quantum world of the very small, and it also fails in the world of the very, very large or very, very fast. That's where Einstein's relativity rules.

According to Einstein, observed time depends on the speed of a particle or a twin or a spaceship relative to events in an observer's frame of reference. This is a difficult concept. It's counterintuitive. No one "gets" it the first time. You have to think about it, again and again. It's worth doing. So reread this chapter. Relativity describes the universe we inhabit. It's fun to have it under your hat.

Should we give up Newton's laws of physics? NO. His laws work well in most of everyday science and technology. Need a refresher? They're listed on page 15.

DILATION means stretching, the opposite of **CONTRACTION**. Time dilation and length contraction are the scientific terms for what you perceive happening to your twin who is traveling at close to the speed of light: Her clocks slow down, and her length is contracted or shortened. (Okay, if you're a guy, that's a fraternal twin in the rocket.)

Relative Gravity

34

[General relativity,] which many physicists believe to be the most perfect and aesthetically beautiful creation in the history of physics, perhaps in all of science, has replaced Newton's theory of universal gravitation. It has cleared up some anomalies in the behavior of planetary orbits...led to significant predictions ...become the basis for all of modern cosmology, including the expanding universe; and is at the present moment...again at the center of scientific interest.
—Jeremy Bernstein, American physicist, *Einstein*

What you think of as gravity—the heaviness you feel standing on Earth or the force that accelerates a plummeting stone—is not real because it disappears in some reference frames (i.e., in any free-float frame). That doesn't mean there's no such thing as gravity; rather, the real gravity must be something that can't be transformed away with a change of reference frame.
—Richard Wolfson, American physicist, *Simply Einstein: Relativity Demystified*

A teacher puts gravity to a potentially painful test in a cartoon titled "The Soundness of Newton's Laws."

Gravity. The very word is awesome. It comes from the Latin *gravis*, which means "heavy." Newton's Law of Universal Gravitation *is* heavy. At the dawn of the twentieth century, it is probably the most revered theory in all of science.

Everyone knows the story of Isaac Newton sitting under the apple tree. And everyone knows that gravity is a force that keeps your feet on the ground and the Moon in the sky. In Albert Einstein's day, no one questions it.

No one, that is, except Einstein. He doesn't seem to be concerned by what others think. He is looking for scientific truths, and he doesn't worry about where he might find them.

And, for Einstein, there is something bothersome about Newton's Law of Universal Gravitation. It doesn't exactly fit with the Special Theory of Relativity. **Newton said that gravity is a force that acts at a distance instantaneously.** Instantaneous is faster than the speed of light. But that can't be. Einstein's special relativity says nothing can go faster than light. There is a problem here. If gravity's influence moves faster than light, then Maxwell is wrong and so is special relativity. Einstein doesn't think that is so.

Most people don't realize that Newton found gravity frustrating. He couldn't explain the "action at a distance" aspect of gravity. How do the Earth and the Moon attract each other through the vacuum of space? In a letter to the Reverend Richard Bentley, he called that property of gravity an "absurdity." Newton provided the mathematics to measure what seems to be gravity's pull, but he couldn't figure out how it happens. He said he was leaving that problem for future generations to solve.

Newton couldn't explain something else, nor could Galileo. Einstein believes it might be the key to understanding gravitation. It has to do with acceleration.

When Galileo dropped a variety of weights, he discovered that they all fell at the same rate. Why? Aristotle had been sure that heavy objects fall faster than light ones. Galileo proved that isn't so. But why doesn't gravity make a lead weight fall faster than a tennis ball?

Galileo said, "I feign no hypotheses," which means he couldn't answer that question, and he wasn't going to guess. Newton repeated the words in Latin, *Hypotheses non fingo*. It became a famous phrase but a maddening one. If Galileo and Newton couldn't explain why **objects all fall at the same rate in a gravitational field**, can anyone?

Einstein is going to try. He is working on a new explanation of gravity that will get rid of instantaneous action. He will do that by building on Galileo's observation of falling objects. At the same time, he will expand relativity.

Einstein knows that special relativity correctly describes motion only when gravity is not a major player. He wants to

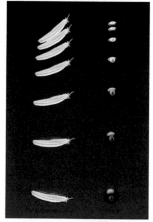

Do a lightweight feather and a heavy ball fall at the same rate? Not if air gets in the way. But in a vacuum or on the Moon (where there is no atmosphere), they certainly do.

Aristotle lived in Greece in the fourth century B.C.E.; Galileo (1564–1642) lived in Renaissance Italy. We're not sure Galileo actually went up the Tower of Pisa—it might have been a thought experiment—but he was experimenting with dropped objects elsewhere.

If you (or a cat) fall from a height, you won't feel a "pull" of gravity; you'll feel weightless, but you will be aware that you are accelerating toward Earth. This acrobatic cat lands on a soft pillow, but if a free fall is stopped by a sidewalk—crunch and splatter—cat, or person, is in big trouble.

enlarge relativity so it will explain the pushes and pulls of the universe, and gravity is giving him difficulties.

If gravity is to be understood and special relativity broadened, Einstein believes **gravitational acceleration** might be a key piece of the puzzle. The former patent clerk spends years thinking about this. He thinks about it in Zurich, then in Prague, then back in Zurich. He is determined to solve the puzzle.

In 1907, while struggling to understand acceleration in a gravitational field, Einstein gets what he says is "the happiest thought of my life." It comes after he talks to a man who fell from a roof. (It is a happy day for the man, too. He falls into a pile of rubbish and isn't hurt.) What the man tells Einstein is important. He says that he did not experience any feeling that, according to Newton's theory, gravitation should have given him. He didn't feel a violent pull to the Earth.

He just felt as if he were in a weightless free fall.

Einstein concludes, "For an observer falling freely from the roof of a house there exists—at least in his immediate surroundings—no gravitational field." To an audience, he puts it in plain words: "If a man falls freely, he will not feel his [own] weight." (Today we're familiar with the sensation of weightlessness in spaceships.)

Einstein used the term *difform motion* instead of acceleration. It means the same thing.

Some years later a *New York Times* reporter will embellish the story. Was there actually a falling man? Today no one is sure. But Einstein understood that the weightlessness a person feels in a long fall can be explained by acceleration.

But how can he picture that acceleration? On a summer hiking vacation in 1913, he figures out a way to do it. The Einsteins (Mileva, Albert, and Hans Albert) and the Curies (Marie, Pierre, Irène, and Eve) are on the slopes of the Alps, south of Lake Como, Italy. The children realize that Einstein is unaware of the spectacular scenery; his mind is elsewhere. Marie Curie's biographer, Rosalynd Pflaum, writes, "One day, the three young people howled with laughter when Einstein suddenly stopped dead, seized Marie's arm, and demanded, peering intently at her: 'You understand, what I need to know is exactly what happens to the passengers in an elevator when it falls into emptiness.'"

That elevator becomes a vehicle for one of Einstein's most famous thought experiments. If there is a connection between acceleration and gravitation, a falling elevator should help him find it. In his mind, Einstein imagines an

Did the central point come to him, as legend has it, from talking to a housepainter who had fallen off a roof and reported feeling weightless during the fall?...[H]e called that 1908 insight the "happiest thought of my life"— the idea that there is no such thing as gravitation, only free-fall. By thus giving up gravitation, Einstein won back gravitation as...a warp in the geometry of space. His 1915 and still standard geometric theory of gravitation can be summarized, we know today, in a single, simple sentence: "Space tells matter how to move and matter tells space how to curve."
—John Archibald Wheeler, in an essay on Einstein

elevator in a very tall building. A man standing inside the elevator takes out his keys and drops them; they crash at his feet. That's gravity, the man thinks to himself.

Now, Einstein cuts the elevator cable (there is no safety mechanism as there is with elevators today). The elevator car begins to fall within its very long shaft. The man, his keys, and some coins from his pocket are all falling but not to the floor. Man, keys, and money float freely inside the

Einstein put a man with keys in an elevator. We put a kitty and a ball in an elevator (top picture). The cat's paws stay on the floor. Cut the elevator rope and kitty, elevator, and ball all go into free fall.

elevator box. They seem weightless. To an outside observer with X-ray vision, they are all falling at the same rate. It's called a free fall.

Like the man who fell from the roof, the man in the elevator can't feel his fall. He is floating. Einstein realizes that as the man accelerates, so does the elevator. There is no relative acceleration between them. For the elevator-floater **gravity does not exist**. No experiment he can do will detect it.

Here's Mr. Kitty in a rocket-powered elevator car in interstellar space. Firing the rocket accelerates the ship. In the top picture gravitation and/or acceleration put Kitty on the red cushion. In the bottom picture gravitation/acceleration causes him to hit his head.

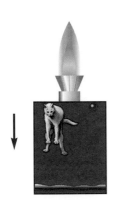

You can easily feel acceleration in a car. Speed up, and you'll be pulled back against the seat. Slow down, and you'll lean forward.

With this mind experiment, Einstein is doing the same thing Galileo did when he dropped balls from the Leaning Tower of Pisa. Everything Galileo dropped fell at the same rate. The same is true of Einstein's passenger, his keys, and the elevator car. As Galileo discovered, objects all fall at the same rate—that is, with the same acceleration—in a gravitational field. The passenger feels weightless; he and the elevator are in a free fall. If the ground weren't in the way, they would keep falling.

Einstein goes on with the experiment. He puts a rocket underneath the elevator. Then he ignites it. The rocket pushes upward, causing the elevator to reverse direction and accelerate. As soon as that happens, the man and his money and keys hit the floor. Without a window (giving him a frame of reference), the man can't tell if it was acceleration or gravitation that stopped his free fall. (If he puts the rocket on top of the elevator, and its downward thrust causes the car to fall faster, the man will hit his head on the ceiling. The point is the same.)

Einstein wonders what will happen if he takes the windowless elevator car into space, far from any planet or body with a gravitational field. Will acceleration still feel just like gravity's pull?

So, in his mind, he sends the box-like elevator car and its passenger into deep space. Again, his passenger floats freely; so does the elevator. Once more, Einstein ignites a rocket beneath the elevator car; the elevator speeds up, and money, keys, and man hit the floor. In the elevator, the effect of this acceleration is exactly the same as the effect of gravitation.

Einstein realizes: The person in the elevator and an outside observer have different points of view. A physicist might call these points of view "frames" or "coordinate systems." The falling elevator only exists in a small piece of space (and time), and so physicists call it a *local* frame. Special relativity doesn't work in all local frames. It works in special frames called "local inertial frames."

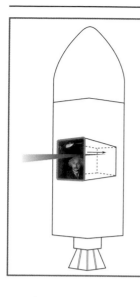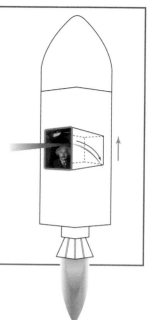

Accelerated Learning

Einstein did another thought experiment on acceleration. He imagined himself in a floating spaceship with one window (left). A beam of light entered that window and, going in a straight line, hit the opposite wall.

Then he turned on the rockets, and the spaceship accelerated (right). The beam of light hit the opposite wall at a point slightly lower than the window. (The same thing would happen to a ball thrown in through the window.)

The conclusion: The effect of acceleration is exactly the same as the effect of gravitation. Therefore, light bends in the presence of gravity or acceleration.

These insights lead Einstein to a new theory of gravitation. He calls it **general relativity (GR)**. GR is built on ideas that will overturn long-held ways of looking at the universe. Here they are:

- First idea: **In a closed box in free float, you can't tell whether you are near Earth or in intergalactic space.** (Think of that box as a local inertial frame.)
- Second idea: **A local inertial frame is always available.** An inertial frame of reference can be set up anywhere. Just get into a windowless box and let it fall freely (as a satellite does within Earth's gravitational field). **This important idea allows special relativity to be used locally anywhere in the universe**—even in a gravitational field—at least for the few seconds of free fall before you hit something. (It's the floor that stops your fall, not some mysterious force.)
- Third idea: Think of the universe as a series of very small local inertial frames, one connected to the next. Do that, and special relativity works everywhere. Einstein's expanded theory, **general relativity, allows us to stitch a patchwork of local frames into a quilt that spans the whole universe.** A patchwork of local frames requires not only

Part of what makes Einstein so astonishing is that his ideas were almost all pure head work—based on a lot of reading, thinking, and number crunching and not on experimentation. (Others would prove his theories.) When he got an idea, he worked it out in his mind and then wrote it down mathematically. When someone once asked him where his laboratory was, Einstein held up his fountain pen. Actually, his lab was under his hat.

Many scientists continue to explain things with Newton's gravitational theory and equations because they work correctly in our local Earth world. But they know (and so do you) that Newton's theory predicts incorrect results in the rest of the universe at very fast speeds.

stitches in space but also stitches in time. It means a clock changes ever so slightly as it eases from one local frame to the next.

• Fourth idea (biggest of all to Einstein's contemporaries): **Newton's theory of gravity needs revising. Einstein has a better explanation. Every free particle does nothing but move in a straight line at a constant speed in a local inertial frame.** One local frame is connected to the next frame. In that next frame, the particle also moves straight, but now in a slightly different direction. **It is the universal patchwork of local frames that turns special relativity into general relativity.**

In other words, Newton's force is an approximation. Einstein's explanation—a free particle moves just as straight as it possibly can across the collection of stitched-together local frames—is more accurate.

Einstein is blowing away Newtonian gravity. He's upsetting the scientific applecart. If acceleration is the same as gravitation, then gravitation isn't instantaneous action at a distance. The great Isaac Newton was wrong about that. (Einstein actually apologizes to Newton—well, to the memory of Newton—who is one of his heroes.)

If it were possible to embrace the Universe with a single free-float (inertial) frame, then special relativity would describe that universe and general relativity would not be needed. But general relativity is needed precisely because typically inertial frames are inertial in only a limited region of space and time. Inertial frames are local.

—Edwin F. Taylor and John Archibald Wheeler, *Exploring Black Holes: Introduction to General Relativity*

Einstein will spend 10 years working on this new theory of gravitation: general relativity. He knows a set of loose ideas isn't enough. He needs to lay them out in a consistent theory expressed mathematically. His Olympia Academy friend Marcel Grossmann is now a math professor at

Zurich Polytechnic, or the ETH; Einstein is now a physics professor at that same well-regarded institution. Grossmann was the one who gave Einstein his detailed class notes when they were both students there. Now, Grossmann rescues him again, helping Einstein understand some of the non-Euclidean math he needs. "Grossmann, you must help me or else I'll go crazy," Einstein pleads. To physicist Arnold Sommerfeld he writes, "I have become imbued with great respect for mathematics.... Compared to this problem the original relativity theory was child's play."

Marcel Grossmann studied hard at the ETH while Einstein skipped classes. Later, Albert regretted that he hadn't paid attention, especially in Professor Minkowski's class.

Be a Skeptic, Find Out, and Think for Yourself

Let's sit and watch a TV show. Our hero, Astronaut Samantha, is floating inside a space shuttle. The narrator explains the crew's weightlessness. "They are beyond the pull of gravity," he says. "They have escaped the g-force," he adds. You can snicker. The announcer doesn't have a clue. He's got it all wrong.

Once seen as a fanciful, futuristic endeavor (above), space tourism has arrived.

Assuming that the space shuttle is somewhere between Earth and the Moon, it is well within Earth's gravitational field. After all, Earth's gravity keeps the Moon in orbit. You calculate Earth's gravitational pull from the center of the Earth, which is about 6,400 kilometers (almost 4,000 miles) below the surface. Space shuttles orbit about 325 kilometers (202 miles) above Earth, which doesn't change the gravitational equation by much.

Samantha and her fellow astronauts seem to be floating, but they are actually in a free fall in a closed box. They don't feel as if they are falling because the shuttle and everything around them is also falling. But when the rockets on the space shuttle turn on, and the shuttle accelerates,

The Moon is in a free fall, circling the Earth (right) like a marble on a curved track. (There's more on warped space in the next chapter.) To leave Earth orbit, an object has to accelerate beyond orbital velocity. The NASA illustration above depicts a future spacecraft designed to carry four astronauts to the Moon.

then the astronauts feel what they can assume is gravity (measured in g-forces) pulling on them. The effects of acceleration cannot be distinguished from the effects of gravity; they are equivalent.

Princeton physicist Robert H. Dicke (1916–1997) was bothered that everyone was hailing general relativity but no one had tested it rigorously. He set out to do so. In one of a series of experiments, he asked *Apollo* astronauts to place three mirrors on the Moon. From Earth, Dicke aimed a laser beam at the reflectors. The time it took the beam to travel there, bounce off the mirrors, and return to Earth gave a precise measurement of the distance from Earth to the Moon ($d = vt$, or distance equals velocity multiplied by time). The experiment also provided measurements of time that proved general relativity. Dicke tried one elegant experiment after another, expecting to find a flaw in Einstein's theory, but each experiment just confirmed it. Read more about Dicke in chapter 44.

At this very time—while Einstein is working on general relativity—German artists, philosophers, scientists, and politicians are all striving to make imperial Germany the most advanced of all cultures. Einstein's 1905 papers have created a buzz in the world's physics community, especially in Germany. He is now seen as a leading physicist.

With the idea of German supremacy in mind, two men board a train in Berlin in 1913. One is Max Planck: elegant, lean, bespectacled, the originator of quantum theory. The other man is rotund Walther Nernst: witty and brilliant, he is a physical chemist. They are on their way to Zurich to ask Einstein to join them at the University of Berlin. They know that he has renounced his German citizenship; they know he is not a conformist. Still, they are prepared to offer him inducements that are probably beyond his dreams (and this man is a dreamer). If Einstein accepts, he will be elected as the youngest member of the Prussian Academy of Sciences, he will become a full professor with no teaching obligations, he will lecture only when he wishes to do so, and his salary will be at the top of the profession.

Still, Albert needs to think about it. His boyhood experiences in Germany have left him wary of its academia. He believes the two men view him "as if I were a prize hen or a rare postage stamp." What he can't foresee in 1913 is what is going to happen in Germany. The nation that sees itself as the world's cultural center will soon hand itself over to ruthless, shortsighted leaders. None of the three, not

General Rubbish?

General relativity didn't impress everyone. Oliver Heaviside (1850–1925), an electrical engineer and mathematician from England, called the theory "drivel" (foolishness).

In America, Columbia University astronomer Charles Lane Poor (1866–1951) said Einstein was a victim of a worldwide mental disturbance caused by the Great War. Perhaps more important, Poor wrote a paper disputing the results of a 1919 solar eclipse observation designed to prove general relativity (more on this in the next chapter).

Thomas Jefferson Jackson See (1866–1962), an astronomy professor at the University of Chicago, called Einstein "a confusionist." See announced, "The Einstein theory is a fallacy. The theory that the 'ether' does not exist, and that gravity is not a force but a property of space, can only be described as a crazy vagary, a disgrace to our age."

GPS: <u>G</u>o <u>P</u>ick a <u>S</u>pot!

If you've been asking yourself, How does general relativity affect me?, one way is through GPS, a satellite-based navigation system that was first launched for military use in 1978. It keeps planes, hikers, bikers, and many of us on course. The next time you decide to trek in a thick jungle, tackle a mountain peak, or wander down a back road, take a GPS device with you: If you get lost, rescuers will know exactly where to find you.

When the first GPS satellite was launched (this is an artist's view of one), some military minds thought general relativity (GR) was unnecessary. They believed clocks run at the same rate everywhere. They turned off the GR program, and the satellite's measurements went askew.

The Global Positioning System is made up of 24 satellites orbiting about 19,300 kilometers (12,000 miles) above Earth. Each satellite weighs about 1 ton, is 5 meters (17 feet) across, travels at a speed of roughly 11,250 kilometers (7,000 miles) per hour, completes an orbit in about 12 hours, and carries an atomic clock (with backup clocks). Those clocks allow the satellite to transmit a unique, timed binary code (a sequence of 1's and 0's) that tracks the satellite in spacetime. That code is picked up by receivers on Earth and matched with a similar code sequence inside the receiver. If you're holding a GPS receiver, the transmitted code will pinpoint your location.

How does general relativity (GR) fit in? Clocks at different altitudes and different relative speeds run at different rates, when you calculate using GR. A GR-programmed GPS receiver measures the time it takes a signal to go from the satellite to the receiver. That time (t) is then multiplied by the velocity (v) to get the distance ($d = vt$). With that information, the receiver can determine your exact position in relation to spacetime's four dimensions: longitude, latitude, altitude, and clock time. You (or a plane in flight) can be found where those four dimensions intersect. Thank you, Albert Einstein.

Planck, not Nernst, not Einstein, has any notion of what is ahead; these are all men of goodwill.

Einstein listens. With his trimmed mustache, dark curly hair, and quick wit, he exudes an easy self-confidence. He tells the two men that he needs a night to think about their offer. Planck and Nernst have planned to take a sightseeing train around Zurich the next morning. Einstein says he will meet them when they return. If he has a white flower tucked in his lapel it means he will stay in Zurich. A red flower means he has chosen Berlin.

As the train pulls into the station the two men look down the station platform. They see a young man with a red flower.

Warps in Spacetime

We use relativity to explore the boundaries of Nature. Special relativity describes the very fast. General relativity— the Theory of Gravitation—describes matter and motion near massive objects: stars, galaxies, black holes. General relativity also describes the Universe as a whole.
—Edwin F. Taylor and John Archibald Wheeler, *Exploring Black Holes: Introduction to General Relativity*

If we wished to, we could use Einstein's laws rather than Newton's in everyday life. The two give almost precisely the same predictions for all physical effects, since everyday life entails relative speeds that are very small compared to the speed of light. Einstein's and Newton's predictions begin to diverge strongly only at relative speeds approaching the speed of light. Then and only then must one abandon Newton's predictions and adhere strictly to Einstein's.
—Kip S. Thorne, American physicist, *Black Holes and Time Warps: Einstein's Outrageous Legacy*

We exist in a locally curved region of space and time created by the mass of the earth. The tug we feel downward as we stand by our beds in the morning is the sensation of our daily slide down that well in spacetime, a slope heading inward toward the center of the earth.
—Thomas Levenson, American science writer and documentary filmmaker, *Einstein in Berlin*

By 1915, Einstein sees spacetime as a quilt made up of local environments. Each patch in the quilt is a place where photons and other particles travel in straight lines, which means inertia and special relativity can rule. How do you get from those local environments to a universal system? *The stitching between the local patches creates jogs (or transitions) that allow for changes of direction and speed.* Those jogs and the patches are so very, very small that they seem to create continuous curves.

Where does gravity fit into the picture? Gravitation is

nature's response to the curvature.

Here's a visual picture: Imagine a rubber sheet (right). You're holding one end; Einstein has the other end. Now stretch that sheet until it is taut, like a trampoline.

What happens to the stretched rubber if you lay something heavy on it? Try a bowling ball. It causes an indentation, or dimple, or warp, or curve in the sheet. Near the ball, the curve is steep; far from the ball, the curve is slight.

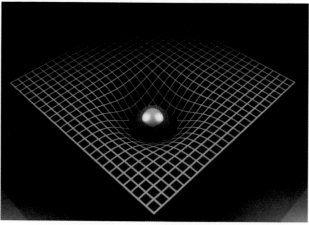

The more massive the object, the deeper the curved gravitational well it forms in four-dimensional spacetime (represented here as a flexible two-dimensional grid).

Now picture that stretched rubber sheet filling space like a fabric of foam. What are the bowling balls? They're the stars and planets. But anything—a rock, a particle, a spaceship, or you—will warp and stretch the fabric it travels through. Near a large mass (a huge star, for instance), the dent is very deep. Einstein also shows that as space stretches, so does time. That makes the fabric four-dimensional spacetime.

What happens if you put an extraordinarily heavy object on the sheet? It will superstretch spacetime's fabric, creating a deep hole—a black hole (left). (Details on black holes are coming.)

Keep in mind: Spacetime isn't a sheet of rubber, and it isn't foam, but it is four-dimensional. Four-dimensional means that it includes time as well as the three customary dimensions: length, width, and height.

Yes, this takes imagination for our Earth-trained minds to envision; no drawing of a rubber sheet can give us a picture that includes time. But scientific thought has always been hugely creative. We live in a universe with at least four dimensions; it's worth the effort to try to imagine it.

So, to repeat: Mass makes curves (warps) in the

Remember, special relativity says those three familiar dimensions aren't absolute; they can be measured differently, depending on the object being measured and the person doing the measuring.

Today, scientists talk about dimensions that go way beyond Einstein's fourth dimension. They say there may be 10 or more dimensions, and problems that seem beyond calculation in our three-dimensional spatial world, or even in the fourth dimension of spacetime, become clear in additional dimensions.

We still talk of gravity as a force, because it works for us to do so, but, as you now know, it is a warping of the fabric of spacetime.

Physicists at the ETH (Einstein's alma mater) in Zurich are still working to describe how nature works in precise, mathematical terms, using both theory and experimentation. This is an apparatus for a gravitational field experiment.

patchwork-quilt fabric that is spacetime, and then anything passing along the curve is pulled into it. A bowling ball, naturally, will make a bigger indentation than a pebble—which means it will have more gravitational pull. And the Moon or Earth? They make huge dents.

Toss a marble on the rubber sheet near the bowling ball. What happens to the marble? Yes, it's pulled toward the massive object. Toss an asteroid into spacetime near the Moon, and it will fall onto the Moon's surface.

Gravitation is not a mysterious push or pull; it's the result of moving objects following the shortest possible path in curved space (those dents). Each object moves as straight as it possibly can in a local frame, across one tiny patch in the spacetime quilt, jogging slightly as it moves from patch to patch. The particle does not feel the jog; spacetime is a continuum, and those patches are incredibly small. Much of spacetime is essentially flat, but where there is mass, there are curves.

Einstein is sure that gravitation, along with the other laws of physics, applies to all natural things in the universe. So when he asks himself, Does gravity influence light? the answer has to be yes. If energy and mass are equivalent—as $E = mc^2$ says they are—then electromagnetism (light) must respond to curved space just as everything else does.

Einstein realizes that a warp in space (a gravitational field perhaps created by a star or planet) will affect a light beam as well as anything else. **Any form of matter or energy—including light itself—must bend its path as it passes a warp.**

Back in the eighteenth century, Newton guessed that gravity would affect light, but he couldn't figure out why. He wrote out questions (he called them queries) for future scientists to answer. His first query was, "Do not bodies act upon light at a distance, and by their action bend its rays?" (Read that again to be sure you have it.) Newton was a very great scientist, which means he knew how to ask questions, and he also knew when he didn't have all the answers.

Einstein answers Newton's question. Massive objects in

Why London Bridge Is Falling Down

We are not aware of gravity when we are in a free fall (skydiving before the parachute opens), but we don't often fall freely. Earth's surface, or a bedroom floor, or the deck of a ship get in the way and stop us; our reaction to those barriers is measured as weight. But if nothing gets in the way, the natural thing for us to do is to float freely as we fall down the slope (the curvature of spacetime).

What causes the curvature? Einstein says that spacetime is filled with slopes (or wells) created by massive objects like Earth. We're all sliding into one of those wells, and we're heading toward Earth's center. That's what gives us the sensation of gravity.

Why is gravity's "pull" different in different parts of the universe? Local gravity reflects

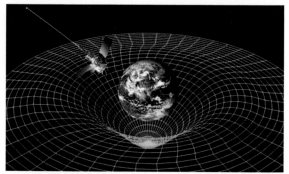

The Gravity Probe B satellite, depicted here by an illustrator, began orbiting Earth in April 2004 to collect data on two unverified effects of Einstein's Theory of Gravitation: the geodetic effect (how much Earth warps local spacetime) and frame-dragging (how much Earth's rotation drags local spacetime around with it).

the steepness of the slope we're falling down. On the Moon, the slope (or curvature) is less steep than on Earth. Changes in the steepness of the slope make for curvature in spacetime.

Newton thought of space as flat. When it occurred to Einstein that spacetime might curve, his whole view of the universe had to change. How can particles and light travel in straight lines following the Law of Inertia if the universe is curved?

Each object moves as straight as it can in a local frame, across one tiny patch in the spacetime quilt, jogging slightly as it moves from patch to patch. But the particle does not feel the jog. Spacetime is a continuum; the patches get smaller and smaller.

We don't have much experience with curvature because we live on the surface of Earth, which is nearly flat locally. On Earth's flat surface, the shortest distance between two nearby points is usually what Euclid told us: a straight line. But for longer distances, the shortest possible paths are curves, like the great circles that pilots use to travel the shortest routes around the curved surface of our Earth.

---

spacetime do bend light. **Then anything that passes through spacetime responds to its warps.**

Einstein's equations confirm his theory. They show light bending in response to curved space. But can math actually predict what happens in the real world? Einstein is sure that it can. He is convinced that **starlight responds to gravitation and bends when it passes near a massive object.** He studies a distant star and figures out exactly how much its rays will bend when they pass the edge of the Sun. He realizes the bend is big enough to observe, but only during the brief moment of a total solar eclipse.

Einstein's equations answer Newton's query about light bending and more. In relativity theory, where time and space are woven together, not only light bends in the presence of gravity—so does time. To repeat: *Time* bends in the presence of gravity. Before the twentieth century, no one had ever grasped that idea.

Here's a solar eclipse photographed in July 1991. An image of the Moon was superimposed over the series to more clearly show its position.

Normally the Sun's rays are so intense that you can't see stars that appear close to the Sun. But during a solar eclipse, the Moon covers the disk of the Sun, allowing us to see stars that aren't normally visible.

And a solar eclipse is coming! For about two minutes on August 21, 1914, the Sun will darken. The eclipse should be especially visible from Siberia in northeastern Russia. Einstein wants to test his new theory of gravity. This would be a good time and place to do it.

Einstein knows that Newton's Theory of Gravitation is a foundation of science. Almost everyone believes Newton is beyond questioning. Max Planck warns Einstein about trying to overturn Newton's gravity. "In the first place you won't succeed, and even if you do, no one will believe you," he says. But even Max Planck, whom he reveres, can't keep Einstein from attempting to prove his theory. Science depends on proofs. And Planck does add, "If you are successful, you will be called the next Copernicus."

Einstein is sure of himself and persuasive. He convinces some members of the German physics establishment that general relativity is worth a test. When a wealthy industrialist agrees to pay for an expedition, it's a go. German astronomer Erwin Freundlich and two assistants make plans to study the predicted eclipse. Einstein writes to his friend Michele Besso, "I am now completely satisfied and no longer doubt the correctness of the whole system, no matter whether the solar eclipse observation succeeds or not."

On June 28, 1914, Austro-Hungarian archduke Franz Ferdinand and his wife are on a state visit to Sarajevo, and (the story has been told again and again) a Serbian terrorist shoots twice and kills them both. Austro-Hungary responds to the archduke's murder by declaring war on Serbia. Russia sides with Serbia. Germany doesn't. Tensions rise.

On July 19, the three German scientists have set up a camp near Kiev, in Russian territory, preparing to photograph the eclipse. Their Russian hosts are suspicious. Are the Germans really enemy spies pretending to be scientists? Their cameras are seized, and, briefly, the three are thrown in jail.

On August 1, Germany declares war on Russia. Now there is no chance of German scientists making it to Siberia. It doesn't matter: Clouds cover the Siberian sky at the moment of the eclipse.

Einstein has had a lucky break. His calculation of the deflection of starlight was wrong, he soon realizes. He had forgotten to include the curvature of both space and time. Since his is a theory of spacetime, it's an embarrassing mistake. Einstein does the math again, and he comes up with a different result.

Einstein is now settled in Berlin. His marriage to Mileva is in trouble. He is spending time with his cousin Elsa, a cheery, motherly woman, who has fallen in love with him. "Life is better here than I anticipated," he writes to a friend, "except for a certain discipline with regard to clothes and so on, that I

Within weeks of arriving in Berlin, Albert's marriage to Mileva was over. Cruelly, he rejected her pleas to stay together. However, everyone expected him to win a Nobel Prize, and Einstein agreed in advance to give the prize money to Mileva. When he finally received it, he did so.

Big-Picture Physics

Special relativity (SR) works well in everyday, Earth-bound situations. It also does fine in experiments where particle physicists send protons, etc., careening at high speeds. Gravity acts like a force. No problem.

But some experiments "at the edge" made Einstein rethink the domain of special relativity: Starlight bends near the Sun, objects at rest dropped side by side both head toward the center of Earth and thus approach one another, and photons change energy as they rise in a gravitational field. No gravitational *force* can explain those happenings.

So Einstein had to rethink SR to deal with gravity. So do the rest of us. We know that SR always works in an inertial frame. (Just let

yourself fall, and you're in an inertial frame.) In free-fall frames, SR works perfectly.

Now consider events that span groups of galaxies, or black holes, or the entire solar system. Whoops. SR fails. You need GR (general relativity). It provides the math to stitch together local inertial frames in order to span big regions of spacetime.

With GR, scientists have created "a new expanded edge." They can deal with long-distance space travel, global positioning, the life cycle of stars, and the universe as a whole. Thanks to GR, the size of the scientific stage (where we can make observations, explanations, and predictions) has grown and grown and grown. It's an exciting place to be.

A September 1914 cartoon titled "Down with the Monster!" rallies French citizens to support the fight against German aggression at the beginning of World War I.

have to submit to on the orders of a few old gents." Later, he will say that the Prussian Academy "granted me enviable living and working conditions during the best years of my life."

But, even in an academic cloister, Einstein can't escape the war. At first, everyone thinks the fighting will be over in weeks; then it turns ferocious. In October of 1914, 93 of Germany's leading artists, writers, scholars, and scientists (including Max Planck) sign a manifesto in support of the war. They say Germany has no responsibility for the hostilities and that it is France and Britain, both allies of Russia, who are at fault.

Einstein is horrified. He and a friend write an antiwar manifesto; only two other people will sign it. On all sides, this begins as a popular war. To his friend Paul Ehrenfest, Albert writes that all Europe seems insane and that "it is difficult to accept the fact that one belongs to that species that boasts of its freedom of will." Meanwhile he keeps working, publishing his paper on general relativity in 1916.

In 1917, with no end to this devastating war in sight, one copy of that paper is sneaked out of Germany, through the

American soldiers on the front lines in Alsace, France, celebrate victory upon hearing news of the war's end on November 11, 1918.

neutral Netherlands, to Great Britain. England's leading astrophysicist, Arthur Stanley Eddington (1882–1944), reads the paper and is captivated by its insights. He spreads word among his colleagues. **Eddington wants to test general relativity.** But can he test the work of a German scientist while England is at war with Germany?

He can't. Then, finally, on the eleventh day of the eleventh month in 1918, the "Great War" (World War I) is over. Eddington and a few other scientists are eager to find out if Einstein's theory is valid.

But some British officials are calling for a boycott of German science. The Royal Society and the Royal Astronomical Society ignore the objections; they sponsor what will later be called one of the grandest experiments in the history of science. By testing "the 'enemy' theory," Arthur Eddington writes later, "our national observatory kept alive the finest traditions of science; and the lesson is perhaps still needed in the world today." English astronomers intend to show that science is an international venture, beyond politics. A solar eclipse is predicted for May 29, 1919. The British astronomers will use that opportunity to measure starlight near the Sun and see if its rays bend—as Einstein predicts.

Sir Arthur Stanley Eddington was Britain's leading astronomer, an early champion of Einstein's ideas, and a man who didn't like to share the limelight. When someone said only three people understood general relativity, Eddington asked, "Who's the third person?" Today, relativity is widely understood.

They set out in two ships; one group (led by Eddington) sails to the island of Principe near West Africa, and others go to Sobral in northern Brazil. Even if clouds hide the eclipse in one region, the other might be cloud free. The ships carry the latest scientific instruments. On the morning of May 29, it pours on the African coast, but in the early afternoon the skies clear enough so that six photographs can be made. In Brazil, seven useful photographs are taken.

One Thing Is Certain

After his return to England from Principe, Sir Arthur Eddington wrote a long poem about that adventure. Here are a few stanzas:

Five minutes, not a moment left to waste,
Five minutes, for the picture to be traced—
The Stars are shining, and coronal light
Streams from the Orb of Darkness—Oh make haste!

For in and out, above, about, below
'Tis nothing but a magic Shadow show
Played in a Box, whose Candle is the Sun
Round which we phantom figures come and go.

Oh leave the Wise our measures to collate.
One thing at least is certain, LIGHT has WEIGHT
One thing is certain, and the rest debate—
Light-rays, when near the Sun, DO NOT GO STRAIGHT.

These negative (left) and positive prints from the same solar eclipse photo come from the 1919 report by Eddington in search of evidence that the Sun's gravity causes light to bend, an effect predicted by Einstein's Theory of Gravitation.

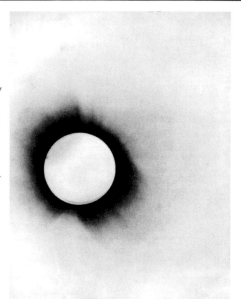 

The photographs must be processed and measurements and other data checked. It will take months of careful work. On September 2, Einstein writes to a friend, "[S]o far nothing precise has been published about the expedition's measurements so that even I know nothing about them." A month later, he and a few others get hints of what is to come.

Then, on November 6, fellows of the two royal societies gather in a joint meeting. Later, the distinguished British mathematician and philosopher Alfred North Whitehead will describe the scene:

> *The whole atmosphere of tense interest was exactly that of the Greek drama.... There was dramatic quality in the very staging—the traditional ceremonial, and in the background the picture of Newton to remind us that the greatest of scientific generalizations was now, after more than two centuries, to receive its first modification...*
> *[and] a great adventure in thought had at length come safe to shore.*

J. J. Thomson (remember him?) is chairman. He rises and

Einstein was discussing some problems with me in his study when he suddenly interrupted his explanation and handed me a [telegram].... It was the news from Eddington confirming the deviation of light rays near the sun that had been observed during the eclipse. I exclaimed enthusiastically, "How wonderful, this is almost what you calculated." He was quite unperturbed. "I knew that the theory was correct. Did you doubt it?"

—Ilse Rosenthal-Schneider, former student and lifelong friend of Einstein's,
  in *Some Strangeness in the Proportion*

announces, "This is the most important result obtained in connection with the theory of gravitation since Newton's day [and] one of the highest achievements of human thought." He continues, "It is not the discovery of an outlying island but of a whole continent of new scientific ideas."

The next day the rest of the world hears the news. **"Lights All Askew In The Heavens: Einstein's Theory Triumphs,"** blares the *New York Times* (right). **"Revolution in Science. New Theory of the Universe. Newtonian Ideas Overthrown,"** proclaims the London *Times*.

Einstein said light would bend as it passed through spacetime near the Sun. The scientists measured. Light bent!

Before this experiment, the general public has mostly been unaware of the 40-year-old, shaggy-headed, German-Swiss-Jewish physicist (eventually to be an American). Suddenly, this man with his wry sense of humor, rumpled clothes, and equation-writing skills is a world celebrity. And science has been transformed.

Does It Change?
Or Is It Changeless?

Why all the fuss, then? Because relativity has made us think again about where we live, which is in time and space. Because Einstein pulled aside the curtains and revealed that, in describing nature, space and time are woven together, and matter and energy are one.
—Brian L. Silver (d. 1997), *The Ascent of Science*

The quality that dominated [Einstein's] personality was a very great and genuine modesty. When anybody contradicted him he thought it over and if he found he was wrong he was delighted, because he felt that he had escaped from an error and that now he knew better than before.
—Otto Frisch (1904–1979), Austrian-British physicist

A t the beginning of the twentieth century, just about every major scientist believed that the universe was infinite and had existed forever. There was no solid evidence to back this up, but no one had proved otherwise, either. It was, as the phrase goes, "accepted wisdom." Science said the universe was and always would be basically the same and unchanging. It was static. Heavenly bodies might come and go, but the overall structure of the universe would stay the same.

The celebrated Scottish geologist James Hutton (1726–1797) said, "We find no vestige of a beginning, no prospect of an end."

This concept went back to the ancient Greeks. Heraclitus of Ephesus wrote around 500 B.C.E., "This cosmos, the same for all, was made by neither god nor man, but was, is, and always will be: an ever-living fire, kindling and extinguishing according to measure."

Archytas of Tarentum (a Greek colony on the heel of the boot of Italy) was active in the early fourth century B.C.E. He was a famous mathematician, and he may have built the world's first flying machine.

Archytas of Tarentum, a follower of the Pythagoreans and a friend of Plato (some say he was Plato's math teacher), pondered the whole universe. He wondered: Is the universe infinite (without a beginning or end)? Or is it finite (with boundaries)? Then he attempted to answer his question with a thought experiment.

The idea that our universe might have companions isn't new. In *An Original Theory or New Hypothesis of the Universe* (1750), English astronomer Thomas Wright (1711–1786) envisioned an "endless Immensity" (infinite space) with "an unlimited Plenum of Creations" (multiple universes). Each sphere represents a universe with "the most perfect of Beings" (God) at the center.

"If the universe is finite, it must have an edge," Archytas reasoned. In his mind, he went to the edge of the finite universe and threw an imaginary spear. "There is nothing in the void to turn the spear around, so it will move on until it lands. But the spot where it lands will be beyond the point which you say is the edge." Having created a new edge, Archytas threw the spear again—and again. With each throw, the universe

What Is Imaginary Time?

Was Archytas right? Is the universe infinite? Physicists Stephen Hawking at Cambridge in England and James Hartle at the University of California at Santa Barbara don't think so. They have proposed that space and imaginary time, married together, is an entity that is finite but without boundaries or edges.

Imaginary time? Hawking says consider ordinary time by picturing a horizontal line. On the left is the past, on the right the future. Now add a vertical line; that's imaginary time. We don't normally experience the dimension where imaginary time exists, but it is an important hypothesis in modern physics.

Philosophers and astronomers aren't the only ones who search for answers. Writing in the 1860s, the decade before Einstein was born, Jules Verne pointed a fictional cannon into space, blasted men onto the Moon's surface, and, in superb science fiction, inspired a generation of thinkers. This engraving is from the original 1865 edition of Verne's book *From the Earth to the Moon*.

became larger. "The universe cannot have an edge, and must therefore be infinite," he concluded.

Like almost all his colleagues early in the twentieth century, Albert Einstein believes that the universe is infinite, eternal, and unchanging. He also believes that the universe is *homogeneous* (basically the same everywhere) and *isotropic* (the same in all directions), and, therefore, that Earth cannot be central to the universe. Einstein is pushing Copernicus to the limit.

When Einstein and others apply the theory of relativity to this isotropic universe, they discover a big problem. It turns out that the universe—with relativity—is unstable. Something is wrong: It can't be unchanging (static). According to Einstein's gravitational formulas (and Newton's, too), every object in the universe is pulled toward every other object. Unless countered by another force, this attraction will eventually bring everything together into a big crunch, which would mean "Good-bye, universe."

Today's scientists agree that the universe is basically homogeneous and isotropic. They *don't* agree with the static picture.

Einstein studies his own theory. He works the math again and again. He can't believe what it is telling him. Finally, he finds a way to keep the universe from collapsing. He adds

Einstein's breakthrough idea that gravity is a distortion of spacetime opened the door for topologists (mathematicians who specialize in surface geometry) to explore the phenomenon. This computer-generated strip of spacetime has two twists.

something new to relativity: It is an unchanging number, a constant, which Einstein calls the "cosmological constant." It represents a very tiny antigravity force and is symbolized by a letter from the Greek alphabet: lambda, λ. (More details on this to come.)

Our Universe Had a Birthday

The laws of physics are fixed; they don't change. If you jump off a ledge, you will fall toward the Earth. Not just once in a while—always. You will never fall the other way. Gravity makes it so.

But you're not a law of physics. You will change. You won't be quite the same tomorrow as you are today.

Do the stars change? Yes. Like you, they are born, they evolve, and they die.

What about the universe as a whole? According to physicist Stephen Hawking (writing in the twenty-first century), "All the evidence seems to indicate, that the universe has not existed forever, but that it had a beginning, about [14 billion] years ago. This is probably

The loops, dips, hills, and banked turns of roller coasters are designed with the fixed laws of physics in mind—especially acceleration (changes in speed and direction).

the most remarkable discovery of modern cosmology." More on this evidence ahead.

Einstein has transformed the world. He now has people seeing time and space in a relative way. He has turned gravity into a spatial thing with warps and weaves and time interwoven. But astronomers are about to bring the heavens into magnified view. Einstein's theories will be put to the test by observations and experimentation. Is the universe static? The stars themselves will answer that question. How about the lambda force? Keep it in your head. Some surprises are coming.

Expanding Times

There is nothing so far removed from us
as to be beyond our reach
or so hidden that we cannot discover it.
—René Descartes (1596–1650), French philosopher and
mathematician

Equipped with his five senses, man explores the universe...and calls the adventure Science.
—Edwin Powell Hubble (1889–1953), American astronomer, *The Nature of Science and Other Lectures*

You can observe a lot just by watching.
—Yogi Berra, American baseball player

I n 1917, a giant 100-inch (2.5-meter) light-gathering telescope—the world's largest—is placed on Mount Wilson in California. It's a high spot where the skies are clear. Then an astronomer as good as Tycho Brahe puts his eye to the lens. His name is Edwin Hubble. He has been an amateur boxing champ, a high school teacher, and a Rhodes scholar. Hubble looks deep into the sky and sees sights never witnessed before (at least not from this planet). And he knows how to interpret the surprises he sees.

Hubble already realizes that our solar system is a tiny player in a big picture. An American astronomer named Harlow Shapley (1885–1972) has determined that we're part of a galaxy of stars—the Milky Way

Rhodes scholars are chosen for their potential to better the world in the field of their choice. They receive a fully paid, two-year scholarship to the University of Oxford in England.

Built in 1917, the Hooker telescope at the Mount Wilson Observatory was the world's largest for the next three decades. It allowed Edwin Hubble to discover (in 1923) that other galaxies exist beyond our Milky Way.

galaxy—and we're not even in its center. We're way off in the suburbs.

This has been difficult for most people to believe. For aeons, we had thought of Earth as the center of everything. Then a Polish church canon, Nicolaus Copernicus, came along and said we had it wrong. It is the Sun that is in the center, said Copernicus. After that, almost everyone believed our solar system was the core of the universe, until, in 1918, Shapley told us that isn't so.

And now, in 1925, Hubble figures out that ours isn't the only galaxy. That is a shocker. (Today's cosmologists are telling us ours may not be the only universe.)

Hubble sees bright, fuzzy "clouds" in space and realizes that they are distant galaxies similar to our own. Once he finds a few galaxies, he keeps finding more and more of them. He becomes aware that there is a vast array of galaxies in the heavens. Suddenly, astronomers (and others who are paying attention) understand that the universe is beyond-belief gigantic—and that our Earth-home is just a speck compared to the totality.

[G]alaxies are as common as blades of grass in a meadow.
—Kenneth F. Weaver and J. P. Blair, *National Geographic* magazine (May 1974)

Does that make our planet less interesting? Not at all. Now that we know the universe is vast and full of surprises, the relationship between Earth and the universe becomes more fascinating—and awesome.

A Dutch astronomer, Willem de Sitter (1872–1934), looks at Einstein's equations and points out that they imply the universe is expanding. Einstein doesn't believe it. An expanding universe? The picture of the universe that most scientists accept is static. In it, stars have fixed movements. There is no expansion or contraction. That static universe is a legacy from Aristotle. Just about all the great thinkers who followed him, including Albert Einstein, accept it.

But Einstein does realize that a static universe has a problem. According to Newton's theory of gravitation and

Tycho Brahe, an amazing Dane, was one of the world's greatest astronomers in the time before telescopes. For more about him, read chapter 4 of *Newton at the Center*.

Edwin Hubble examines a negative photo of a galaxy (light is dark, dark is light). Prior to the giant Hooker telescope, other galaxies appeared as fuzzy smudges of light and were mistakenly thought to be spiral gas clouds within our own galaxy.

Annie J. Cannon (1863–1941) was a great astronomer in the tradition of Tycho Brahe—an observer and classifier. But she had a big advantage over Brahe: telescopes. Cannon classified one-quarter of a million stars, the work of a lifetime. Her nine volumes of charts, published by the Harvard Observatory, are still used today.

Einstein's own general relativity (GR), the heavenly bodies attract one another, which means eventually the static universe will collapse into itself. But that doesn't seem to be happening. A few physicists are saying that the universe is not static, that it is expanding. That would mean that the universe has a story to tell and that it is a tale of change. Nothing in previous science suggests that is true. It doesn't seem to make sense. How do you explain gravity in an expanding universe? In a letter to Willem de Sitter, Einstein says, "This circumstance of an expanding Universe is irritating.... To admit such possibilities seems senseless to me."

So, in order to make his equations work in a static universe (and ward off collapse), Einstein hypothesizes an antigravity force. While gravity pulls stars and planets inward, this gravitational repelling effect pushes them apart—creating a balance that keeps the universe static and stable. Einstein adds his antigravity force, coded as a **cosmological constant**, **lambda**, to his equations. This force

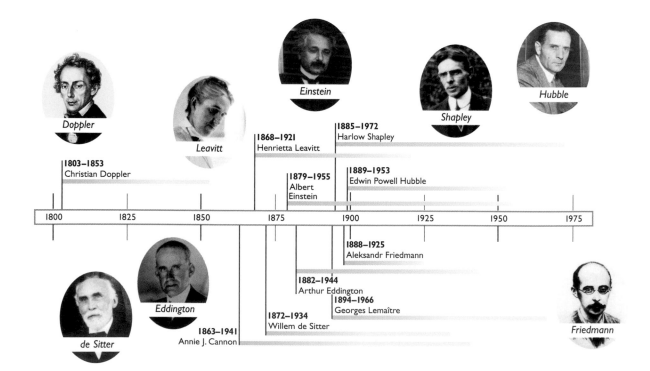

Doppler

1803–1853
Christian Doppler

Leavitt

1868–1921
Henrietta Leavitt

Einstein

1879–1955
Albert
Einstein

1885–1972
Harlow Shapley

Shapley

1889–1953
Edwin Powell Hubble

Hubble

1800   1825   1850   1875   1900   1925   1950   1975

1888–1925
Aleksandr Friedmann

1882–1944
Arthur Eddington

Eddington

1894–1966
Georges Lemaître

1872–1934
Willem de Sitter

de Sitter

1863–1941
Annie J. Cannon

Friedmann

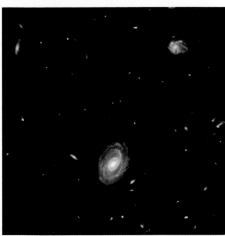

How far can we see? In 2004, the Hubble telescope set a record with this view of galaxies, some of them 13 billion light-years away—and thus 13 billion years old.

is important in the vast cosmos but unnoticed over short distances. It seems to do the trick.

But ideas fly off in unforeseen directions; there's no telling in advance where they may land. Einstein's theories are fertile. Aleksandr Friedmann (1888–1925), a young professor in St. Petersburg, Russia, hears about general relativity, studies it, and takes it seriously (he has the math skills to do so). While Einstein is trying to adapt general relativity to fit his belief in a static universe, Friedmann goes with what the GR equations are telling him: that the universe must be expanding. In 1922, he writes a paper saying just that. How does Einstein react? He says Friedmann has blundered.

Another actor appears on the cosmological stage. He is a shy, plump Belgian priest; his name is Georges Lemaître (1894–1966). "There is no conflict between science and religion," says Lemaître, who was a soldier in the Belgian army during World War I and then became a civil engineer, a mathematician, and a cleric. "Once you realize that the Bible does not purport to be a textbook of science, the old controversy between religion and science vanishes," he says.

Did God play a role in Big Bang creation? Scientists can't answer that question. God is not a provable theorem. That doesn't mean the God theorem is not true or that eventually it might be proved by a new discipline different from our current science.

Lemaître studies physics at the University of Cambridge in England under Arthur Eddington. Then he spends two years at MIT (the Massachusetts Institute of Technology), where he learns about the work of Edwin Hubble.

Back in Belgium, as a professor at

Aleksandr Friedmann's mother was a piano teacher and his father a composer. Aleksandr said he became a scientist because he had no musical talent. In World War I, he joined the Russian air corps (airplanes were in their infancy) and developed bombing procedures. After the war, he became a professor at St. Petersburg's Academy of Sciences. One of his brightest students was exuberant George Gamow. In 1925, 37-year-old Friedmann died of typhoid fever.

Lemaître (above) took the universe back to a beginning point. What happened when that point exploded? In the beginning, did relativity and quantum mechanics dance to the same music? What was the stuff of creation? Where is the universe heading? Billions of years from now, might space and time stop expanding, shrink, and finally crunch back into a singularity? Or will the universe continue expanding forever? These are the questions that physicists ask.

The Big Bang Connection

Hubble's discovery that the universe is expanding turns out to be one of the most powerful ideas of all time. Why? Because, if the universe is expanding, it must have expanded from something much, much smaller.

That confirmed what Georges Lemaître had already discovered when he took Einstein's equations, ran time backward, and imagined galaxies and stars scrunching together into a very dense, subatomic speck. That kernel, said Lemaître, burst forth in a blazing hot creation. As the universe expanded, it cooled.

Now, wrote Lemaître, "Standing on a well-chilled cinder [he means Earth], we see the slow fading of the suns, and we try to recall the vanished brilliance of the origin of the worlds."

Here's one of many ways that science illustrators have imagined the Big Bang.

Universal expansion can be compared to raisins rising in bread dough. The dough takes the raisins with it. The dough is the universe; the raisins are galaxies.

the University of Louvain, Lemaître thinks about general relativity and about the discovery of galaxies. He links observation and theory. Lemaître doesn't seem to have read Friedmann's paper, but he, too, comes to the conclusion that the universe must be expanding. If the universe is expanding, he asks himself, what is it expanding from? The galaxies are now billions and billions of light-years apart, but in the past they must have been much closer.

Lemaître does a thought experiment: He puts the universe's tape on rewind. He focuses on stars and galaxies and sees them getting closer and closer to one another. The universe is contracting. He watches stars and galaxies scrunching together in a cosmic smashup.

Lemaître takes his mathematical calculations back, back, back—to a time billions of years ago when all the matter of the universe was squeezed into an infinitely small pinpoint: a "cosmic egg," the "primordial atom." Finally, he can go back no further. And then, "we had fireworks of unimaginable beauty," says Lemaître. "There was the explosion followed by the filling of the heavens with smoke. We come too late to do more than visualize the splendor of creation's birthday."

It is "a day without [a] yesterday." The beginning of time and space, the bursting forth of the universe, the release of all

Shifting Red, Shifting Blue—What's Going On?

Even with a great telescope, you can't actually see the stars move—at least not the way you can see a motorcycle zoom down the road. Why not? Because stars are so many light-years away that their motion is hardly discernable. (The apparent motion we see in the night sky is caused by Earth's rotation.) So how did Edwin Hubble know the stars are moving?

He knew because he could see something called a **redshift**. And, if you want to understand space talk or space travel, you need to know what a redshift is. So here's an explanation.

Let's start with Isaac Newton, who split visible light into a spectrum of colors by refracting (bending) it through a prism. The spectrum ranged from red at one end to blue at the other. Why? Because red wavelengths are longer than blue ones and so refract at different angles.

The top diagram shows the Earth and a galaxy relatively at rest. The spectrum of light from the galaxy has a dark absorption line in the center. If the galaxy is moving away from Earth (center), the wavelength stretches, and the line shifts toward the red. The opposite occurs if the galaxy is moving toward Earth (bottom).

Hubble had to understand wavelengths and color—and a few other pieces of the puzzle—before he could figure out that stars move.

He knew, that in 1842, Christian Doppler (1803–1853), an Austrian mathematician/physicist, discovered a principle now called the Doppler effect. It works with all types of waves, including sound and light. Imagine standing by a road as a motorcycle approaches. Coming toward you, the sound of the motorcycle is high-pitched; leaving you, it is lower-pitched. (You can hear this for yourself. It's easy with an ambulance siren.) The change in pitch—the Doppler effect—is caused by the scrunching of sound waves from an approaching source and the stretching out of sound waves from a source moving away from you.

Light waves also show a Doppler effect. The effect is too small to see if you're watching the headlight on an approaching or receding motorcycle, but it is important to astronomers tracking the light of a star across vast distances of space. Christian Doppler figured out that, as a star moves away, its light waves would seem longer (no surprise, since they are getting farther away). Those longer waves mean more red and less blue. So a star whose spectrum is shifted to longer wavelengths must be moving away from us. Its light is shifting to the red part of the spectrum. The spectrum of a star heading in our direction will show a shift to shorter wavelengths—a blueshift.

Doppler also understood that the greater the shift, the faster the star's movement relative to us. With that insight, he found a way to tell both the direction and speed of the stars. But he had a problem. Telescopes of his time couldn't see any of this. Doppler was sure there would soon be more powerful telescopes. In 1842, he wrote of the effect he had found:

It is almost to be accepted with certainty that this will in the not too distant future offer astronomers a welcome means to determine the movements and distances of such stars which, because of their unmeasurable distances from us . . . until this moment hardly presented the hope of such measurements.

Hubble and Spitzer in Orbit

Astrophysicist Lyman Spitzer, Jr. (1914–1997), who was head of Princeton's observatory and astrophysics department, believed that a telescope orbiting above Earth's atmosphere would give much clearer images than any placed on our planet's surface. That was in 1946, more than a decade before the first Earth-made object (a little satellite named Sputnik) was launched into space. No one paid much attention to Spitzer's idea, but he kept lobbying for it.

In 1977, Congress authorized funds. It took eight years to build, and it cost $1.5 billion, but finally, in April 1990, Spitzer's brainchild, a 16-meter (52-foot) telescope, entered orbit 569 kilometers (353 miles) above Earth. To name it after pioneering Edwin Hubble seemed the right decision. During its first 16 years on the job, the Hubble Space Telescope has taken 750,000 images of 24,000 celestial objects.

Spitzer was described as a man of "incredible discipline, diligence, and politeness." He was analyzing data from the Hubble Space Telescope on the day he died.

In 2003, NASA launched an infrared telescope named the Spitzer Space Telescope.

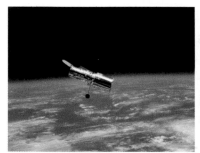  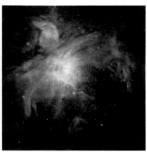

Data from the Hubble (left) and Spitzer (center) telescopes produced this false-color image of the Orion Nebula (right), which includes combined infrared, visible light, and ultraviolet frequencies.

The name "Big Bang" wasn't meant seriously. British astronomer Fred Hoyle (1915–2001) didn't like the whole idea and used the term derisively. But the name stuck.

the energy and mass that is ever to be. It is a creation, a beginning; it will come to be called the Big Bang. Out of a birthing fireball emerges all the potential of the world to come.

Lemaître doesn't have it quite right, although he is close. It wasn't an explosion as we think of explosions. Things didn't fly out into space. There was no space. There was no time. Every place that exists today began in a dense, smaller-than-an-atom seed—space and time were born together, and space expanded. According to this Belgian priest (and he was right), the newborn universe was unimaginably hot; it seethed and swirled with high-energy particles.

In 1927, Lemaître writes a paper describing his expanding model of the universe. **Lemaître has come up with an idea that will, eventually, define modern cosmology.** But, at first,

The gravitational attraction between the galaxies would bring them all together if they were not rushing away from each other. The universe can't stand still.

—John D. Barrow, *The Origin of the Universe*

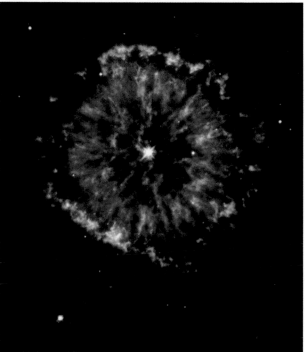

that idea doesn't make much of a bang in the scientific community. Hardly anyone pays attention. The expanding universe and the Big Bang are just unproved theories. Einstein rejects them; he still thinks of the orb of stars as fixed and changeless in its actions. "Your calculations are correct, but your physics is abominable," says Einstein to Lemaître.

Einstein doesn't know that theory and technology are about to come together—as they did when Galileo's telescope helped prove Copernicus's ideas.

In 1929, Edwin Hubble, using that giant Mount Wilson telescope, actually sees that the galaxies are moving apart and at an astonishingly fast pace. Hubble finds that **galaxies are moving apart at speeds that increase in direct proportion to their distance from an observer.** That will later be known as Hubble's law, and it is exactly what one would expect if the beginning of everything was a big bang: Farther away means higher velocity.

What about the static universe? Does Einstein still believe in it?

He is about to go to California to see what Hubble has seen. Einstein is in for a surprise.

In the eighteenth century, Germany's Immanuel Kant and England's William Herschel suggested that some of the hazy nebulas in the heavens are "island universes," or galaxies. Their telescopes weren't powerful enough to tell the difference, but ours are plenty sharp. That's a spectacular planetary nebula, above, and an infrared/ultraviolet view of the Andromeda galaxy, upper left. Andromeda is moving toward us because we're so *attractive* (it's that gravity thing).

An Expanding Universe

On January 29, 1931, Albert Einstein and
Edwin Hubble seat themselves in a fancy
Pierce-Arrow touring car. A chauffeur drives
them up a long, winding, dirt road to the
top of Mount Wilson, near Pasadena,
California. There, Hubble shows Einstein the
world's largest telescope, the Hooker, with its
100-inch mirror (right). Einstein, now 51, acts like a little
boy with a new toy. He is excited and awed. His second
wife, Elsa, is less impressed. When told the astronomers are
using the huge telescope to determine the shape of the
universe, she says, "My husband does that on the back of an
old envelope."

Scott Teare (above), a
physicist who specializes in
optics, has just given the
100-inch mirror from the
Hooker telescope a shiny
new aluminum face for
collecting starlight. The job
has become a yearly labor of
love for what Teare calls a
"wonderful piece of history."

But Hubble knows things that her husband doesn't know.
These two titans are about to share their knowledge:
Einstein the theorist and Hubble the experimenter.

Hubble arrived on astronomy's stage just as advanced
telescopes with great light-gathering power became a reality.

The telescopes at the Mount Wilson Observatory perch atop a long, winding, mountain road outside Pasadena, California. The two towers on the left observe the Sun, especially its magnetic field and sunspots. The domed telescopes peer farther, at stars and galaxies. The smaller 60-inch scope is retired from professional use but still gives the visiting public a powerful peek at the night sky.

At the same time, knowledge of the physics of the heavens had taken great leaps.

It is now nearly 90 years after Doppler's 1842 insight, so Hubble understands that when he looks through his telescope and sees a shift to the red or blue portion of the visible spectrum it is caused by the motion of the star relative to the motion of the observer (who is moving with Earth). The faster a star is moving away from us (relative to Earth), the more its light will redshift.

Using the world's best telescope, Hubble soon realizes that **virtually all the light he sees coming from galaxies is redshifted.** That's a huge surprise. Hubble puzzles over this

Long = Red, Short = Blue

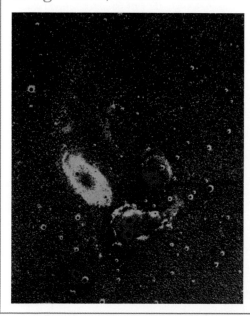

Imagine someone throwing balls at you at a constant rate. You catch one ball every second. Now, as he moves away from you (or you from him), it takes longer for each ball to reach you so you no longer get one per second. As he moves toward you, you get slightly more than one per second. In either case, the rate at which he throws the balls hasn't changed, but the rate at which you catch them has. Same thing with light waves. Some are longer and some shorter because of the time it takes them to reach us, and those longer and shorter times make light shift to the red or blue part of the spectrum.

In this false-color image, the four small red galaxies (two are lumped together in the center, interacting) have a greater redshift than the large, pale, doughnut-shaped galaxy. That means Stephan's Quartet, as they are collectively called, is receding from Earth faster and is farther away.

The image in the background is a falsely colored, visible-light view of two chain-like clusters of galaxies merging. The red swooshes represent radio emissions from four galaxies doing a "complicated dance," as described by the investigators.

and figures it out: **Redshifts mean the galaxies are all moving away from us.**

It is no surprise to learn that stars move. But just about everyone believes that heavenly movement has to do with orbits around a bigger star or planet.

Hubble is seeing something else. He is observing whole galaxies moving, but not in orbits around one another. The redshifts give him a tool with which to figure out the speed at which the galaxies move. They are moving very fast—away from us and from one another.

Expansion? Not in My Backyard

In the Woody Allen movie *Annie Hall*, nine-year-old Alvy Singer is depressed and has stopped doing his homework. Why? Alvy explains to his mother and the doctor that there's no point to homework: "The universe is expanding," he says, "...and if it's expanding, someday it will break apart and that would be the end of everything!"

"What has the universe got to do with it?" Alvy's mom asks. "You're here in Brooklyn! Brooklyn is not expanding!"

She's right! Brooklyn isn't expanding, nor is Earth, nor our solar system, nor our galaxy. All those collections of matter are held together by local gravitational or chemical interactions, which are stronger than the force of expansion.

The expansion of the universe happens in the space between the clusters of galaxies. Outside the great clusters of galaxies, universal expansion tops local gravitational pull. So it's the galaxy clusters that are speeding away from one another. Think of a dust mite sitting on an expanding balloon; the balloon gets bigger, and so does the space between mites, but the dust mite doesn't grow.

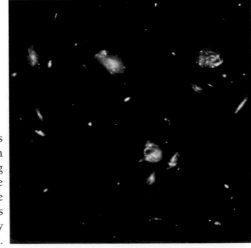

To be clear: Galaxies are not flying through space; they are being carried along as space itself expands. The galaxies themselves don't grow bigger; they just get farther apart.

When Hubble found the universe was flying apart, he forced cosmologists to consider how the universe was created and how it would end. Before, those had been questions for theologians and philosophers, not scientists.

English astronomer John Goodricke (1764–1786) discovered the first Cepheid (Delta Cephei) in 1784. Henrietta Leavitt (right) single-handedly cataloged some 2,000 pulsating stars out of 20,000 known. She plotted their period (time between pulses) and luminosity (actual brightness) on a graph.

Now that Hubble realizes there are many galaxies in the heavens, he wants to pinpoint some and figure out how far they are from Earth. How can he do this?

He knows that in 1912, an American astronomer named Henrietta Leavitt (1868–1921) discovered something important about a kind of rare star—a Cepheid (SEE-fee-id) star. Cepheids stand out in the heavens like lightning bugs on a dark night. They blink! Which means they brighten and dim. The time between blinks is called a "period." Leavitt discovered that a Cepheid's period is directly related to its luminosity (true brightness); the longer the period (the slower the blinking), the greater the luminosity of a Cepheid.

Distance from Earth is the measurement Hubble is after, and Cepheids solve his problem. Their regularity makes them dependable. Hubble's colleague (and sometimes rival) at Mount Wilson, Harlow Shapley, has pioneered the use of Cepheids as marker stars. If you know the luminosity, the period, and the distance of a marker star, you can put that data in a formula to calculate the unknown distance of other Cepheids. Hubble can discover how far away a galaxy is by measuring its Cepheids. He finds that the galaxies are much, much farther from us

LUMINOSITY is the amount of energy that a star kicks out—regardless of how far away it is. It's the actual (or intrinsic) brightness of the star. But a super-luminous star that's extremely far away can appear dim to us. How bright a star *appears* to us (measured in units of magnitude) is called its **apparent brightness**. The brightness decreases by the square of the object's distance—valuable information to astronomers.

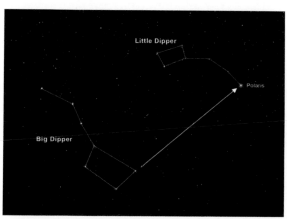

Polaris, the North Star (marked with arrow), was the first Cepheid used as a marker star. Polaris had long served as a guide for travelers making their way on Earth before it became a beacon for astronomers.

You'd think that a face-on spiral galaxy as pretty as this pinwheel would have a proper name, but it's known only in technical terms, as NGC 1309. Astronomers pegged its distance at 100 million light-years thanks to blinking Cepheid stars within its midst. For a few weeks in 2002, the bright light from a resident supernova, known as a Type Ia event, reached Earth and provided a clue to the expansion of the universe. The "how" is coming up, in chapter 46.

If the universe is static, then there is no before and no beginning. Everything remains the same. The expanding universe theory gives the cosmos a history. Now we know our universe has a story to tell. Stars come and go. So do their orbiting planets. And every instant in cosmic time is unique.

and from one another than anyone has imagined.

The tools to understand the skies are piling up, but no one before Edwin Hubble has put the pieces of the puzzle together. Newton connected the falling apple to the Moon in the sky. **Hubble comes up with a new, expansive picture of the universe: It is astonishingly vast, it is without a center, and it is not static.**

He discovers something else that is important: **The farther away a galaxy is, the faster it moves relative to us.**

Lemaître and Friedmann figured out that the universe is expanding by studying Einstein's equations. Hubble, without knowing of their work, actually sees and measures that expansion. For thousands of years, humans have imagined an unchanging universe. Hubble buries that concept.

Ah, if only we could hear their conversation as Hubble shows Einstein around and explains what the telescope sees. What we know is that Hubble's discovery convinces Einstein that he is wrong about a static universe.

Edwin Hubble (with the pipe) and Walter Adams, director of the Mount Wilson Observatory, invite Albert Einstein to peer through the eyepiece of the Hooker telescope. The 51-year-old Einstein went farther—leaping on the structural supports to examine the state-of-the-art telescope more closely. The January 1931 visit came two years after Hubble confirmed that the universe is expanding.

Einstein now calls Friedmann's calculations "both correct and clarifying." Of Lemaître's ideas he says, "very, very beautiful."

As for his cosmological constant—the antigravity force he came up with to keep the universe static—he now calls it the greatest scientific "blunder" of his life. Einstein won't live long enough to know that the constant, lambda, is not quite the blunder he thinks it is.

Lambda Returns

Through most of the twentieth century, physicists agreed with Einstein: Lambda, the cosmological constant, was a blunder.

But quiet Georges Lemaître didn't think so. He was fascinated by lambda. Einstein had pictured it in geometric terms; Lemaître said lambda is a form of matter that exerts a repulsive gravitational effect but is unaffected by other matter.

Long after both Einstein and Lemaître were dead, lambda was resurrected to possibly explain a new mystery in spacetime. Lambda might (or might not) be involved in dark energy (more on this later).

Today, some astrophysicists hypothesize that antigravity has existed from the first instant after the Big Bang. Einstein was called prescient (farseeing) for finding lambda; hardly anyone remembers Lemaître's belief in it.

And as for a static universe? That idea never comes back.

A Luminous Indian

Like all Indian students newly arrived in England, Chandra had to contend with loneliness, homesickness, bland food, and the difficulty of keeping to a vegetarian diet, not to mention…the racial prejudice that was as prevalent in Cambridge as anywhere in England. Fiercely proud of his Indian heritage, Chandra strongly objected to any hint of discrimination.
—Arthur I. Miller, American professor of the history and philosophy of science, *Empire of the Stars*

The life of a star of small mass must be essentially different from that of a star of large mass…. For a star of small mass the natural white dwarf stage is an initial step towards complete extinction. A star of large mass cannot pass into the white dwarf stage and one is left speculating on other possibilities.
—Subrahmanyan Chandrasekhar (1910–1995), Indian-American astrophysicist, *The Observatory* 57 (1934)

His real name is Subrahmanyan Chandrasekhar (soo-brah-MON-yon chan-drah-SAY-kar). But everyone knows him as Chandra, which is appropriate because the word means "moon" or "luminous" in Sanskrit (the ancient language of his native India). Chandra will become a shining light in a twentieth-century science that combines the disciplines of physics and astronomy: astrophysics.

Cambridge University's loss was the University of Chicago's gain in 1936 as Chandrasekhar, a rising star at age 26, accepted a post at Yerkes Observatory.

He is a Brahman. In caste-divided Hindu India, the Brahmans are families at the top of the social heap. Chandra's family has a tradition of scholarship and high achievement: His grandfather was a math professor, an uncle will win a Nobel Prize in physics, and his mother is acclaimed for her translations of literature.

Even in that family, Chandra is a star. From an early age, he excels in math and astronomy. In 1927, when he is 17 and

From the Indian Subcontinent

Chandra was born in the city of Lahore in 1910. It was then in British India but is now part of Pakistan.

After India won independence from Britain, in 1947, Pakistan split off to form a separate nation of mostly Muslims. The region that stayed as India was mostly Hindu. The split led to terrible violence between Muslims and Hindus. Seven and one-half million Muslim refugees fled India for Pakistan. Ten million Hindus left Pakistan for India.

Long before that, when Chandra was eight, his family moved to Madras in southern India, a cultural and business center with beautiful Hindu temples and a famous beach on the Bay of Bengal.

In 1996, Madras was renamed Chennai. Today, Chennai is known as the automobile capital of India and is also a major center for outsourced jobs from the Western world.

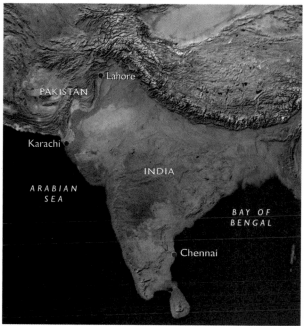

This satellite photograph tells you that most of the Indian subcontinent is arid.

a student at Presidency College in Madras, he discovers a book written the previous year by the eminent British scientist Arthur Eddington, *The Internal Constitution of the Stars*. Eddington is Great Britain's leading astrophysicist, known for his charm and wit as well as his intellect. (Remember, it was Eddington who led the eclipse-tracking expedition that verified Einstein's general relativity.) In his book, Eddington describes something that is baffling astronomers of the day. It is a kind of star called a "white dwarf."

"White dwarfs are probably very abundant," writes Eddington. "Only three are definitely known, but they are all within a small distance of the sun." Astronomers can see those three white dwarfs through their telescopes. They have measured the mass and circumference of the closest one, Sirius B, and found to their astonishment that 1 cubic inch of Sirius B weighs 1 ton. That enormous density is unlike anything found on Earth, so most astronomers just don't

When a Sun-sized star burns up most of its nuclear energy, gravitation will cause it to contract, eventually becoming a small, dense body—a **WHITE DWARF**.

That figure, 1.4 Suns (or solar masses), means almost one and one-half times more massive than our Sun. Remember that mass isn't about size—it's about the amount of matter. A white dwarf is a massive object that's very small compared to our Sun.

Our Sun is stable (for the moment) because outward pressure from nuclear burning balances the inward gravitational pressure. But once it burns off its massive layers of hydrogen, that inward pressure will decrease, and it will swell into a red giant. (The somewhat mysterious object below, astronomers suspect, is a red giant–white dwarf pair.) Red giants turn stored helium into carbon and other heavier elements. When its helium is gone, our red giant Sun won't be hot enough to burn the carbon, so the core will contract, releasing a huge envelope of energy. The Sun will then expand beyond Earth orbit and blow off its outer layers, leaving a dense core about the size of Earth: a white dwarf.

be? The answer is in their atoms, but knowledge of the atom has not gone far enough for Chandra to know where to look. In 1930, neither he nor anyone else can take the next step. Still, he is sure his calculations are right.

When serious-minded young Chandra arrives in England, he is eager to meet the man he idolizes, a man who began life as a working-class boy and has made it to fame and a knighthood. The star of British astrophysics is now Sir Arthur Eddington. No one in astrophysics can match his prestige.

Chandra tells Eddington of his discovery: that all stars do not become white dwarfs. For the next five years, with Eddington's knowledge and encouragement, Chandra works on his theory, fine-tuning his calculations, with Eddington following his progress in detail. At last, Eddington arranges for Chandra to present his conclusions on white dwarfs at a meeting of the Royal Astronomical Society on January 11, 1935.

A hundred distinguished astronomers are seated in the tiered lecture hall. This is no minor subject: The fate of stars involves the fate of the universe. The ambitious young scholar gives a detailed talk on his finding: **White dwarfs have a maximum mass. Giant stars cannot end their lives as white dwarfs.** This totally new idea means rethinking much of astrophysics. The audience listens. Then Eddington speaks. He laughs.

Standing tall and gun-barrel straight with wire-rimmed spectacles clipped on his thin nose and a gold watch hanging from his vest pocket, the distinguished Cambridge

don says, "There should be a law of Nature to prevent a star from behaving in this absurd way!" He ridicules the results of his young colleague. If you had been in the audience, whom would you have believed: England's greatest astrophysicist or an unknown student from India?

Eddington is not finished; every chance he gets, he derides Chandra's idea. He says the math is correct, but it has nothing to do with the stars. Chandra has used both quantum theory and relativity. Eddington says they are diametrically opposed. "I do not regard the offspring of such a union as born in lawful wedlock," he says. Eddington calls Chandra's theory "stellar buffoonery."

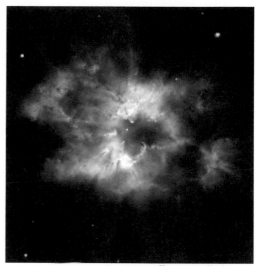

Our Sun will eventually be a white dwarf star surrounded by a cocoon of blown-off gas. The tiny white dwarf at the center of the nebula above is one of the hottest ones we know. White dwarfs keep cooling until, after billions of years, they can cool no more. Then they become black dwarfs, astronomers conjecture, though no one has actually spotted a black dwarf. (It's not easy; they're black.) Perhaps the universe isn't old enough for these dark stars to have formed, but no one is sure.

> **Throughout its normal lifetime, a star is an uneasy battleground between the force of gravity and the energy of nuclear fusion. Gravity tries to compress the star into a tiny ball, while the heat and light of its nuclear furnace…tries to blow it apart.**
> —Charles Seife, *Alpha and Omega: The Search for the Beginning and End of the Universe*

In Denmark, Niels Bohr considers the conflict and realizes that Chandra is right, but Bohr is busy with other matters, and it is Eddington who holds sway in astrophysics. (Read *Empire of the Stars*, by Arthur I. Miller, for details on this anguishing story.) Chandra leaves England in 1936, moving to the Yerkes Observatory at the University of Chicago. He says he doesn't want to spend his life fighting Eddington. He won't come back to white dwarf research for more than one-quarter of a century.

The question remains: Is there a limit to the mass of a star that can decay into a white dwarf? Chandra says no star larger than 1.4 times the Sun's mass can become a white dwarf. Is this Chandrasekhar limit valid? If it is, then what happens to giant stars? What is their destiny?

Explosive? And How!

[T]he real reason that Eddington was so opposed to Chandra's theory was that it entirely undermined his fundamental theory, which he had been honing with obsessive care for seven years.

—Arthur I. Miller, American professor of the history and philosophy of science, *Empire of the Stars*

Prostrating myself, I have observed the appearance of a guest star; on the star there was a slightly iridescent yellow color.... This shows that a Plentiful One is Lord and that the country has a Great Worth. I request that this be given to the Bureau of Historiography to be preserved.

—Zhou Keming, Chinese astronomer, describing a supernova in 1006

The thing that is funny (if you can use that word) about the Eddington-Chandra dispute is that Arthur Eddington understands both relativity and astronomy better than almost anyone of his time. He even points out that a small, dense star will create a deeper well in spacetime than a bigger but less dense star.

Eddington realizes that when a star uses up all its fuel, if it still has a mass greater than what will one day be called the Chandrasekhar limit (1.4 Suns), it will "go on radiating and radiating and contracting and contracting until, I suppose, it gets down to a few kilometers in radius, when gravity becomes strong enough to hold in the radiation, and the star can at last find peace." In other words, it will be so small and dense that it will create a chasm in spacetime from which nothing can escape, not even light. The star will end up, Eddington adds, "nowhere."

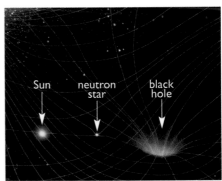

Our Sun makes barely a dimple in spacetime, according to this illustration, while a smaller but much more massive neutron star forms a slightly toothier dent. A yawning chasm created by a black hole extends for light-years.

This neutron star is a mere 16.8 miles (27 kilometers) across at best, smaller than the tiny Martian moon Phobos. Though extremely hot, it's 100 million times dimmer than the faintest stars you can see with your unaided eyes. The Hubble telescope spotted it with its Faint Object Camera, an instrument designed to stretch the limits of visibility.

Eddington's instincts are taking him in the right direction—toward what we now know as a black hole—and yet he doesn't have the imagination or the conviction to go with those instincts. He is sure that nothing like a black hole could actually happen. It seems too weird. Much later, Chandra will say that Eddington's "enormous physical insight" made him understand that the "existence of a limiting mass implies black holes." But for Eddington, the black hole idea is unthinkable. Too bad for him. "If [Eddington] had accepted that," says Chandra, "he would have been 40 years ahead of anybody else."

Chandra would have been there, too. And so would the world of astrophysics. As it happens, that world has to wait those 40 years to learn about black holes. It doesn't have to wait as long for **neutron stars**.

"In lecture after lecture during the 1930s, and article after published article," Kip Thorne will write in *Black Holes and*

The plural of supernova is either *supernovae* (Latin-style) or *supernovas* (English-style), which is the one I've chosen to use.

Time Warps, "[Fritz Zwicky] trumpeted the concept of a neutron star—a concept that he, Zwicky, had invented to explain the origins of the most energetic phenomena seen by astronomers: supernovae, and cosmic rays."

Like a black hole, a neutron star is the corpse of a giant star, the leftover after an enormous explosion. There is no way for Zwicky to see one in 1933, when he and his German colleague at Caltech, Walter Baade, first propose the neutron star's existence. But Zwicky does a lot of stargazing using the Mount Wilson telescope, and he figures out (correctly) that some unusually bright stars must collapse and become super-dense. That process of collapse produces a big blast; from Earth it looks as if a new star has appeared in the heavens.

Zwicky and Baade coin the name **supernova** to describe these short-lived beacons. They realize that a supernova is the result of the rare but spectacularly violent explosion of a massive star. (We now know that the star is blasting its outer layers away from its core, and that sends neutrinos, electromagnetic waves, and a whole lot of atomic nuclei and other debris off into its galaxy and beyond.)

In 1572, astronomer Tycho Brahe witnessed a supernova explosion. Here's the messy aftermath, more than 400 years later, courtesy of the X-ray eyes of the Chandra telescope. High-energy areas are colored blue; the glowing rim is a shock wave created as the debris expands.

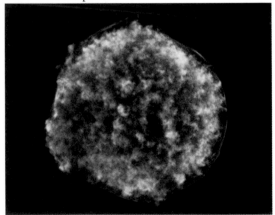

For a month or so, a supernova is a lightbulb in the sky where none has been before, and then it fades. The visible radiation from a supernova can be as bright as all the stars in our galaxy put together. In 1572, the Danish astronomer Tycho Brahe saw a supernova and couldn't believe what his eyes were showing him (not only because it was bright, but also because at that time it was assumed that the heavens never change).

Zwicky is trying to make sense of supernovas at the same time that Ernest Rutherford at the University of Cambridge realizes something is missing from the scientific picture of the atom's nucleus. Rutherford knows that the nucleus includes positively charged protons; he guesses that there may *also* be a neutral particle in the nucleus. In 1932, James

Chadwick, at Cambridge with Rutherford, finds that missing particle with no charge and calls it a neutron. For particle physicists this is big news; it answers lots of questions about the atom. (See chapter 15 if you've forgotten the details.)

Most astronomers don't pay attention to Chadwick's discovery. After all, a neutron is a tiny nuclear particle; it doesn't seem related to their work. But Fritz Zwicky is a maverick with a history of thinking differently. He hypothesizes that, given the intensity of a supernova explosion, a star's atoms must undergo change. What happens to its electrons, protons, and neutrons? He conjectures that the force of the eruption might send the negative electrons whamming into the positive protons, turning them into neutrons. That would leave the star's core stripped of its electron cloud but with a double dose of neutrons. It would become a neutron star.

"What made the idea so daring was that Baade and Zwicky had thought of understanding some of the largest known objects, stars, in terms of a tiny part of the nucleus of the atom," Arthur Miller would comment later.

In 1934, Zwicky and Baade write, "Supernovae represent the transition from ordinary stars into neutron stars." The problem is, no one pays attention. Zwicky is a nonconformist—you could call him a weirdo, or crazy, or a genius—with a talent for insulting people. Born in Bulgaria to Swiss parents, he graduated from Einstein's old university in Zurich, the ETH, and came to Caltech in 1925 at the very time that Hubble and others at the Mount Wilson Observatory were discovering wonders in the heavens.

Zwicky concludes that when a massive star explodes into a supernova, all that is left is an orb about the size of Manhattan. But that small orb has the density of an atom's nucleus. Arthur Miller explains, "On Earth a teaspoonful of white dwarf matter would weigh more than 6 tons. The same tiny amount of neutron star matter would weigh a billion tons—probably enough to take it plunging

Neutrons don't repel one another, so they are close together in a neutron star; there is almost none of the empty space found in atoms and no electric charge. As a result, a neutron star has enormous density, essentially the density of a nucleus. It's 100 million times as dense as a white dwarf.

Chroniclers at the Abbey of St. Gall in Switzerland described the supernova of May 1, 1006, as "a star of unusual magnitude, shimmering brightly... in the extreme south, beyond all the constellations." Frank Winkler wrote in the New York Times: "It would have been bright enough to read by (for those few who could read in the 11th century) ... [and] remained visible for at least two and a half years, according to Chinese records." A thousand years later, the remnant of the explosion (below) now spans 70 light-years—and growing.

Some Neutron Stars Are Pulsars

In 1967, Antony Hewish and Jocelyn Bell discovered a star that emitted beeps: periodic bursts of radio waves. Hewish recalled the moment in his Nobel Prize speech:

One day around the middle of August 1967 Jocelyn showed me a record indicating fluctuating signals....[We] first thought that the signals might be electrical interference.... [It] was not until November 28th that we obtained the first evidence that our mysterious source was emitting regular pulses of radiation at intervals of just greater than one second. I could not believe that any natural source would radiate in this fashion and I immediately consulted astronomical colleagues at other observatories to enquire whether they had any equipment in operation which might possibly generate electrical interference....Still skeptical, I arranged a device to display accurate time marks at one second intervals....To my astonishment the readings fell in a regular pattern....[T]he pulsed source kept time to better than 1 part in [1 million]....Having found no satisfactory [earthly] explanation for the pulses we now began to believe that they could only be generated by some source far beyond the solar system.

Hewish and Bell actually nicknamed the pulses LGMs—"Little Green Men"—until they found a natural explanation for them. The source was a star, which they named a "pulsating radio star," or *pulsar*.

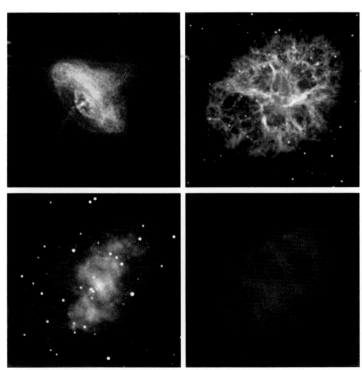

Astronomers soon figured out that a neutron star can have a very intense magnetic field millions of times stronger than anything that can be produced on Earth. Charged particles pour out of the star's magnetic poles, producing powerful beams of radiation (usually radio waves, but they can be X rays or light). As the pulsar spins, its rays are like the beam of a turning lighthouse beacon. If the beam sweeps past Earth, astronomers record a pulse, or bleep, of radiation.

There might be at least 1 million active pulsars in our Milky Way galaxy alone. Not all neutron stars are pulsars, and only a small fraction of pulsars sweep their beam past our observational instruments on Earth.

A pulsar resides at the center of the Crab Nebula, seen here in false-color images taken at X-ray (top left), visible-light (top right), infrared (bottom left), and radio frequencies.

through Earth." Picture that small, dense star as if it were a whirling figure skater, rotating faster and faster, and then add enormous attractive power.

For the next 20 years, Zwicky and Baade collect evidence to back up their theory. Hardly anyone takes them seriously.

Meanwhile, in 1932, a Russian physicist, Lev Landau, has come to the same conclusion that Chandra reached earlier: There is an upper limit to the size of a star that can become a white dwarf. (Landau is unaware of Chandra's work, and his methods are different.) Landau says that if a star is heavier than 1.4 Suns, its core will keep collapsing past the white dwarf stage. He makes sure others know of his ideas. Landau wants the world to pay attention. His life is at stake. Russia's brutal dictator Joseph Stalin is killing intellectuals.

Landau has other theories: One of them is about what makes stars shine. He says it is because of *nuclear burning*, which is another term for fusion. He is right, although he hasn't figured out the details. Stalin isn't impressed; Landau is sent to a Soviet *gulag*. (He survives that prison, but it shatters him emotionally.)

Tiger! Tiger! Burning Bright (Wrote Poet William Blake)

A poet can make of burning whatever he wishes. But to a computer user, burn means to add data to a computer disk. To a chemist, burning is combustion. (Burn the toast black, and you'll see a chemical reaction that leaves carbon from the original carbohydrates, which means "carbon-plus-water.")

When a physicist says burn, she is talking of a nuclear change. In nuclear burning, the nuclei of atoms are fused together, becoming more massive nuclei of a different element. Stars gradually combine hydrogen atoms (with one proton) into helium atoms (with two protons) and then helium into heavier elements—oxygen and carbon. A supernova explosion can create very heavy elements—gold and uranium—on the spot.

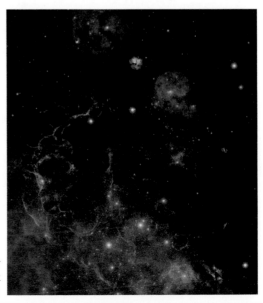

The beautiful Vela supernova remnant is roughly 11,000 years old. Who on Earth saw an extra-bright star appear, dazzle, and then fade in the night sky?

A former Leningrad classmate of Landau, tall, blond, wildly creative George Gamow, takes the next step in the starry drama. He focuses on fusion. Then Dutch-Austrian-German Fritz Houtermans and Welsh Robert d'Escourt Atkinson, inspired by Gamow's work, write a paper describing how the fusion of hydrogen nuclei into helium nuclei allows a star to shine for billions of years. Later, Houtermans writes of a walk "with a pretty girl" (who will become his wife). "As soon as it grew dark the stars came out, one after another, in all their splendor. 'Don't they shine beautifully?' cried my companion. But I simply stuck my chest out and said proudly: 'I've known since yesterday why it is that they shine.'" (His girlfriend laughed.)

Houtermans has only part of the star-shine picture. This is the late 1930s, and the world is in chaos. War is coming, and many physicists are desperate to escape from Europe.

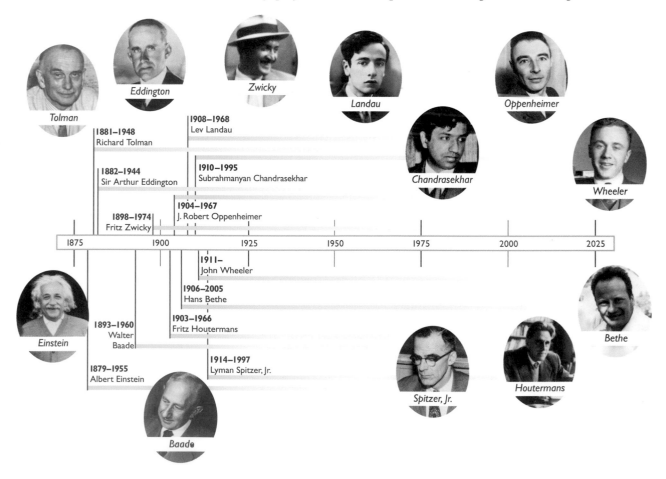

Tolman

Eddington

Zwicky

Landau

Oppenheimer

Chandrasekhar

Wheeler

1908–1968
Lev Landau

1881–1948
Richard Tolman

1910–1995
Subrahmanyan Chandrasekhar

1882–1944
Sir Arthur Eddington

1904–1967
J. Robert Oppenheimer

1898–1974
Fritz Zwicky

1875   1900   1925   1950   1975   2000   2025

1911–
John Wheeler

1906–2005
Hans Bethe

1903–1966
Fritz Houtermans

1893–1960
Walter Baade

1914–1997
Lyman Spitzer, Jr.

1879–1955
Albert Einstein

Einstein

Baade

Spitzer, Jr.

Houtermans

Bethe

In Washington, D.C., where things are relatively peaceful, astronomers and physicists meet to consider the stars and try to figure out why and how they shine. The astronomers realize that they have much to learn from the physicists and vice versa. Hans Bethe, who attends the meeting, is obsessed with the stars and with the fusion process. In 1939, Bethe publishes what has been called an "epoch-making paper" on the subject.

While all this is going on, Chandra is on the sidelines. He's writing books, he's a popular teacher, he's editing the *Astrophysical Journal*, and he is going on with his studies of the stars. He was hurt too badly by Eddington's rebuke to continue investigating white dwarfs. Still, he is aware that no one has come up with an answer to the question he posed back when he was a student at Cambridge. What happens to a collapsing star that has a mass greater than the limit of 1.4 Suns?

Elsewhere J. Robert Oppenheimer considers that question, teaming with Caltech's expert on general relativity, Richard Tolman. They read Lev Landau's paper. There is no way to contact Landau, who is in a Soviet gulag. In 1939, Oppenheimer does what Tolman suggests. Instead of using Newton's equations, as Landau has done, he uses general relativity's equations to consider

Lev Landau was sent to a Siberian gulag in 1938. Those slave labor camps were scattered throughout the Soviet Union. Here, prisoners build a canal in Uzbekistan in 1939. Novelist Aleksandr Solzhenitsyn described the brutal camps in *The Gulag Archipelago*.

the fate of massive stars. Oppenheimer and graduate student George Volkoff study stars that have reached old age and have burned most of their hydrogen. In those stars, the thermal (heat) pressure, which pushes out, can no longer balance the gravitational pressure, pushing inward.

This is what happens: The star begins to contract, becoming more and more compressed. The compression raises the temperature of the star, and, with its hydrogen mostly gone, its *helium* begins to burn, turning into carbon and heavier elements. When the amount of helium falls enough, the process repeats itself: contraction, raised

Chandra in the Sky

In 1999, NASA launched an X-ray observatory into space and held a competition to name it. Tyrel Johnson (a high school student in Laclede, Idaho) and Jatila van der Veen (a physics teacher in Camarillo, California) suggested calling it the Chandra X-Ray Observatory. That name was picked from 6,000 entries.

Taking pictures in the X-ray wavelength of the electromagnetic spectrum (rather than in visible light), the Chandra space telescope, operated by the Smithsonian Astrophysical Observatory at Cambridge, Massachusetts, was soon providing a never-seen-before view of the heavens. Subrahmanyan Chandrasekhar would have been pleased.

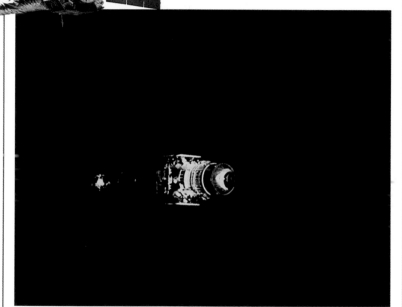

The Chandra X-Ray Observatory is a full-fledged spacecraft with thrusters, solar panels, and antennas to transmit data. The X-ray energy its telescope collects shows us a universe that's more active and violent than we imagined.

temperature, burning to heavier elements. Finally, at its greatest heat, iron nuclei form. **Iron, the most stable element, does not burn. Added energy is necessary to create nuclei heavier than iron.**

This is where the supermassive stars separate from their less massive cousins. Stars like our midsized Sun can't find added energy. They cool down and become white dwarfs.

But in a massive star, the scenario is different. When the fuel finally runs out—which can happen in a fraction of a second—the star caves in, creating enough force and pressure to heat its core to still-higher temperatures. It becomes so hot that it explodes. That blast, even in our vast universe, is hard to miss.

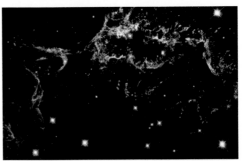

It is a supernova explosion, and all the products of the long cooking will get spewed into space. **It brings enough energy to create heavier-than-iron nuclei.** The swirling debris of this celestial wham-bang collects and condenses, eventually becoming new stars, planets, and more. **The heavy elements on our planet and in our bodies come from stellar explosions. We are made of star stuff.**

The Cassiopeia A supernova remnant is 10,000 light-years away, and so light from the exploding star took 10,000 years to reach Earth, finally arriving in 1667. Supernovas seem rare to us, but that's because the universe is so large. Our tiny part of it, the Milky Way galaxy, sees a supernova about once per century. Those stellar bombs go off about once per *second* in the greater universe.

By 1939, supernovas have become cutting-edge science. A few years earlier, when Zwicky connected neutron stars and explosive supernovas, no one paid attention. Now they do. Chandra's paper is rediscovered. Robert Millikan, Caltech's Nobel Prize–winning experimental physicist, says that supernovas provide "the birth cries of the elements." The stars and elements are falling into place. And another big insight is on its way.

Oppenheimer and his physics colleagues in the United States and around the globe figure out that **some stars that go past the white dwarf barrier** *also* **go past the neutron star stage.** Those stars become so small and so dense that they disappear from view, creating **deep gravitational wells in spacetime from which nothing can escape**, not even light. **Those wells will be called black holes.** The substance of a star has plunged to the center of a black hole, creating what is known as a singularity. No one has directly observed a **singularity**, because no light escapes from inside an outer boundary of the black hole that is now called the "event horizon."

The really massive explosion that leads to a black hole is called a *hypernova.*

These are big ideas, but, for a while, they aren't going anywhere. Research on the heavens is closing down. The world is at war.

Singular Black Holes

In 1964…black holes were thought to be…holes in space, down which things can fall, out of which nothing can emerge. [But] one calculation after another, by more than a hundred physicists using Einstein's general relativity equations, had changed that picture. Now…black holes were regarded as…dynamical objects: A black hole should be able to spin, and as it spins it should create a tornado-like swirling motion in the curved spacetime around itself. Stored in that swirl should be enormous energies that nature might tap and use.
—Kip S. Thorne, American physicist, *Black Holes and Time Warps: Einstein's Outrageous Legacy*

Being torn apart at the center of a black hole is bad enough. But according to some calculations, you will not even make it to the center alive: Your atoms will be scrambled by violent, chaotic tidal forces some distance from the center—especially if you fall into a young black hole.
—Edwin F. Taylor and John Archibald Wheeler, *Exploring Black Holes: An Introduction to General Relativity*

When *supermassive* stars die, they don't stop (they can't stop) at the neutron star stage. But where do they end up? In the early decades of the twentieth century, no one knows.

You can't bring a star into the laboratory, so J. Robert Oppenheimer and Hartland Snyder do a thought experiment to try and answer the question, What happens when a really massive star collapses?

Thanks to $E = mc^2$, Oppenheimer knows that energy can change into mass. He figures out that the humongous energy (and pressure) unleashed when a gigantic star crunches into itself will turn into additional mass and density—which will add *more mass* to the dying star and therefore increase its rate of contraction, until you have a runaway collapse that

No one has ever seen a singularity, but general relativity predicts it is an incredibly small point. General relativity says the singularity has infinite density; quantum mechanics says that is impossible. Take your pick. Physicists have been unable to work themselves out of that dilemma.

makes the star so dense that it seems to swallow itself. **All the mass of the huge star—plus the added mass from the compression itself—ends up as a very tiny, very compact singularity at the base of a pit or well in spacetime.** (It is 1939, and the term "black hole" has not yet been coined, but that's what we're talking about.)

Oppenheimer and Snyder write a paper about their conjecture. They believe that a singularity must create a scrunch point in spacetime with extraordinary gravitational pull. Anything that gets too near the singularity is churned and tugged and finally falls across its horizon into a hole. General relativity tells Oppenheimer and Snyder that the hole is an infinitely deep well, and it gulps down everything in its vicinity—including stars, gas, and photons. All that feasting just adds to its mass. (While

Big galaxies have few young stars, according to data from the Galaxy Evolution Explorer orbiting telescope. Korean astronomer Sukyoung K. Yi believes the presence of supermassive black holes (that's one illustrated, above) stifles star birth.

A Singular Definition and *the* Problem in Today's Astrophysics

We think every black hole has a singularity at its center. What exactly is a singularity? No one knows in detail. Answering that question is a quest for the twenty-first century.

However, we do have a starting definition. Caution: It is difficult. So, don't fret if you don't get it at first (some people spend years on this).

A singularity is an infinitesimally small point in spacetime where gravity is so strong that matter is infinitely dense.

Infinitesimally small? Infinitely dense?

Suggestion: Think about that word *infinite*.

Now keep in mind: Mass curves spacetime. Start with that idea, and then consider this: At a singularity, general relativity says that the curvature of spacetime becomes infinite.

And that's the problem. General relativity predicts this infinite density, but quantum mechanics (thanks to Heisenberg) says that nothing, not even a single electron, can be crushed into an infinitesimal point! So we have a contradiction between two great physics theories that, so far, has not been resolved. This is one of the most important challenges facing science today. Solving it will almost certainly lead to a new scientific paradigm.

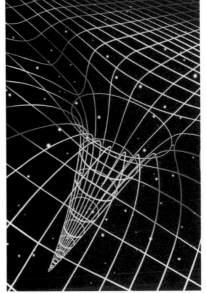

Here's another attempt to picture in two dimensions a black hole's four-dimensional curvature of spacetime.

the singularity itself is tiny, the entire structure inside the horizon can be huge, along with a large surrounding, spinning disk.)

Gravity pits in spacetime! Bizarre!

In September 1939, when the Oppenheimer/Snyder paper is published, hardly anyone takes it seriously. The world is choosing sides in preparation for war. The following year, Germany, Italy, and Japan sign a military agreement calling themselves an "axis" and agreeing to go to war together. Then, on December 7, 1941, Pearl Harbor is attacked by Japan, and the United States joins what has become World War II. In June 1942, "Oppie" is asked to direct scientific research for the Manhattan Project and is soon busy making nuclear bombs. He sets aside his study of dead stars that transform themselves into powerful, dense objects.

It is the late 1950s before the subject is picked up again, this time by Princeton professor John Archibald Wheeler, who starts thinking seriously about those **singularities** and the structure that embraces them.

Wheeler looks more businessman than physics professor. Solidly built, he wears a coat and tie and has courtly manners. But the man is full of surprises: He keeps firecrackers in a chest near his desk, and when he has something to celebrate—usually a new physics insight— he ignites them, sometimes in academic hallways.

Originally a nuclear physicist, Wheeler studied with Niels Bohr in Copenhagen, and, in 1939, he and Bohr wrote a classic paper on the theory of nuclear fission. A member of the Manhattan Project, Wheeler worked on the reactors that created plutonium for one of the nuclear bombs.

After the war, Wheeler turns his attention to general relativity. Einstein is living in Princeton, which is handy for Wheeler, who, like Einstein, is searching for nothing less than the universe's big secret: How do quantum theory and relativity work in harmony? Those two theories must harmonize, but no one knows how. Studying the fate of the largest stars seems a good place to look for answers to questions about the origin and destiny of the universe.

Here's John Archibald Wheeler in Copenhagen, where Niels Bohr was his mentor. Later, Wheeler was a mentor to generations of young physicists in the United States.

Wheeler keeps a box of aphorisms on his desk (some are his, some are from others). One of his says, "We will first understand how simple the universe is when we understand how strange time is."

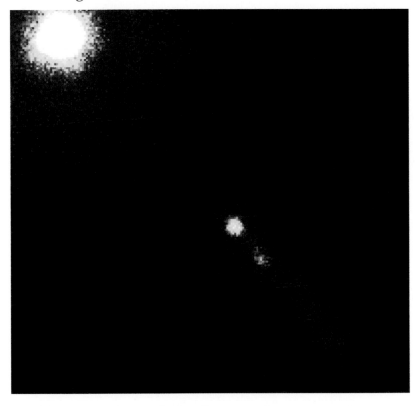

The universe came into being in a Big Bang, before which, Einstein's theory instructs us, there was no before.
—John Archibald Wheeler

What causes violent jets of gas to blast away from quasars at close to the speed of light? Why are these gas clouds lumpy instead of in a continuous stream? And what makes them slow down? The Chandra X-ray telescope is helping scientists investigate these mysteries and more.

The universe has no lack of strange things. In 1962, Dutch-American astronomer Maarten Schmidt uses the big telescope on Palomar Mountain in California to study an object in space called 3C 273. It is the most distant object known (at the time) and unusually bright. It has a star-like appearance, which means it must be a (relatively) small object, and so its brilliance is baffling. When he spots it, Schmidt goes home "in a state of disbelief" and tells his wife, "Something incredible happened today."

At first, neither Schmidt nor any other astronomer can explain this super-bright object. What can it be? It stays around; that means it isn't a supernova. Whatever it is, it needs a name. Astronomer Hong-yee Chiu comes up with *quasar*, short for quasi-stellar radio source.

Quasi means "resembling to some degree." A **QUASI-STELLAR OBJECT** (an updated term for quasi-stellar radio source) is not really a star, but it's sort of like one.

Putting Our Feet Down on the Moon

In May 1961, President John F. Kennedy announced that we would put a man on the Moon before the end of the decade. We were early. And we put two men there: astronauts Neil Armstrong and Edwin "Buzz" Aldrin.

Along with astronaut Michael Collins, they arrived in Moon orbit on July 19, 1969. The next day, Collins stayed in the *Apollo 11* space capsule while Armstrong and Aldrin descended in a small lunar lander named the Eagle. As Armstrong stepped onto the Moon's surface, he said the famous words: "That's one small step for a man, one giant leap for mankind."

The three-man *Apollo 11* crew (left) arrived at the Moon four days after their July 16, 1969, liftoff (second photo). While Michael Collins orbited overhead in the command module, Buzz Aldrin became the second man to set foot on the dusty surface (third photo), after Neil Armstrong's famous "giant leap for mankind."

That giant leap celebrated the marriage of science and technology that has helped define the modern world and has changed the lives of most of Earth's inhabitants. Where would we go next?

Buzz Aldrin, addressing Congress, said, "[S]ince we came in peace for all mankind, those footprints belong also to all people of the world.... The Apollo lesson is that national goals can be met where there is a strong enough will to do so.... The first step on the Moon was...a step toward our sister planets and ultimately toward the stars."

This is what Michael Collins had to say:

As we turned, the Earth and the Moon alternately appeared in our windows. We had our choice. We could look toward the Moon, toward Mars, toward our future in space—toward the new Indies—or we could look back toward the Earth, our home, with its problems spawned over more than a millennium of human occupancy.

We looked both ways. We saw both, and I think that is what our nation must do.

Soon, other quasars are spotted; their redshifts show that they are very far away. Astronomers realize that when they see a quasar they're looking at the universe as it was in the far, far distant past. Quasars appear star-like, they are small, but each sends off more energy than 100 giant galaxies. The nuclear reactions that power suns can't explain them, which means they aren't stars. So what does explain the extraordinary amounts of energy a quasar releases? No one is quite sure.

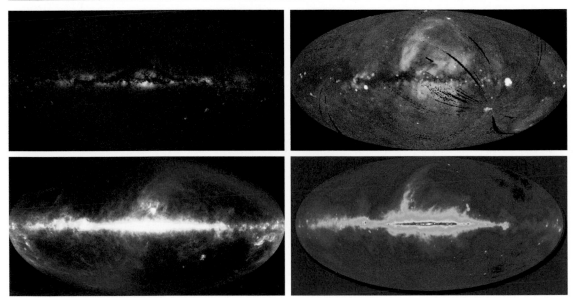

Quasars are just part of the growing excitement that new technology is bringing to the field. Telescopes that "see" radio waves from space, rather than just visible light, have been part of the astronomer's kit of tools since the late 1930s. Big-dish radio telescopes arrive in 1957. X-ray telescopes, carried above Earth's atmosphere by rockets, followed in the mid-1970s. Each frequency of electromagnetic waves records celestial objects in a unique way, adding to the expanding map of the heavens.

Meanwhile, John Wheeler has seized on the name "black hole" to describe the gravitational pits that hold the dense remains of collapsed stars. The name makes sense: Light can't escape from these holes—that's what makes them black.

"Why bother with those turbulent wells?" some physicists ask, believing black holes can never be spotted. Wheeler doesn't agree; he says we are "right within the most interesting expanding-and-collapsing system of all, the universe. To try to analyze its physics would not seem to be a meaningless occupation."

Ultraviolet telescopes, gamma-ray telescopes, and infrared telescopes are added to the astronomer's arsenal. By the 1970s, some of these telescopes are in orbit. A 1982 report on

Here's a good example of what we were missing in the days when telescopes were limited to visible-light frequencies and computers weren't around to digitally color-code the images. Today's new instruments have revealed stunningly diverse views of our Milky Way galaxy in visible-light (top left), sharp-resolution X-ray (top right), infrared (bottom left), and radio frequencies. The horizontal band through each image shows the plane of the galaxy, thick with stars, gas, and dust.

Remember, all electromagnetic waves travel at the same speed in a vacuum, but a single radio wave cycle (crest to crest or valley to valley) can stretch for miles; X-ray waves are less than one-millionth of an inch from crest to crest.

If a star gets close to a black hole, its gas will be attracted to the accretion disk around the hole. The gas spirals inward, and particles rub against one another, creating friction that boosts temperatures and radiation in the form of X rays. At the event horizon, X rays shoot off like fireworks. Since X rays are rare in space (compared to visible and UV light), they can be a sign of something unusual—like a black hole.

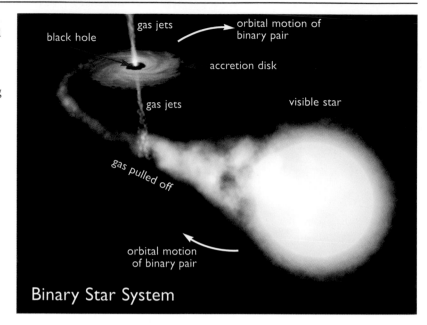

Binary Star System

astrophysics from the National Academies Press (in the U.S.) says, "[T]here have been only two periods in which our view of the Universe has been revolutionized within a single human lifetime. The first occurred three and a half centuries ago at the time of Galileo; the second is now under way."

Interest in black holes grows, although so far nobody has found one in the skies. Scientists looking for them focus on stars orbiting at high speeds around the center of active galaxies. What are they orbiting? There must be a powerful source of gravity at those centers: Could they be black holes with singularities at their cores? The doughnut-shaped area around which stars and other matter swirl is labeled the **accretion disk**. (With today's orbiting telescopes, we can see the swirling stars and the accretion disk. There's a photo on page 366.)

The **edge of the black hole** is called the **event horizon**. Just outside that edge, the black hole's gravitational influence isn't quite strong enough to suck in light rays. That's why objects are visible up to the horizon, but once inside the black hole, there's no trace of them; light can't escape from inside the horizon. (There's a slight exception to this; see box on page 363.)

By the last decades of the twentieth century, a number of physicists are signing up for black hole research, including Stephen Hawking, Kip Thorne, Roger Penrose, and Dennis Sciama. Hawking has ALS (named Lou Gehrig's disease after a famous American baseball player). ALS gradually paralyzes muscles and impairs speech (but not the mind); it seems to sharpen Hawking's focus and determination.

With his 1988 book *A Brief History of Time*, Stephen Hawking popularized high-level theories about cosmology and black holes. But, he cautions, "Any physical theory is always provisional . . . [because] you can never prove it. No matter how many times the results of experiments agree with some theory, you can never be sure that the next time the result will not contradict the theory. On the other hand, you can disprove a theory by finding even a single observation that disagrees with the predictions of the theory."

A Leaky Radiator

Many scientists now believe that black holes are dynamic entities, not just dead-end holes in space from which nothing escapes. Stephen Hawking startled the scientific world when he said some radiation is emitted from some black holes—but very, very slowly. (The rate depends on the mass of the black hole. Small, primordial black holes, if they existed, have already evaporated.)

The leak involves particles and antiparticles created near the event horizon of the black hole. One of the duo falls into the hole while the other escapes and carries a bit of the black hole's energy with it. Those escaping particles are called "Hawking radiation."

How long would it take for a massive black hole to evaporate? A whole lot longer than 13.7 billion years (which is the age of our universe), says Hawking.

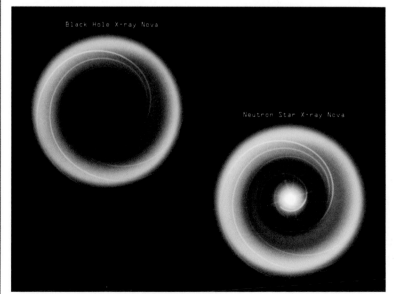

Gas at the center of a black hole is invisible because light can't escape (left illustration). In contrast, gas glows brightly when it hits a super-dense neutron star (right).

Yakov Zeldovich was born in Minsk, now the capital of Belarus, but moved to Russia at age 17. At the time of this photo (ca. 1950), he was a leader in developing thermonuclear weapons for the U.S.S.R.

"The same illness that had chained his body loosened his mind," writes Dennis Overbye in *Lonely Hearts of the Cosmos*. He quotes Hawking: "Before the illness set on I was very bored with life. I drank a fair amount.... It was really a rather pointless existence. When one's expectations are reduced to zero, one really appreciates everything that one does have." According to Overbye, "Hawking was the ideal black hole cosmonaut."

Wheeler (at Princeton) and Hawking (at Cambridge) are sure black holes exist; their physics and their equations tell them so. But how do you find one? Yakov Boris Zeldovich (1914–1987), a small, fiery, no-nonsense physicist, figures that out.

Zeldovich, at a center of cosmology in Moscow, understands that a spinning black hole must attract a swirl of cosmic dust, much like the swirl of water around a drain. That rotating disk (the accretion disk) would, by compression and friction, shoot off X rays. So, if you're searching for a

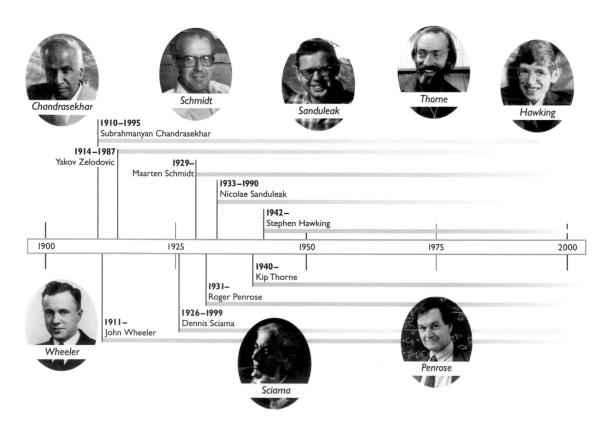

Chandrasekhar
Schmidt
Sanduleak
Thorne
Hawking

1910–1995
Subrahmanyan Chandrasekhar

1914–1987
Yakov Zelodovic

1929–
Maarten Schmidt

1933–1990
Nicolae Sanduleak

1942–
Stephen Hawking

1900   1925   1950   1975   2000

1940–
Kip Thorne

1931–
Roger Penrose

1911–
John Wheeler

1926–1999
Dennis Sciama

Wheeler
Sciama
Penrose

black hole, Zeldovich says, look for a star that seems to be orbiting nothing (which would be a black hole). You can measure the mass of a black hole by the speed and size of the orbit of the orbiting star (just use Kepler's laws). You should also look for a lot of X-ray activity.

Hawking goes to Russia and comes back with new equations and conjectures to aid in the hunt. Then optical (visible light) astronomers find a giant star, Cygnus X-1, which is 6,000 or so light-years away in the constellation Cygnus. Every 5.6 days, it whirls around an invisible object. Physicists figure out the mass of that unseen companion by using the speed and size of Cygnus's orbit. They realize it must be 10 times the mass of the Sun, and that is much too massive to be anything but a black hole.

Still, Stephen Hawking is skeptical. How can anyone be sure? "If it isn't a black hole, it has to be something even more exotic." So he bets Kip Thorne (Wheeler's student, who is now at Caltech) that Cygnus X-1 is not orbiting a black hole. That wager becomes famous. The loser will have to pay off with a subscription to a racy magazine.

Stephen Hawking puts black holes and their equations onto the blackboard that is inside his head. Then he retreats

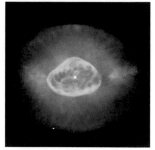

Besides Cygnus X-1 (below), the constellation Cygnus features this beautiful cat's eye of an object. It's a star surrounded by a (digitally colored) green "iris" of hot gas.

Cygnus is the Latin word for "swan." It's also a large constellation in the Northern Hemisphere, between the constellations Hercules and Pegasus.

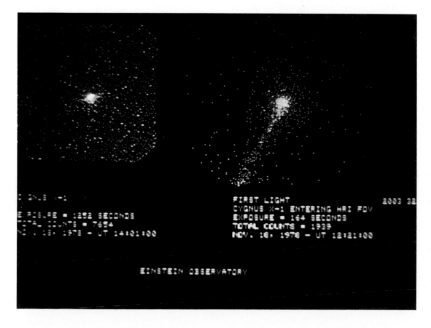

The first space object recorded by the Einstein Observatory spacecraft was Cygnus X-1, a suspected black hole. A blue supergiant companion star orbits Cygnus X-1 once every 5.6 days.

Supermassive Black Holes: A Breed Apart

In 1916, while he was serving as a German solider on the Russian front, Karl Schwarzschild took Einstein's equations for spacetime and applied them to a spherical mass. His solution implied that what we now call black holes might exist, but neither he nor Einstein realized this. J. Robert Oppenheimer showed that pits in space could come from the natural collapse of a massive spherical dust cloud. A few decades later, John Wheeler examined the idea, rejected it, but then was converted. Wheeler studied the process of gravitational collapse of real materials and used the name "black hole."

When astronomers finally pinpointed black holes in the sky, they found something startling: **Some black holes are so large that they could never have been a *single* star.** Kip Thorne explains:

[G]igantic black holes ... were never predicted to exist by any theorist. They are thousands or millions of times heavier than any star that any astronomer has ever seen in the sky, so they cannot possibly be created by the implosion of such stars. Any theorist predicting these gigantic holes would have tarnished his or her scientific reputation. The discovery of these holes was serendipity [sheer luck] in its purest form.

In this illustration, a supermassive black hole has a black, doughnut-shaped dust cloud. X rays can peer inside to the accretion disk (orange glow).

Gigantic black holes are too massive to have been formed by the accidental collision of regular stars or the vacuuming up of stars in a region of reasonable size in a galaxy. These combinations might form a really big black hole, but not one that is *supermassive*. That's big, really big, but it isn't **supermassive**. The term *supermassive* is reserved for **gargantuan black holes believed to exist in the centers of galaxies—including our own.** These are black holes whose mass can range from that of 1 million to 1 billion Suns.

How do those supermassive behemoths form? Astronomers are not yet sure. But the giants have made us reappraise quasars, which are now thought to be supermassive black holes in young galaxies. The quasars, now understood to be a type of active galactic nucleus (AGN), are spinning and acting like giant dynamos, sending out huge amounts of radiation. *Young galaxies?* Aren't quasars/AGNs billions of years old? Yes, the oldest we've spotted is 12 billion light-years away. We're not seeing it as it is today, but as it was 12 billion years ago—when it was a young galaxy in an infant universe.

Which comes first: the supermassive black hole or the galaxy? They probably grow together.

At the core of Galaxy NGC 4261, as observed by the Hubble and other telescopes, is an accretion disk (right) and high-energy gas jets (left).

The universe may
Be as great as they say.
But it wouldn't be missed
If it didn't exist.
—Piet Hein, (1905–1996),
   Danish poet, designer,
   and inventor

into his mind and has a breakthrough thought. It has been called the first successful combination of quantum mechanics and relativity, and it changes much that has been believed about black holes.

Hawking says a black hole is not just a celestial vacuum cleaner; it is a dynamo in space. A black hole can radiate energy, he says. The smaller the black hole, the greater the radiation. He sees black holes (in his mind) eating information, growing fat with it, and coughing some of it back up in a new form. To science writer Dennis Overbye, Hawking explains, "[I]f we send someone off to jump into a black hole neither he nor his constituent atoms will come back, but his mass-energy will come back."

After the Hubble Space Telescope is launched into orbit in 1990, astronomers find high-energy centers in every galaxy they study—including our own Milky Way galaxy. Those dynamic core regions have all the attributes of black holes. Is that what they are? Each of these giant black holes has a mass millions of times the mass of our Sun. Infrared space telescopes, like the Spitzer (launched in 2003), peer right through interstellar gases and dust into the center of a galaxy, finally confirming black holes. We learn that black holes come in a range of sizes, from tiny to supermassive. Our "picture" of the universe is growing.

As for the previously unexplainable **quasars, scientists come to believe that quasar light comes from the accretion disk of a spinning black hole gobbling gas and stellar fragments** from its surroundings. These starry materials do not go quietly to their fate; they rage and storm and shoot off photons as they spiral in toward the event horizon where compression creates sizzling heat and sends off huge jets of radiation. That's why a quasar outshines most galaxies.

Quasars are much too large to have originated as single stars. Some are so far away that their light has taken at least 2 billion years to reach us. So we see these quasars as they

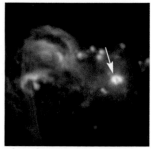

The white arrow points to a supermassive black hole at the exact center of the Milky Way. Red stands for radio energy, green for infrared, and blue for X ray.

Chandrasekhar came back to the world of dense stars in the 1970s, developing a theory of black holes based on general relativity. In 1983, at age 73, he published *The Mathematical Theory of Black Holes*, which is still a classic in the field. That same year he won the Nobel Prize in physics for his earlier work.

Heavenly Hindsight

In July 2005, astronomers using optical, radio, and X-ray telescopes on land and in orbit saw a massive stellar crash that had happened 2 billion years earlier! (Remember, when you look at the stars, you're seeing events that occurred in the past, because of the time it takes for starlight to reach us.)

The 2-billion-year-old collision was a big bang, not the Big Bang. Astronomers were alerted to it when they tracked an unexpected shower of gamma rays, the most energetic form of electromagnetic radiation. The origin of gamma-ray bursts (GRBs) had been a mystery. It now seemed clear that GRBs are products of huge celestial explosions that keep exploding like long fireworks displays. GRBs are messengers of black-hole formation.

A space illustrator interpreted a gamma-ray burst (GRB) recorded by NASA's High-Energy Transient Explorer (HETE). Three days later, the Chandra telescope observed an X-ray afterglow (lower right corner inset), which helped pinpoint the location in the outskirts of a spiral galaxy 2 billion light-years away.

"This is the real deal," said Donald Lamb of the University of Chicago in a *New York Times* article. He was understating the excitement of astrophysicists, who had just tracked a burst of high-energy radiation brighter than 1 million billion Suns.

Physicists expect to find that a supermassive black hole came out of that colossal fender bender. Did gravity waves, ripples in spacetime, come from this heavenly wham-bang? A new detector called LIGO is looking (see the next chapter).

NASA's Swift space telescope can capture gamma-ray bursts (GRBs), as illustrated at left, in X-ray and ultraviolet/optical frequencies at the same time, giving us a varied glimpse of black holes being born. A GRB is released during the hypernova explosion and collapse of a star into a black hole. The inset is Swift's official patch and sticker.

were 2 billion years ago. Quasars are snapshots from the young universe when galaxies were just being formed.

Here is physicists Neil deGrasse Tyson, Charles Liu, and Robert Irion's take on quasars (from a book titled *One Universe*): "Quasars are just early chapters in the lives of the cores of ordinary galaxies. As the central black hole ages and grows larger, the quasar becomes less hyperactive— and the galaxy around it looks more and more like a normal quiet galaxy. If that's true, supermassive black holes should be common.... A growing set of data supports this notion."

Astrophysicists call gigantic black holes "AGNs" (active galactic nuclei). Quasars are one kind of AGN. Are there any nearby? There don't seem to be. In the cores of galaxies fewer than 1 billion light-years away, supermassive black holes lick their lips as they swallow nearby stars. But none match the distant quasars in size and ferocity—which tells us that the universe was more energetic when it was young.

As for Hawking's bet with Kip Thorne? Stephen Hawking finally agrees that he has lost the bet: Cygnus X-1 is orbiting a black hole. Thorne gets his magazine.

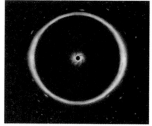

The intense gravitation of a black hole curves the light passing by it, creating a gravitational lens. In this artwork, light from a distant galaxy hidden behind a black hole (center) is distorted into what's called an "Einstein ring." Bent light from a gravitational lens sometimes appears as multiple images.

Here's an artist's view of a supermassive black hole snacking on the remnants of a Sun-like star, based on an event observed by the Galaxy Evolution Explorer. The yellow star (far left) stretched apart into a yellow blob (center) and then broke into pieces, which spiraled toward the core. As the stellar crumbs heated up, they radiated high-energy light before disappearing inside the event horizon.

Gravity Waves?

[A]ccording to general relativity, when an object with mass oscillates, gravitational waves must be emitted.... However, they carry unimaginably small quantities of energy, and are therefore very difficult to detect.

—Mauro Dardo, Italian physics professor, *Nobel Laureates and Twentieth-Century Physics*

[G]ravity waves are generated whenever space is fiercely disturbed. These waves are not really traveling *through* space, say in the manner a light wave propagates; rather, they are an agitation of space itself, an effect that can serve as a powerful probe.

—Marcia Bartusiak, American science writer and author, *Einstein's Unfinished Symphony*

An aerial view of the LIGO Livingston facility, stretching across a Louisiana forest, puts the grand scale of the experiment in perspective.

Deep in a pine forest in Livingston, Louisiana, not far from Baton Rouge and the Mississippi River, engineers have laid out two steel pipes, each 1.2 meters (4 feet) in diameter and 4 kilometers (2.5 miles) long. The pipes come together to form a giant L. Nothing is inside the tubes—literally. The air has been pumped out, leaving a vacuum, like interstellar space.

At the corner of the L, a beam splitter takes an entering laser beam and divides it between the two arms. Mirrored weights hang at the end of each arm and at the joint where they come together. Those weights are free to move horizontally. The precision laser beams bounce back and forth between the mirrored weights many times. Then, because the two arms are identical in length, the two beams can be recombined in such a way that the peak of one wave adds to the valley of the other wave, leading to a cancellation: a zero signal at the detector. That process

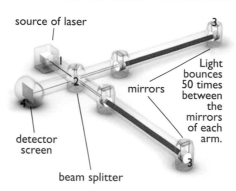

source of laser

Light bounces 50 times between the mirrors of each arm.

mirrors

detector screen

beam splitter

of recombining is called "interference." If one arm should stretch, or change by even the slightest amount, the two signals will no longer exactly cancel, and a photodetector will pick up a small signal, noting a change in the interference pattern.

The steel L is an interferometer similar in principle to the one that Michelson and Morley used in their famous experiment (chapter 4), except this one is giant-sized and uses lasers instead of regular beams of light. As for the stretch or change this device is designed to track? It should be caused by a gravitational wave: a ripple in spacetime.

This is LIGO (rhymes with "I go"), or the Laser Interferometer Gravitational-Wave Observatory, and it includes an identical interferometer 3,200 kilometers (about 2,000 miles) away at the Hanford Nuclear Reservation near Richland, Washington. Why two sites? It greatly reduces the possibility of errors caused by local disturbances; a signal detected at both locations will not be due to a truck passing near one of them.

Both sites in Louisiana and Washington are under the wings of the California Institute of Technology (Caltech), the Massachusetts Institute of Technology (MIT), and the National Science Foundation (NSF). In addition, the international LIGO Scientific Collaboration (LSC) includes some 40 institutions and gravitational-wave observatories in Italy, Germany, Japan, and Australia.

LIGO splits a laser beam (1) at the corner of the L (2), sending light down to mirrors on each arm (3). If a gravity wave changes the length of an arm, some of the light is diverted to a photodetector (4). Adam Frank wrote in *Astronomy* magazine, "The expected changes in length are... equivalent to seeing Saturn move closer to the Sun by the diameter of a single hydrogen atom."

An interferometer is any device that uses an interference pattern, which is a combination of two or more waves, to make a measurement. Here, the vibration of a membrane in headphones—a sound— causes interference between laser beams. A computer plots and displays the membrane's movement. Goggles protect the researcher's eyes from the lasers.

LIGO is a worldwide scientific collaboration. A French/Italian interferometer, VIRGO, is near Pisa, Italy; GEO600, a British/German effort, is in Hanover, Germany; TAMA is near Tokyo, Japan; and AIGO is close to Perth, Australia.

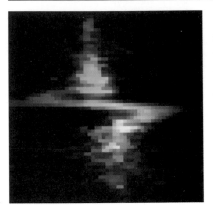

Like EM waves, gravity waves probably can't escape a black hole, so we'll continue to rely instead on indirect evidence such as this zigzag spectrum (above) at the center of galaxy M84. The shift indicates rotational movement of stars consistent with a supermassive black hole. Without a black hole, there'd be no zig or zag—just a vertical line.

PLASMA is a soup of electrons and the ionized atoms from which they came, produced at very high temperatures (as in the stars). The early universe, right after the Big Bang, was a hot plasma. Important point: Gravitational waves can travel through a plasma, but EM waves cannot. So if we want to learn about our cosmic beginnings, tracking gravitational waves is important. Those waves can also tell us what happens in the center of stars, where an envelope of plasma blocks EM waves.

If gravitational waves travel at the speed of light, as Einstein's formulas say they do, then the delay in reception at LIGO detectors at various locations on Earth will give information about the direction from which a gravitational wave arrives.

Why are we searching for gravity waves? And is it worth the effort?

In the past, most of the information we had about the cosmos came to us by way of visible light. The portrait that visible light paints is serene. It is of a universe filled with stars that twinkle and planets that move in predictable orbits.

In the 1950s, we began examining the heavens with radio and X-ray telescopes, followed by microwave, gamma ray, and infrared telescopes. Each of these electromagnetic eyes perceives a different part of the spectrum; they quickly shattered the picture of a tranquil heavens. The universe they bring us is violent, energetic, and fast-changing. Still, we've only been seeing the EM spectrum, and there's more to the universe than that.

Most electromagnetic waves come from regions of relatively weak gravity, like the surfaces of stars. We know that there are celestial regions of intense gravity: near a black hole, in areas where black holes collide, at the birthing of a neutron star, and at the Big Bang itself. We don't know much about those events and regions, and EM waves aren't bringing us detailed information.

That's a big reason for tracking gravity waves. We want to learn about the places where gravitation is so strong that Newton's equations fail and Einstein's are necessary, where matter and spacetime vibrate and swirl at close to the speed of light. In those regions of high gravity, heavy layers of matter block electromagnetic waves, as do plasmas where high temperature drives electrons out of atoms. But such barriers *don't* block gravity waves. If we can track those waves of gravitation, it should give us a very new picture of the universe.

According to general relativity, disturbances in the cosmos must ripple the fabric of spacetime itself. A cosmic collision should cause spacetime's fabric to rock and roil, which will send off large gravitational waves. (Imagine throwing a pebble into a pool. Watch the ripples circle out from the spot where the pebble hits the water. Gravity waves are believed to do a similar thing to space.)

Rock and **ROIL**? No, that's not a mistake. *Roil* means to be turbulent or agitated. It comes from an old French word that describes what happens when you muddy water by disturbing sediment.

LIGO's goal is to directly detect those waves. The search is difficult; by the time gravitational waves reach Earth-bound detectors, they are very slight cosmic shudders—so slight that even our most sensitive measuring equipment hasn't yet been able to discover them. But interferometers, like LIGO, are expected to do so.

Gravity waves ripple outward in all directions from a source, just as water waves do.

So where do we look for sources of gravitational waves? We search the skies for scenes of cosmic violence. When a star burns out and becomes a supernova, it does so in a blazing explosion. When two black holes collapse into each other, the force of that mating makes the universe tremble. When a black hole collides with a neutron star, it creates violent gamma-ray bursts. As for the greatest of all cosmic events—the Big Bang—if we can find a remnant of that blast, we will not only track gravitational waves, we'll have direct information about what happened in the universe's first second. Scientists are hoping that LIGO will detect gravity waves from these awesome events—even billions of years after they happened.

Slight cosmic shudders? Yes, very slight. According to Einstein's often-tested field equations, if two black holes with the mass of 10 Suns are 1 billion light-years away, and they wham into each other, the gravitational waves that reach Earth will displace the oceans by just 10 times the diameter of an atomic nucleus!

Neanderthals and *Homo sapiens sapiens* (yes, *sapiens* is doubled in our official name) were both walking the Earth close to 170,000 years ago when a brilliant star, now known as Sanduleak, exploded in the Large Magellanic Cloud, a galaxy that is our neighbor. Sanduleak was a supernova—a dying star. Stars like Sanduleak do not go gently. They rage, expand, and burst forth in dramatic, luminous death throes.

Explosions in the heavens are not unusual, but those that

we humans have been able to see with our naked eyes are rare. Besides, it takes time for the action to reach us. The hominids and woolly mammoths on Earth when Sanduleak blew apart didn't see it happen. The Large Magellanic Cloud is in our celestial neighborhood, and yet it is still 170,000 light-years away. So that's how long we earthlings had to wait before we could see Sanduleak detonate, which we finally did in 1987.

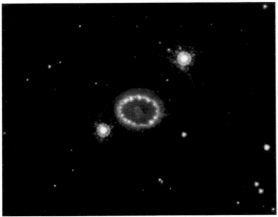

Sanduleak got its name 17 years earlier, in 1970, when the Romanian-American astronomer Nicolae Sanduleak (1933–1990) listed it as a blue supergiant. He didn't realize he'd be lucky enough to see it blow up. Here's what's left, as seen through Chandra's X-ray eyes.

In that year, light from Sanduleak's wham-bang made it to Earth and could be seen with unaided eyes (if you were looking and knew the skies). By then, the globe was peppered with telescopes, and back-porch astronomers and observatories could turn their scopes on that long-gone supernova and see far more than any human had seen in the aeons before there were telescopes. Astronomers took precise measurements and made notes of the electromagnetic waves that had at last reached our planet from the ancient supernova.

Now, what about gravity waves? According to Einstein's theory, when Sanduleak blew itself apart, it must have created waves of gravity. Science writer Marcia Bartusiak describes what else happened: "A fraction of a second before the detonation, the core of the star had been suddenly compressed into a compact ball some 10 miles wide, an incredibly dense mass in which a thimbleful of matter weighs up to 500 million tons (roughly the combined weight of all humanity)."

In that awesome dying process—collapsing, contracting, giving birth to a dense neutron star—the supernova must have shaken spacetime itself and shaken it

Light beams are continually absorbed by cosmic debris—stars, gas clouds, and microscopic dust particles—as they roam the universe. Gravity waves, on the other hand, travel right through such obstacles freely, since they interact with matter so weakly. Thus, the gravity wave sky is expected to be vastly different than the one currently viewed by astronomers.... In addition, [gravity waves] will at last offer proof of Einstein's momentous mental achievement...that space-time is indeed a physical entity in its own right.
—Marcia Bartusiak, *Einstein's Unfinished Symphony*

When a Black Hole Collides

On July 9, 2005, NASA's High Energy Transient Explorer satellite (known as HETE, pronounced HET-ee) and the orbiting Chandra X-ray telescope, along with Hubble and some other satellites worldwide, watched what they believe was a neutron star whamming into a black hole. It was a mammoth collision. Astronomers had theorized this kind of titanic event, but this time they actually saw it happen.

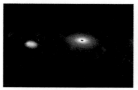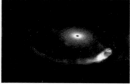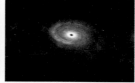

A NASA illustrator imagines what happens when a black hole attracts, stretches, crumbles, and devours a neutron star. We observe the feast as a short but intense gamma-ray burst (far right), followed by a brief afterglow of lower-frequency energy waves (as pictured on page 368). The new generation of gravitational-wave detectors—LIGO and its cousins—should be able to detect gravity waves from such events when those waves reach Earth.

As the neutron star's matter was sucked into the black hole, it detonated a colossal explosion, sending off jets of plasma (a soup of electrons and ionized atoms) and bursts of gamma rays. The collision had happened 2 billion years earlier. It had taken all that time for the information to travel—*at the speed of light*—from a faraway galaxy to our planet. That we can locate and identify these events is one of the wonders of modern astronomy.

"This is very good news for LIGO," said Dr. Stan Whitcomb of Caltech in a *New York Times* article, because such events should generate relatively strong gravitational waves.

hard. "The resulting ripples would have rushed from the dying star as if a giant cosmic pebble had been dropped into a spacetime pond," writes Bartusiak.

We didn't spot gravity waves from Sanduleak, but we're pretty sure they must have been formed. Gravitational waves agitate space itself on their journey through spacetime. But gravity's strength falls off quickly with distance; matter on Earth jiggles only slightly as the gravitational wave passes over it. Still, on reaching Earth, the **gravity waves** generated by Sanduleak must have **stretched and squeezed** (by a very tiny amount) **mountains and seas and land** before continuing on their journey through spacetime.

We missed that show because our gravitational-wave detectors were not yet ready. But that's the kind of event we now hope to measure.

This trio of spacecraft, collectively named LISA (Laser Interferometer Space Antenna), will trail behind Earth as they circle the Sun, as shown in this NASA illustration. By staying a few million miles apart in an equilateral triangle and beaming lasers at one another, they'll be able to measure gravity waves too weak for the much smaller, Earth-bound LIGOs to detect.

Along with gravity waves, there should exist the corresponding particle, called a graviton. Quantum theory tells us that waves are particles and particles are waves. So if gravity waves exist, so must gravitons. And many scientists are quite sure they do exist. The quantization of Albert Einstein's theory of general relativity says gravity waves must be out there. The challenge is to find them. (But even when we find gravity waves, we will not have pinned down gravitons, just as detecting radio waves does not explicitly detect photons of the corresponding energy.)

LIGO is designed to sense motion as small as one-thousandth of a Fermi (less than one-trillionth of the diameter of a human hair). That's impressive, but the Laser Interferometer Space Antenna (LISA), currently planned for launch in 2015, will do even better.

Three spacecraft will form the corners of a huge equilateral triangle, with laser beams from each spacecraft pointed at the other two. LISA's arms (the distance between spacecraft) will be *5 million* kilometers long, compared to LIGO-Louisiana's 4-kilometer arms. LISA's interferometer will be able to measure a change in length 1,000 times smaller than the diameter of a proton. LISA will be sensitive to gravitational waves of much lower frequencies than Earth-bound detectors, perhaps tracing emissions from two neutron stars near the beginning of their mutual death spiral or from eruptions in the early universe whose frequencies we can't predict.

We now map the universe by observing its electromagnetic energy. Light waves show us one thing, radio waves something else. Physicists are working to detect the gravitational world. That's the important reason for LIGO. It will give us a new and different picture of the universe.

Is there really a point to this?
Yes.
Einstein's theory of general relativity predicted gravity waves in 1916. They have yet to be confirmed directly. ("The worship of Einstein, it's the only reason we're here,"

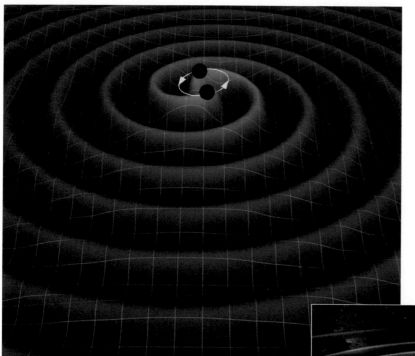

Two orbiting stars (below) or black holes (left) create a pattern of ripples in spacetime that, upon reaching Earth, can in theory be measured to reveal the distance and direction of the source and the shape of the masses that made them. The white dwarf stars belong to an interesting binary system called J0806. Because they lose orbital energy as they generate gravity waves (one of Einstein's predictions), they've locked each other in an ever-tightening death spiral. Eventually, they'll unite.

said an MIT scientist at LIGO, who may have had his tongue in his cheek.)

There has been indirect confirmation. In 1993, Princeton astrophysicists Russell Alan Hulse and Joseph Hooton Taylor Jr. won a Nobel Prize for finding two neutron stars orbiting each other and losing energy in just the amount predicted by the release of gravity waves.

If the LIGO scientists can detect and track gravitational waves from stellar quakes and cosmic ferocities—even billions of years after they happen—the information that brings should take us one more step up the ladder of understanding the cosmos. EM waves can only bring us back to a time a few hundred thousand years after the Big Bang. If there are gravitational waves from the birth of the universe, scientists believe they will tell us much about the first nanosecond after the Big Bang, about the actions of large bodies in the cosmos, about black holes, and about the future of the universe.

Here are some questions LIGO is expected to address:

- Do gravitational waves exist?
- Do they move at the speed of light?
- Do they displace the matter they travel through?
- What can they tell us about black holes and cosmic events such as supernovas and the Big Bang?

May the Interaction Be with You

The Four Interactions

Here they are, listed from strongest to weakest:

1. The **strong** nuclear force holds particles inside an atomic nucleus. The particle that mediates this force is called a meson, symbolized by π in the illustration below.
2. **Electromagnetism** keeps atoms intact and joins them into molecules. Opposite charges attract; like charges repel.
3. The **weak** nuclear force controls radioactive decay. The trigger is the exchange of a particle called a W boson. The ν (Greek *nu*) stands for neutrino. On the right side of the illustration, that's e⁻, an electron, leaving the nucleus.
4. **Gravity** deals with matter in bulk.

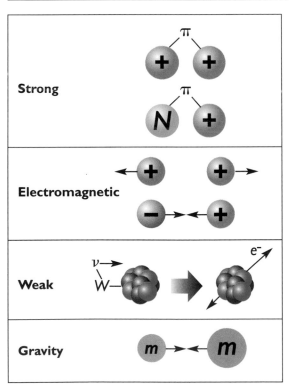

Strong

Electromagnetic

Weak

Gravity

Here are the four fundamental interactions (or forces) listed from strongest (top) to weakest. (In the top box N stands for neutron. In the bottom box *m* stands for mass.)

Finally, more than 2,400 years after Aristotle, we are able to talk of four forces that seem to be the handles that turn the universe. We had to look inside the nucleus of an atom before we could find two of them.

Two old pals—**gravitation and electromagnetism**—are joined by two named (without much imagination) the **strong nuclear force** and the **weak nuclear force**. While the term *forces* is widely used, if you're really savvy you'll call them *interactions*, because they tell us how matter interacts with other matter.

Each interaction is carried by tiny particles: electromagnetism by photons, the strong nuclear force by **mesons (with some exceptions),** the weak nuclear force by particles labeled simply W and Z, and gravity by gravitons. (Gravitons have never actually been detected.)

Gravitation is caused by mass

You'll notice that I use *force* and *interaction* interchangeably. Purists have reason to growl; they are happiest with interaction alone. You can decide for yourself how you want to use those words.

curving spacetime and also responding to that curvature. "Spacetime tells mass how to move; mass tells spacetime how to curve" is a favorite saying of the general relativity physicist John Wheeler.

Gravity seems powerful because it only attracts (it never repels); it is cumulative, which means it adds up—if you have lots of mass, you have lots of gravity; and the gravitational attraction is believed to extend forever, to infinity. Yet gravity is the weakest of the four interactions—by far (see box). Each time you pick up a nail, you are overpowering the gravitational pull of the entire Earth.

James Clerk Maxwell's equations united electricity and magnetism into a single force: **electromagnetism**, which is much stronger than gravity. Electromagnetism holds electrons in an atom and holds atoms together in molecules. But it doesn't add up the way gravity does: Electricity's positive and negative charges and magnetism's north and south poles can cancel each other out.

Radio waves, light, X rays, and gamma rays are different frequencies of electromagnetism. Electromagnetism on a nanoscale is the reach of quantum electrodynamics (QED), one of science's big success stories (see chapter 27). QED tells us how charged particles, such as

electrons, interact with photons inside atoms. With that knowledge, we can explain things like the magnetic properties of ultra-cool superconductors and what happens when hot, charged particles collide.

Since similar electric charges repel each other, physicists couldn't understand why the positive protons packed into nuclei didn't blow one another away. There had to be something holding them together. For decades they assumed the existence of what we now call the **nuclear force**, a force of attraction greater than the electric repulsion between protons. This force was carried by particles called mesons. For typical separation between protons in a nucleus, the nuclear force is a hundred or so times as strong as the electric force.

More recently we have understood that protons and neutrons are themselves made of truly elementary particles called quarks. Quarks interact through a force carried by gluons (great name, right?). The gluon forces are the true source of the strong force that holds the nucleus together.

The strong nuclear force is 137 times stronger than electromagnetism, but it only reaches within the nucleus. *It's a very short-range force.* If its range were greater, it would pull on other nuclei and scrunch atoms out

A Gravitational Tutorial

1. Isaac Newton said that gravity is a force acting instantaneously at a distance.

2. Einstein said gravity might seem like a force acting at a distance, but it is actually objects responding to the local curvature of spacetime. A large mass, like the Sun or Earth, creates a deep dent (curvature) in the spacetime near it. (That curvature is more gradual at greater distance from the mass.) A small mass caught in that deep curvature, like a satellite or a moon, responds by moving straight.

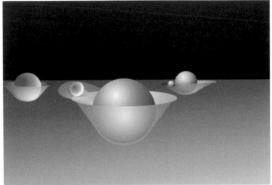

Masses warp the fabric of spacetime. The warps can cause two masses to fall around each other—to orbit.

 3. How can it go straight in curved spacetime? Look at a small enough region of Einstein's curved space, and it will appear almost flat. A satellite follows a straight path across that little bit of space. Think of a celestial patchwork quilt with squares that get smaller and smaller. The path of a particle is straight across each local tiny patch but jogs slightly as it moves from patch to patch, leading to an overall curved trajectory.

4. The new physics of gravitational lensing (see page 426) is showing us that the universe does indeed seem to be filled with a changing, rippling continuum of spacetime. The lensing itself, however, can take place even within static centers of gravitational attraction.

5. Ripples in spacetime can move through space as gravitational waves.

6. Waves are associated with particles, like the photons that carry electromagnetism, so physicists have hypothesized the existence of gravitons—particles that carry gravity. Gravitons have not been detected directly in part because the gravitational force is very, very weak.

7. Spacetime curvature is caused by mass (see number 2); $E = mc^2$ says that energy and mass are interchangeable. So, Earth moves around the Sun not because of a gravitational force from the Sun but because it is following a straight path in the curved spacetime created by the mass of the Sun.

Gravity waves don't travel instantly from one point to another. They fan out in all directions from a source.

of existence. Atomic nuclei are the size that they are, and no smaller, because of the range of the strong nuclear force.

The **weak nuclear force** (100 billion times weaker than electromagnetism, but stronger than gravity) has a range even smaller than the size of an atom's nucleus. Enrico Fermi discovered it in 1933 when he tried to understand why some atoms give off electrons and change into other elements—a process he called beta decay, later recognized as one form of radioactivity.

Besides radioactivity, the weak force is involved in nuclear fusion. The Sun gets its power from the weak nuclear force when it converts hydrogen into helium (not bad for a "weak" force).

In 1967, Americans Steven Weinberg and Sheldon Glashow, Pakistani Abdus Salam, and other physicists mathematically combined the weak force and electromagnetism into the **electroweak force**. So now, most physicists say there are only three interactions.

An important idea here is that all these interactions (even gravity) share the particle-wave duality that is fundamental to the universe.

How do the interactions relate to one another? To figure that out, scientists are trying to understand the earliest moments of the universe, when they are quite sure the forces were united. Putting gravity into a Theory of Everything (T.O.E.) or a Grand Unified Theory (G.U.T.) is one of current science's biggest quests.

Is there a superforce that includes gravitation? Most scientists think there is and that finding it will be the clue that solves a mystery. If we can figure out the secrets of the superforce, will intergalactic travel become part of the everyday world? Stay tuned.

There's no such thing as an intergalactic spacecraft that can travel near the speed of light to another star, but artists have no trouble imagining such a journey.

A Singular BANG with a Background

Things in the cosmos do not happen at random, but every event is governed by one of a small number of natural laws—laws that we can discover in our laboratories. Everything we see in the sky, like everything on earth, happens in a rational, orderly way.

—James Trefil, American physicist, *The Dark Side of the Universe*

The biggest misunderstanding about the Big Bang is that it began as a lump of matter somewhere in the void of space. It was not just matter that was created during the Big Bang. It was space and time that were created.

—Stephen Hawking, British physicist, quoted in *Stephen Hawking's Universe* by John Boslough

[I]n 1965, scientists had their first direct glimpse of the face of creation. The cosmic background radiation, the faint hiss of microwaves from all regions of the sky, is a snapshot of the infancy of our universe.

—Charles Seife, American science writer, *Alpha and Omega: The Search for the Beginning and End of the Universe*

It is 1948, just after World War II, when George Gamow, the Russian-American physicist at George Washington University, and Ralph Alpher, at Johns Hopkins University, write an important paper. Then they get Hans Bethe, at Cornell, to agree to add his name to it.

The names on the paper—Alpher, Bethe, and Gamow—become a pun on the first letters of the Greek alphabet: alpha, beta, and gamma (α, β, and γ).

Gamow, known as a jokester, releases the paper on April Fools' Day; maybe that's why no one

Not a Myth

Pick up almost any serious science book about the origins of the universe, and it will have a sentence like this one from *One Universe* (by physicists Neil deGrasse Tyson, Charles Liu, and Robert Irion): "Of all the stories yet devised about the birth of the universe, the Big Bang is best supported by scientific evidence."

Supported by scientific evidence? How can there be evidence of an event that happened 13.7 billion years ago?

It seems there is. This chapter and the next tell a bit of the story.

takes it too seriously. As it happens, the trio is heading in a productive direction. They are among the first to begin an analysis that will allow science to picture the first moments after the universe's creation.

They have the background for it. Gamow was a student of Aleksandr Friedmann at the University of St. Petersburg (in Russia), which means he was trained by one of the first scientists to understand that the universe is expanding. Alpher, while working as a graduate student under Gamow (in the United States), investigated the way atoms can be built from simple particles. And Bethe (also in the United States) has come up with a theory on the steps needed for energy and chemical elements to be created in the stars.

Research for the Manhattan Project (building nuclear bombs) has added immensely to their understanding of fusion and the way particles combine.

Given that start, these physicists begin to paint a portrait of the universe at the moment of the Big Bang.

As scientific sleuths they are astonishingly farsighted, providing a foundation for later physicists to build on.

They understand that the newborn universe must have been awesomely hot—a seething, dense, gassy plasma of tiny subatomic particles (quarks and electrons) so scorching and fast moving that particles wham into one another and nothing holds together.

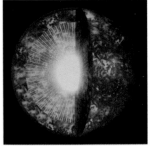

The ancients believed the Sun was eternal. But in 1853, a scientist named Hermann von Helmholtz (1821–1894) realized that some kind of burning is going on in stars, and that they won't last forever. Hans Bethe (1906–2005), a German-American physicist at Cornell University, clarified the burning process—nuclear fusion—and won a Nobel Prize for it in 1967.

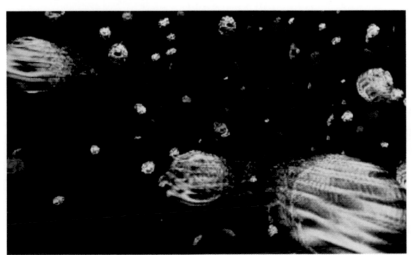

At its birth, the universe was trillions of degrees hot. Here's a color-coded inventory of ingredients in the plasma "soup" less than one-millionth of a second after the Big Bang: quarks (red, green, dark blue), antiquarks (light blue, pink, yellow), electrons and bosons (gray), and gluons (other colors). One second later, the temperature cooled to 10 billion degrees Celsius (still much hotter than the hottest star today), and the quarks combined into protons and neutrons.

A **BARYON** is a particle of matter, like a neutron or a proton, built from three quarks.

The three quarks in a baryon are held together by the strong nuclear force. The fuzzy cloud stands for the exchange of virtual gluons, which mediates the strong nuclear force.

neutron

proton

About 98 percent of the normal matter in the universe is still hydrogen and helium. Those two elements (along with a trace of lithium) were created in the Big Bang. Then came beryllium and boron, produced by cosmic rays. Heavy elements, from carbon to uranium, were manufactured by nuclear processes inside stars, and then in stellar explosions, they were sent off to be used by other stars.

deuterium
nucleus

But gases cool as they expand, and, right away, the universe starts expanding. Quarks combine in threesomes, becoming neutrons and protons (known as baryons). The interaction between the baryons and electrons creates photons (EM radiation) and a blinding light. (The book of Genesis, the first book of the Bible, describes it this way: "And God said, Let there be light; and there was light.")

Here's what Alpher and Gamow say in their famous paper: "Various nuclear species must have originated...as a consequence of a continuous building-up process arrested by a rapid expansion and cooling of the primordial matter."

They've got the basic idea. When they say "nuclear species," the physicists start with the simplest nucleus, that of hydrogen, the proton on which all other nuclei build.

As the temperature falls, protons and neutrons stick together as deuterium nuclei (a hydrogen isotope with one neutron and one proton). Then a hydrogen explosion (think of a hydrogen bomb) fuses protons and neutrons, creating helium nuclei (two protons and two neutrons). That explosion produces most of the helium that will ever exist.

Key to Diagrams

up quark

down quark

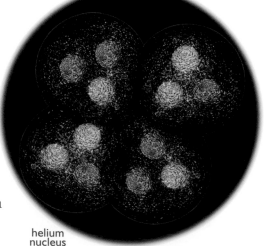
helium
nucleus

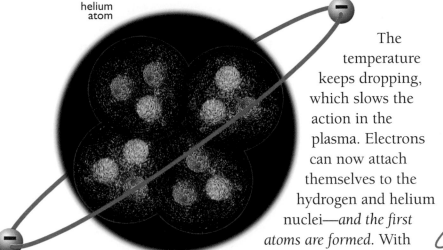

helium atom

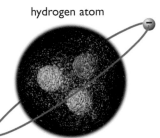

hydrogen atom

The temperature keeps dropping, which slows the action in the plasma. Electrons can now attach themselves to the hydrogen and helium nuclei—*and the first atoms are formed.* With electrons engaged—most are no longer zooming freely—photons are able to move in straight lines over long distances. The universe becomes transparent, as it is today. (This time of transition happens about 380,000 years after the Big Bang.)

But no stars shine. Slowly, very slowly, tiny variations in the universe's density begin to take shape as clusters of matter. Because of the lack of starlight, the universe is dark. (Later, this time will be called the "Dark Age.") The clusters keep condensing, contracting, and heating up until—thanks to nuclear processes—stars fire up. When stars ignite, the Dark Age is over.

Scientists call the time when atoms were first formed "recombination." But there was nothing "re" about it. It was the first time that electrons paired with nuclei (left and below).

Using Hubble telescope data, scientists plotted the density of normal matter in the universe (left, brighter means denser) and, for the first time, dark matter (page 390). Meanwhile, the Spitzer telescope detected an infrared glow that might come from the time when the very first stars formed (right). The gray spots are places where foreground stars were digitally removed.

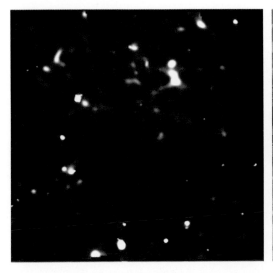

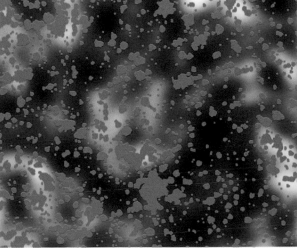

"Every minute of every day, the Earth is bombarded with a barrage of photons.... Most of these come from the Sun and the stars: they were emitted anywhere from today to a few thousand years ago," writes Michael D. Lemonick in *Echo of the Big Bang*. He explains that the photons from the cosmic microwave background—the photons that look like snow on some TV sets— are billions of years old, the oldest radiation in the universe.

"Is THAT IT? Is THAT THE BIG BANG?"

Bored with TV? Tune an antennaed (non-cable) TV to a channel that your set doesn't receive, and some of the static you see will be due to cosmic microwaves. You can watch the young cosmos not long after its birth!

In the 1940s, Alpher, Bethe, and Gamow know some but not all of this. They keep working hard to try to understand the early universe. Alpher writes a second prophetic paper in 1948, this one with a Johns Hopkins colleague, Robert Herman. Alpher and Herman suggest that when the universe leaves the dense plasma phase and photons are able to move freely, those photons stay around. They can't leave the universe; there is no other place for them to go.

Alpher and Herman believe photons from the primordial stew must still surround us! Just as embers glow long after a campfire has died out, some of that original radiation from the Big Bang must still be glowing.

Starting more than 13 billion years ago as short-wavelength, high-frequency, frantic gamma waves, the cosmic photons must have stretched with the universe as it expanded. Traveling across the electromagnetic (EM) spectrum, the photons became X rays, then ultraviolet light, then visible light, then infrared, and then long-wavelength, low-frequency microwaves. Alpher and Herman say those microwaves should now be glowing at us from all directions in the cosmos. They even figure out that their temperature today must be close to absolute zero, the coldest temperature possible.

If those cosmic microwave photons can be found, they will be an echo from the universe's infancy and *clear evidence of the Big Bang.*

But, in 1948, no one pays much attention to this theory. The technology of the time doesn't seem refined enough to find celestial microwaves. Gamow suggests that radio receivers could track the microwaves. But no one shows interest. Actual proof of the Big Bang! That seems like an impossible daydream. The Alpher/Bethe/Gamow and the Alpher/Herman papers are mostly forgotten.

Then, in the early 1960s, Princeton physicist Robert H. Dicke (1916–1997) comes up with the same insight: He says there must be microwaves in the cosmos that are leftovers from the Big Bang era.

Like Alpher, Herman, and Gamow before him, Dicke realizes that ancient photons—released from the hot plasma more than 13 billion years ago—must now be streaming (they move straight!) in the cold microwave region of the universal EM spectrum. Finding them will help prove Big Bang theory, Dicke says. He and some other Princeton researchers decide to set up experiments to track that cosmic microwave radiation. One of them, P. James E. Peebles, writes a paper about the predicted microwave remnants of the post–Big Bang plasma era.

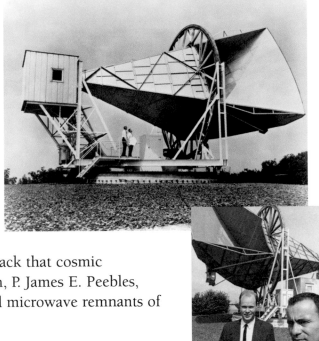

Meanwhile, half an hour's drive from Princeton, two young radio astronomers, Arno A. Penzias and Robert W. Wilson, are working at New Jersey's Bell Telephone Laboratories. Just out of graduate school, they are refitting a three-story, horn-shaped radio receiver designed to pick up microwave satellite signals.

The Nobel Prize for finding the CMB went to Wilson (left) and Penzias (in 1978), which left out Dicke, Peebles, Alpher, Herman, and Gamow. The two scientists removed "white dialectic material"—pigeon droppings—from their horn antennae (top) to make sure they collected "clean" data.

Penzias and Wilson are turning the receiver into a radio telescope; they plan to study the Milky Way's halo. But there is unexpected noise in their device—a bit like the static you get on an old radio—and it is messing up their work. They fine-tune their radio-wave receiver; they turn its direction; they clean off pigeon poop; they build a cooling shield. They do everything they can think of to make that hiss go away. No matter what they do, it stays. It is the same day and night and throughout the seasons (which means the Earth's rotation doesn't affect it, nor does Earth's orbit around the Sun). Talk about frustration! Penzias and Wilson are determined to find the source of that bothersome signal.

The word *radio* can be confusing because radio waves aren't sound waves. They're low-frequency electromagnetic waves (see page 23). Your bedside radio and big-dish receivers transform EM radio waves into sound.

Back to Jim Peebles at Princeton. He, too, is frustrated.

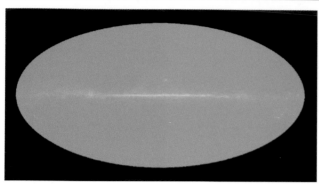

No matter which way astrophysicists point their radio telescopes, the cosmic microwave background (CMB) is there; it speaks to us from a realm more distant than the oldest stars. Here's a map of the microwave sky, ca. 1965.

Big Bang theory tells us: There is no edge and no center to the universe. Matter is distributed uniformly (on a large scale). To use a scientific word, the universe is *isotropic*. The CMB is also isotropic as measured to an accuracy of about one-half of a percent.

He can't get his paper published, and he can't get those cosmic microwaves out of his mind. In a speech at Johns Hopkins University in Baltimore, Maryland, Peebles talks about his problem.

Someone in the audience is a friend of a friend of Arno Penzias at Bell Labs. In a phone conversation, Penzias mentions the unexplained hiss in the radio waves to his friend, and that friend tells him about Peebles's talk. Penzias calls Dicke, and the Princeton physicists are soon on their way to Bell Labs. They all quickly realize that the signals Penzias and Wilson have been hearing on the radio telescope are those that Dicke and Peebles predicted. Without knowing it, Penzias and Wilson have been listening to microwaves left over from the universe's infancy.

What began as a photon sea within the hot plasma of the newborn cosmos (remember, that plasma was the whole universe—except for dark matter) has become cool radiation streaming through today's expanding universe. Penzias and Wilson have found a souvenir of the Big Bang!

The Penzias and Wilson experiment is soon repeated at observatories around the world. Then, in 1989, NASA launches an orbiting telescope: the Cosmic Background Explorer (COBE). Its goal is to study and measure those nearly 14-billion-year-old microwaves with a precision unavailable to land-based telescopes. John C. Mather (of NASA's Goddard Space Flight Center) and George F. Smoot (at Lawrence Berkeley National Laboratory) lead a team of more than 1,000 scientists, engineers, and technicians.

COBE's results are spectacular. The satellite not only measures the temperature of the microwaves (at 2.73 degrees above absolute zero on the Kelvin scale, it is very close to the 1948 Alpher/Herman prediction), COBE also finds that the cosmic microwave background has slight—very, very slight—spots where the temperature varies.

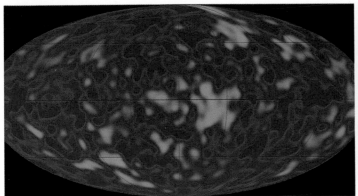

Those irregularities are thought to be the seeds from which stars and galaxies grew. "What we have found is evidence for the birth of the universe and its evolution," says George Smoot, announcing the findings in 1992. COBE's map of the skies, showing splotchy temperature variations (pinpointing future galaxies) is called the first "baby picture" of the universe. Smoot says, "It's like looking at an embryo that's a few hours old."

In 2006, Mather and Smoot win the Nobel Prize in physics for their discovery; they split 10 million Swedish kronor (about $1.37 million). The Royal Swedish Academy says, "The COBE results provided increased support for the Big Bang scenario."

By 2006, COBE has been replaced in space by WMAP, the Wilkinson Microwave Anisotropy Probe, another NASA satellite (this one is launched in 2001). Orbiting 1 million miles above Earth, WMAP is soon sending around-the-sky pictures (an Earth-bound telescope can't take pictures from

Compare these digitally colored maps of the CMB with the first CMB map on the opposite page. Data from the COBE spacecraft (above) of the 1990s and its early twenty-first-century successor, WMAP (below), sharpened our view. Why are the images ovals? Putting curved four-dimensional spacetime—or the curved three-dimensional Earth—on a two-dimensional page is impossible. It can't be done without distortion. So we pick the projection (there are many) that distorts the least amount for our purposes. These are called elliptical Mollweide projections.

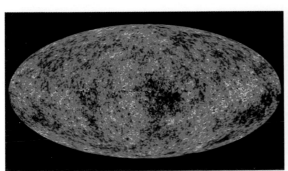

all directions because Earth gets in the way). Data from WMAP tells us the age of the universe: 13.7 billion years. It shows us that the geometry of the universe is essentially flat. And it informs us that what we think of as ordinary matter (baryonic, atom-producing matter) makes up only 4 percent of the universe. The rest is dark matter (20 percent), which does not emit or absorb light, and a mysterious dark energy (76 percent). (You'll find some thoughts on dark energy in chapter 46.)

WMAP keeps sending information on the very early superhot plasma of electrons, baryons, and photons. The photons carry the interactions that keep the cosmic soup hot. That roiling soup is under tremendous pressure, which means that it resists concentration or clumping. (Think of trying to compress a balloon between your hands; it pushes back.) Along with the hot plasma, there's cold dark matter. It stays separate from the electrons, baryons, and photons, although it occupies the same expanding space—the only

Why Is the CMB Still Around?

Why are we still surrounded by the cosmic microwave background (CMB), the leftover radiation from the Big Bang?

To understand, you have to realize that the Big Bang was not like a bomb explosion. No matter how big a bomb is, the explosion is still localized: Its energy spreads to surrounding regions.

But there was nothing local about the Big Bang; there was no "other" space; there was nowhere for the radiation to spread. As the universe expanded, all of space expanded, taking that radiation with it. So the CMB just kept stretching, along with spacetime, and now it still surrounds us.

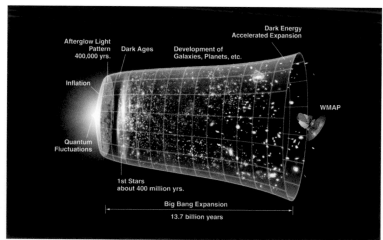

This chronology of the universe includes everything (as far as we know), starting with the Big Bang (far left). Our understanding of the fast-expansion period called inflation (coming up in the next chapter) is being fine-tuned by data from the WMAP spacecraft (far right), which maps the CMB (that green-blue disk here and on the previous page).

space there is. Dark matter is made of particles so small that they hardly interact with other particles except in the earliest, very dense universe. After expanding more than 1 billion times, that dark matter has cooled to near absolute zero.

Why Is Dark Matter Cold?

The quick answer is: No one is sure. This is a field being intensely investigated right now. No one has ever knowingly seen or touched dark matter, and so no one has taken its temperature.

Scientists deduce it must be cold from subtle premises about galaxy formation. Using those deductions, they have build substantial theories. The next step is to search for solid proof.

Here is some current thinking on dark matter: We live in an expanding universe, and expansion cools. We can deduce that the temperature of the expanding universe must have begun to drop right away. Physicists believe that, very quickly, the photon-baryon plasma and dark matter stopped interacting with each other. Because of their different interactions, the plasma cooled at one rate and dark matter at a faster rate.

At first that didn't make a big difference, but over the aeons it has. The universe has expanded more than 10 billion times since dark matter and plasma last interacted. Today, dark matter is believed to be very cold—close to absolute zero.

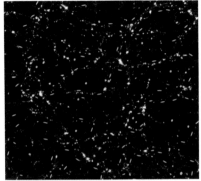

This computer model maps the presence of invisible dark matter (red). Its position is plotted by measuring how the matter bends light from distant galaxies (blue).

It's that cold dark matter that is first to thicken into the concentrations that will become galaxies. The plasma cools more slowly. It takes about 400,000 years for its temperature to drop enough so that electrons can combine with baryons to form atoms. When that happens, photons are released to fly through the universe's spacetime (and eventually reach WMAP's detectors). The concentrations of dark matter now attract the slowpoke atoms. From the clumps of dark matter, with added atomic matter, come galaxies, stars, and all the processes that will allow a small planet in a small solar system to harbor life and the technology that leads to WMAP.

Charles Bennett, the Johns Hopkins University physicist heading WMAP, says, "It amazes me that we can say anything about what transpired within the first trillionth of a second of the universe, but we can. We have never before been able to understand the infant universe with such precision."

This chapter and others near the end of this book deal with cosmology and today's quantum physics. These cutting-edge sciences hold ideas now being tested by very sophisticated emerging technologies. You can expect new knowledge to change some of the conclusions described here. So keep informed: This is your world.

Inflation? This Chapter Is *Not* About Economics!

Ten or twenty billion years ago, something happened—
the Big Bang, the event that began our universe.
Why it happened is the greatest mystery we know. That
it happened is reasonably clear.

—Carl Sagan (1934–1996), American astronomer, *Cosmos*.
When Sagan wrote this, we didn't have an exact date for
the Big Bang; now we do: about 13.7 billion years ago.

Antimatter can pop into existence out of thin air. If gamma-ray photons have sufficiently high energy, they can transform themselves into electron-positron pairs, thus converting all of their seriously large energy into a small amount of matter, in a process whose energy side fulfills Einstein's famous equation $E = mc^2$.

—Neil deGrasse Tyson and Donald Goldsmith, *Origins: Fourteen Billion Years of Cosmic Evolution*

I want to emphasize that science is not merely a collection of facts, but is instead an ongoing detective story, in which scientists passionately search for clues in the hope of unraveling the mysteries of the universe.

—Alan H. Guth, American physicist, *The Inflationary Universe*

A tiny acorn holds all the information necessary to create a huge oak.
And a kernel, much smaller than a subatomic particle, once held all the ingredients needed to produce everything that is in today's observed universe (which has a radius of almost 14 billion light-years).

Now, an acorn is nourished by soil and moisture from a world that already exists. But the cosmic kernel was different. It was all there was.

"IT WAS A LOT EASIER TO KEEP AN EYE ON THINGS BEFORE THE BIG BANG. EVERYTHING WAS ALL IN ONE PLACE THEN."

The kernel was all of space; it was all of time. As it grew bigger, so did spacetime.

From where had that cosmic seed come?

As modern science answers it: from nothing.

"Nothing can be created from nothing," said Lucretius, the Roman poet and philosopher, back in the first century B.C.E. But many of today's scientists don't agree with Lucretius. They don't know why that seed was created, but physicists speculate that **a universe could have come out of NOTHING.** Nothing doesn't mean a vacuum, because a vacuum is empty *space*. It means a state where even space doesn't exist. Nor does time. It really means beginning with nothing.

Physicists speak of a "false vacuum," a special condition during inflation. *False* means temporary, and *vacuum* means the lowest energy, so the false vacuum is a temporary condition that exists during a super-brief period. Our universe might have erupted from a false vacuum.

Cambridge University's Stephen Hawking says this: "Relativity and quantum mechanics allow matter to be created out of *energy* in the form of particle/antiparticle pairs." Hawking continues: "[W]here did the energy come from to create the matter? The answer is that it was *borrowed* from the gravitational energy of the universe." (These are my italics.)

There's an astonishing piece of information in that sentence. Here is the concept restated:

Gravitational fields, a kind of underlying fabric of the universe, are filled with potential energy; that potential can be borrowed to allow other things to happen.

Massachusetts Institute of Technology's Alan H. Guth explains, "[N]o energy is needed to produce a gravitational field, but instead energy is released when a gravitational field is created."

Here's more from Alan Guth:

> [T]he vast cosmos that we see around us could have originated as a vacuum fluctuation—essentially from

In this bubble chamber photo, energy in the form of gamma-ray photons (not seen) created two pairs of matter/antimatter particles (green/red tracks). One electron/positron track is curlicued; the other isn't. Why? A slower particle curves more in the chamber's magnetic field; this is one way to measure a particle's speed.

QUANTUM MECHANICS deals with subatomic particles. Quantum field theory applies quantum rules (chapter 18) to a field—an electromagnetic field or a gravitational field, for example.

Imagine mass at a point in space. The force experienced by the mass at that point, or any point, is one way to describe a gravitational field.

Alan Guth (below) predicted that quantum fluctuations would turn up as ripples in the CMB (cosmic microwave background). The COBE and WMAP satellites (chapter 43) have found and measured those ripples.

POSTDOCTORAL describes research, a researcher, or a university appointment after completion of a Ph.D. (Doctor of Philosophy) degree. Technically, it's an adjective, but "postdoc" (for short) is often used informally as a noun to describe a person working in a research position.

COSMOLOGY is the study of how the universe was created and how it has evolved. Astrophysics deals with the chemical and physical properties of the universe.

nothing at all—because the large positive energy of the masses in the universe can be counterbalanced by a corresponding amount of negative energy in the form of the gravitational field.

This isn't as difficult to understand as it may seem. The first law of thermodynamics says: Energy can't be created; it can just change form. *But there is a loophole.* The first law does allow energy to be created as long as it comes equally in positive and negative amounts and the total adds up to zero. Gravitation, which is negative energy, can pull positive energy (a fabric of photons, electrons, and quarks) out of quantum fields. So Earth and the stars and all the energy around us could come out of quantum fields. The first law allows it.

Now to a startling statement from Guth (who is a modest fellow): "[M]y work...led to a new theory of the origin of essentially all the matter and energy in the universe." (Those are my italics. And, yes, many physicists agree with him.)

Who is this fellow Guth?

In 1978, Alan Guth is a "postdoc" at Cornell University when he learns that Robert Dicke will be giving a lecture there. Guth is a particle physicist and not especially interested in Dicke's research interest, which is cosmology, but he knows about the Nobel Prize awarded a month earlier to Penzias and Wilson for discovering the CMB (cosmic microwave background), and he knows Dicke played a big role in that discovery. The CMB is a hot topic; Guth hopes Dicke will talk about it.

Dicke does more than that; he explains that the standard Big Bang theory is incomplete. Dicke talks about a problem called "flatness."

Physicists have discovered that the universe, overall, is flat. That means parallel lines will never meet. (Yes, the Earth and the stars are anything but flat, but cosmologists are thinking big. Spacetime's curves and dimples exist within a large-scale flat universe.) Classic Big Bang theory cannot explain that flatness.

Later, Guth learns of another challenge known as the

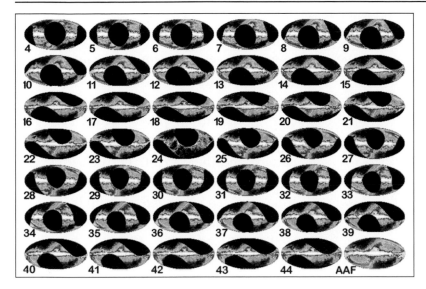

DIRBE (*Diffuse Infrared Background Experiment*), an instrument aboard the COBE satellite, recorded the cosmic *infrared* background (CIB), a higher frequency energy than the cosmic *microwave* background (CMB). The colorized maps show about half of the sky; numbers 4 through 44 each stand for one week's worth of data. As Earth (and COBE) journeyed around the Sun for 41 weeks, DIRBE took in the infrared view from different angles. The last image is an annual average of the data.

"horizon" problem. To a physicist, the distance that light could have traveled to an observer since the beginning of time (13.7 billion years ago) is the **horizon distance**. Since information—traveling from one particle to another—can go no faster than the speed of light, there is no way information can have made it beyond the horizon distance. But some information seems to have done that.

In measuring the CMB, astronomers have found a uniformity of temperature and density (averaged over a huge scale) that tells them there must have been a time when information was exchanged across the whole universe. How did that happen? Some places in the universe are beyond one another's horizons, and they were beyond one another's horizons soon after the Big Bang. So how do you explain the large-scale sameness of the universe? Is it just an amazing coincidence?

Alan Guth is intrigued. He tucks the Big Bang problems away in the back of his mind. Meanwhile, his friend Henry Tye persuades him to work on a bothersome cosmology problem. Tye has been researching magnetic particles thought to have existed in the early universe (but that no one has ever detected). Those particles, called monopoles, have either a north or a south pole but not both. Tye and Guth are soon concentrating on monopoles.

The **HORIZON DISTANCE** is sometimes called the particle horizon because it's the distance photons (the fastest particles) can have traveled to an observer since the creation of the universe.

Did rapid inflation give birth to separate "bubble" universes? This illustration explores the moment before that conceptual event, when several areas of inflation created domes in the early universe.

To learn more, read *The Inflationary Universe* by Alan H. Guth, a first-person account of the development of inflation theory.

In economics, inflation describes what happens when prices go up and the value of money goes down. Germany probably set the world inflation record after World War I: In just a few months, the price of a loaf of bread went from one German mark to millions of marks. People needed wheelbarrows of cash to buy basics. Yet that was a pittance compared to how fast and how much the universe inflated in the split second before the Big Bang.

Important ideas sometimes come not because someone is looking for them but because working hard on one problem can lead naturally to insights on others. Considering magnetic monopoles takes Guth deep into the physics of the early universe and that makes him think about flatness and the horizon. On December 6, 1979, he stays up late working on equations. The next morning he looks at his calculations and writes, "SPECTACULAR REALIZATION."

Guth has done it. He has solved the flatness and horizon riddles. To do that, *he has had to revise Big Bang theory.* He truly has come up with "a new theory of the origin of essentially all the matter and energy in the universe." It will be called the **Theory of Inflation**.

Inflation theory says that for a very, very short time within the first second after the creation, our universe grows exponentially—it doubles, and doubles, and doubles again, and again—approximately once every 10^{-37} seconds. It *inflates* at a rate never since repeated.

Alternate Theories

Cosmologists Paul Steinhardt of Princeton University and Neil Turok of Cambridge have come up with a cyclic model of the universe that also explains the horizon and flatness problems. Some see it as an alternative to inflation theory. By cyclic, they mean that the universe undergoes a large and perhaps infinite succession of big bang/big crunch cycles.

Portuguese cosmologist João Magueijo, in a book titled *Faster Than the Speed of Light*, describes yet another alternative. He says that early in the universe's history, light might have traveled at a faster speed. He calls it the "varying speed of light" theory, or VSL. If VSL can be verified, it would solve the problems of the Big Bang without inflation theory.

These are speculative (unproved) hypotheses intended to clarify Big Bang theory.

This illustration chronicles the history of the universe in eight key stages. From left to right: (1) the Big Bang, (2) the birth of the early universe, (3) the formation of quarks, (4) quarks colliding into protons and neutrons, (5) the emergence of electrons and positrons, (6) neutrons and protons becoming atomic nuclei, (7) the nuclei becoming atoms as electrons join them, and (8) the formation of today's stars and planets and galaxies from those atoms. Steps 1 through 5 happened in about one second.

From something smaller than a subatomic particle, our universe quickly becomes about the size of a marble. It expands more that 1 billion billion billion times.

"The theory of inflation modifies our understanding of just the first tiny fraction of a second of the history of the universe," says Alan Guth. But that fraction of a second makes the universe we know possible.

Just before the inflationary stretch, every part of our then-very-tiny universe must have been in contact with every other part. That's when information was exchanged and a common temperature reached. That explains why the universe as a whole is uniform. This solves the horizon problem.

If the universe started as a Planck-length particle, 1.6 times 10^{-33} centimeters, and, after inflation, it is 1.6 centimeters (the size of a marble), then the multiplication factor is 10^{33}, which is an expansion of more than 1 billion billion billion times.

As for the flatness problem: The inflationary expansion pulls space into flatness. Stretch any bumpy surface far enough, and each small region of that surface will look nearly flat. (These are simplified explanations; mathematical explanations will take you further.)

What happens in our universe during and after the initial inflation? During inflation, the temperature stays more or less constant at 10^{26} kelvins. But after the fraction of a second of inflation, when the inflationary energy is dumped into the existing cosmos, the temperature rises to 10^{29} kelvins. The cosmic kernel erupts with never-to-be-equaled fireworks. **This is the familiar, hot Big Bang.** The infant universe is a dense, scorching fireball. Remember, this explosion doesn't

A temperature of 0 kelvins equals −273.15 degrees Celsius. Water freezes at 273.15 kelvins (0 degrees Celsius) and boils at 373.15 kelvins (100 degrees Celsius).

The Italian writer Dante Alighieri (1265–1321), in his great classic, *The Divine Comedy*, imagined hell as hot enough to melt brimstone. This illumination (above) is titled *Inferno, Purgatory, and Paradise*. Brimstone is the element sulfur (S), which melts at 386 kelvins (113 degrees Celsius or 235 degrees Fahrenheit)—far, far cooler than the early universe.

Some thinkers don't agree that the Big Bang was the beginning of time. They believe there was a pre–Big Bang universe. For thoughts on this, look up loop quantum gravity (LQG) and string theory or M theory.

Alpher, Bethe, and Gamow were right: The newborn universe was a thick brew of basic particles and so hot that even nuclei couldn't form. Physics tells us that if you compress a gas, it heats, and vice versa. So, naturally, as the universe expanded, it cooled. (You know this story from the previous chapter, but it bears some repeating.)

Some 400,000 years after the creation of the universe, electrons began bonding with nuclei to form atoms. As soon as the electrons were busy inside atoms, photons were able to roam around (those same roaming photons that make up today's CMB), and the universe became transparent, which means there was light.

The first stars and galaxies were created from atoms of hydrogen, from some helium, and from a little bit of lithium, all pulled together by gravitation. Heavier elements were created in hypernovas and supernovas and in the fusion process inside stars.

Our Sun and solar system formed about 9 billion years after the Bang (that means about 5 billion years ago). Life on Earth was underway a little over 1 billion years later.

This is all crazy! Do sane people believe it?

Yes.

Big Bang theory—at least what happens after the period of inflation—is accepted by essentially all of today's leading scientists. American scientist Robert M. Hazen writes, "Big bang cosmology has become as close to conventional wisdom as is possible in the science of the whole universe." Steven Weinberg, another American scientist and winner of a Nobel Prize, says, "About the big-bang theory we can be quite confident.... The big-bang theory is not...likely to be blown away by the next round of astronomical observation, but is almost certain to endure as part of any future theory of the universe."

It is a theory that explains a lot, but is there experimental evidence for inflation?

There is.

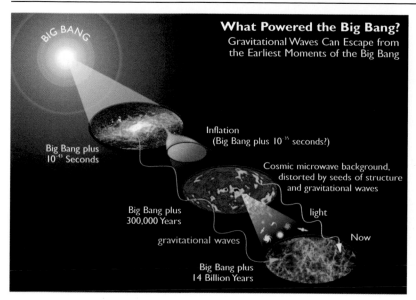

What Powered the Big Bang?
Gravitational Waves Can Escape from
the Earliest Moments of the Big Bang

BIG BANG

Big Bang plus
10^{-43} Seconds

Inflation
(Big Bang plus 10^{-35} seconds?)

Cosmic microwave background,
distorted by seeds of structure
and gravitational waves

Big Bang plus
300,000 Years

light

gravitational waves

Now

Big Bang plus
14 Billion Years

Here's another view of the evolution of the universe, a portrait that's never finished. Recent data from WMAP about the CMB (pink-blue oval) shows clear new evidence of a tremendous growth spurt in the universe's infancy in support of inflation theory. How will our picture change when the first gravitational-wave detectors (chapter 42) weigh in?

Three years of data compiled by the WMAP satellite and released in March 2006 bring a detailed portrait of the early universe about 380,000 years after its creation. It provides hints of what occurred much earlier, even during the time of inflation.

"The observations are spectacular, and the conclusions are stunning," says Brian Greene, a professor of physics at Columbia University. Michael Turner of the University of Chicago calls it "the first smoking gun evidence for inflation."

What's next? We're still working to understand the four interactions. We believe that, at the instant of creation, gravity, electromagnetism, and the strong and weak nuclear forces were all combined into one "superforce." Then gravity divorced its partners. Soon the other interactions also split. Why? How? We don't know.

Finding gravity waves might help us find the superforce, further test inflation theory, and also answer questions about our theory of gravitation (general relativity). General relativity doesn't work in the quantum world—a problem Einstein tried to resolve but couldn't.

There's something else, and it's startling: Inflation theory suggests that the observed universe might be just a small

One reason we're searching for gravitational waves is to learn about the period of inflation. If we can track gravitational waves from the early universe, they should tell us about the structure of space and time during the first seconds after the Big Bang.

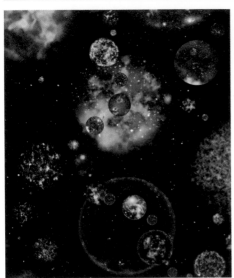

Did inflation allow for the creation of many universes? Maybe. Here's a colorful, visual interpretation of multiverses.

fraction of the entire universe. If inflation happened once, why can't it happen again and again? Astronomer Mario Livio says, "Inflation changed the status of our universe."

Inflation is not likely to have completely stopped, and so pieces of the universe are probably continuing to inflate into new universes. Physicists now talk of a cosmos that is a multiverse: a universe of universes. "People call this idea 'eternal inflation,'" says Livio.

Alan Guth says, "If eternal inflation is correct, then the big bang...was the beginning of our 'pocket universe,' but not the universe as a whole. The full universe existed long before our pocket universe, and will continue to exist for eternity."

> Let your soul stand cool and composed before a million universes.
>
> —Walt Whitman (1819–1892), American poet, *Leaves of Grass*

You're the Center of the Universe

The cosmological principle says that the universe looks just about the same, on average, no matter where an observer is and no matter in which direction he or she is looking. Yes, the universe has lumpy galaxies and empty stretches, but on a very large scale, it all evens out and becomes homogenous.

No matter where you stand, on Earth or on some distant galaxy, you will feel that you are the center of the universe. But there is no actual center—or rather it is equally suitable for any person anywhere in the universe to think that she or he is at the center.

The cosmological principle makes each of us feel as if the universe is centered on us. We believe that this principle applies anywhere and everywhere in the universe.

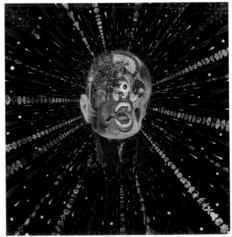

In Fred Tomaselli's *Cyclopticon 2*, a one-eyed cyclops attracts and disperses matter at the center of a (not *the*) cosmos. In Greek mythology, a cyclops was a son of Uranus ("Sky") and Gaia ("Earth").

TOE Be or Not TOE Be

Artist Fred Tomaselli, who was born in California in 1956, now lives in Brooklyn, New York. *Abductor* is one of his dazzling collages made of leaves, pills, magazine cutouts, and painted parts (all sealed under smooth resin). He says, "Painting has traditionally been seen as a window into another reality...." Is today's reality found in the world of science? Is painting a window?

For the last 30 years of his life, Einstein worked to find a basic rule underlying the universe—the T.O.E., or the Theory of Everything.

Einstein knew that the rules for general relativity and quantum theory don't work together. We are missing something vital. Today, that missing piece of information has created a quiet crisis in the scientific world, much like the nineteenth-century conflict between Newton's theory of gravity and James Clerk Maxwell's electromagnetism. It was that challenge that led Einstein to formulate special and then general relativity.

Is finding the T.O.E. the next step? If so, it should help us better understand the creation of the universe and how nature's basic forces affect the tiniest particles. And it may hinge on our understanding of gravity. "The marriage of general relativity and quantum theory," says Nobel Prize–winning physicist Leon Lederman, "is the central problem of contemporary physics."

Entanglement? Locality? Are We Talking Science?

> [T]he most perplexing phenomenon in the bizarre world of the quantum is the effect called entanglement. Two particles that may be very far apart, even millions or billions of miles, are mysteriously linked together.
> —Amir D. Aczel, Israeli mathematician, *Entanglement: The Greatest Mystery in Physics.*

[Quantum mechanics] describes nature as absurd from the point of view of common sense. And it agrees fully with experiment. So I hope you can accept nature as She is—absurd.
—Richard P. Feynman (1918–1988), American physicist, *QED: The Strange Theory of Light and Matter*

A Note to the Reader: This is an abstruse (not easily understood) chapter that is worth the effort. It's about a forward-looking concept that is on the platter in today's scientific cafeteria. Read on if you are interested in a field of physics that may play a big role in future technologies.

Our friend Albert is looking pleased with himself.

Albert Einstein never believes that quantum theory is complete. He keeps looking for a theory—he calls it a unified field theory—that will bring big and little together.

Most of his colleagues, like Niels Bohr, seem to think that relativity explains the world of the very big, that quantum theory explains the world of the very small, and that Newton explains the in-between world. Each has its own domain. The Copenhagen physicists don't see the need for a merger. Quantum mechanics is dazzling and leading to technological wonders; Bohr and his colleagues accept its core of uncertainty and get on with their work.

Einstein doesn't give up. For him, uncertainty is unacceptable. He keeps challenging Bohr. Albert is sure there are undiscovered properties of particles that will tell us exactly (instead of statistically) why particles behave as they do. To try to prove his point—that quantum physics is incomplete and uncertainty invalid—Einstein comes up with "gotcha" situations, one after another. Again and again, Bohr hunkers down with his tribe of physicists, and each time they find proofs to show Einstein that he is wrong about quantum mechanics.

Still, Einstein insists that quantum theory is not the final word. In 1935, Einstein, in Princeton, writes a paper with two colleagues: Boris Podolsky and Nathan Rosen. The three scientists intend to show a flaw in quantum mechanics. Their paper, known as the EPR paradox, is based on mathematical calculations. It will become famous. Einstein, Podolsky, and Rosen (EPR) believe they have found particles that transfer information at a speed faster than light, and they know that is impossible. Einstein actually uses the phrase "spooky action at a distance" to describe what they have discovered.

Using mathematical equations, E, P, and R have come across something known as "entanglement." It's another weird-to-us quantum phenomenon, but it's real. Imagine you have a twin, and you are so tied to your sibling that when you bite your tongue, your twin's tongue, wherever she is, instantaneously begins to throb. Bizarre? Yes. It doesn't happen in the macro world. But entangled particles do something like that (as experiments will prove).

To understand, try thinking like a detective. You're looking for entangled particles: particles that have a mysterious link to one another—no matter where they are in the universe.

Your quest is difficult because your mind is trying to trick you. It has you thinking "classical" instead of "quantum." It has you thinking "common sense" instead of "quantum reality." Even Albert Einstein falls into the trap. Get the story straight, and you will understand an important secret of the

If spacecraft A and spacecraft B carry entangled particles, a message to A can affect the entangled particle on B—a spooky (says Einstein) action at a distance. One future application, according to quantum scientist Christoph Adami, is that "two [atomic] clocks could be started and stopped simply by acting on only one of them." This would allow us to instantly set the time aboard a spacecraft that's light-years away!

Although the EPR paradox didn't disprove quantum uncertainty, it did show something important. More than 70 years after the paper's publication, it is still razor's-edge physics. Even when he was wrong, Einstein's thinking was fertile.

A particle is a **QUANTUM SYSTEM**. An atom is a more complex quantum system than a single particle. So are two or more entangled particles.

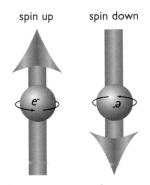

spin up   spin down

Spin is a property of many particles, including many atoms. If two photons or two electrons come from a common source, they might have a total spin of zero. If they do, then each particle will have a spin that is up or down. (Photons have something called *polarization*, which is almost but not quite the same thing.)

The value of the electron spin is constant; it's the *direction* of the electron spin that gets measured as spin up or spin down.

universe and perhaps be able to write unbreakable codes and do computing that is far in advance of anything the world knows in the early 2000s.

You are not only a detective; you are also a physicist (careers that are much alike). So you plan an experiment on a quantum system in a particular state. You know that you can do only one experiment with this system. That's because, this being the quantum world, doing the experiment will put the system in a new state. Given quantum uncertainty, you don't get a second chance with the same particle.

As a detective/physicist, you know that most particles have a property called "spin." When you do an experiment to detect the spin of an electron, quantum mechanics says there are only two possible results: spin up or spin down. Picture two atoms: If one is spin up and the other is spin down, they cancel each other, and the total spin is zero.

As for entangled particles? Well, they are zero-sum teams with two or more players: One is spin up; one is spin down. Suppose an atom with zero total spin spits out two identical electrons. They zoom off in opposite directions, each going one light-year away, which means they are soon two light-years apart. The total spin of the two emitted electrons must still add up to zero, because it was zero before the atom released them. Now, this is where you have to beware of classical thinking and put on your quantum glasses. Until measurement, each electron has the *potential* to spin up *or* down. But once one is up, the other must be down, to keep that zero total.

Knowing that, you decide to measure the spin of one of the entangled particles. You can measure the spin along *any* direction (you decide the direction); you don't know in advance whether the spin will be up or down. It can go either way; *your choice of direction is part of the process* of determining the direction of electron spin. You are a player in this (yes, spooky) process!

As soon as you find the spin, the other particle *instantaneously* spins in the opposite direction (even if it is 1 billion light-years away). It *must* do this, so that the pair of electrons, even though distant from each other, has the same total spin as the original system—namely, zero.

This is where Einstein got off track. He thought one particle was sending a message—information—to the other. He knew that couldn't happen at a speed faster than light, and "instantaneous" is faster than light. So how does the second particle get the message? Clue: This is quantum theory; classical thinking won't do.

Before anyone interacts with one electron of the pair, its spin direction is in a super-position, which means it has the potential to be spin up or spin down. The choice of direction hasn't been made. When you experiment, you help determine that choice. Knowing the spin direction of one particle will tell you the spin of the other particle; it won't change the other spin direction.

Here is an analogy: You write two notes. One reads *Up*, and the other reads *Down*. You give one note to a traveler heading north. He puts it in his pocket; he can't look at it. Neither can the other traveler, who goes south. According to the rules, the travelers cannot send signals to each other. When they are far apart, one of the travelers looks at his note. It reads *Up*; he knows *immediately* that the other traveler's note reads *Down*. There is no error. He hasn't broken any rules. When he read his note, he did not send a signal to the other traveler.

An electron is a tiny magnet, described by an imaginary arrow called its spin, which has a curious quantum-mechanical property. If you pick any direction you wish, and decide to measure the direction of the spin, you will find that it points either along the direction you picked (spin up), or the other way (spin down). Somehow, the spin arrow is simply not allowed to point 30 degrees from the axis along which you aligned your measuring device—or in any other direction except parallel or antiparallel.

—Hans Christian von Baeyer, *Information: The New Language of Science*

Remember, the Uncertainty Principle says *you can't know everything* about a quantum system of particles until you experiment. Once you experiment, *you can know* specifics of the whole system.

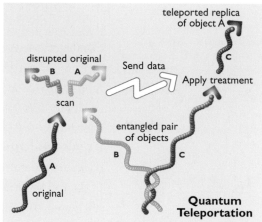

Quantum Teleportation

How does quantum teleportation work? In this scenario, quantum object A (bottom left) turns quantum object C (top right) into a replica of itself without ever meeting it. Instead, A encounters B, which is entangled with C. B sends A's data (its quantum state) to C, which then ends up just like A. The originals, A and B, are lost after the exchange.

It is possible (if you know what you're doing) to "prepare" the spin of an electron in the "spin up" state: A later measurement of the spin *along that same direction* will always give the result "up" (unless something has disturbed the spin in the meantime). Does that tell us there is some certainty in quantum mechanics? Yes. That is what is meant by a quantum state. Engineers can use this knowledge as long as they don't try other experiments on the original system, in which case there is no certainty.

> ## Secret Codes?
>
> Suppose you send a particle, which is part of a coded message. The particle is in a quantum state in which "spin up" means, "The army is moving tonight." Can the enemy intercept your message and find out your military plans? Yes, if they know the code. But by doing so, the system is put into a new state—and that tells you that your message has been intercepted, so you can change your plans.
>
> Actually, it's a bit more complicated than that: Unbreakable codes using entangled particles require more than one spin to carry the message. But you get the idea—no one can tamper with them without your knowing about it.

A Physicist's Uncertainty Joke

Three baseball umpires go into a bar and talk about problems in calling balls and strikes.

The first umpire says, "I calls 'em as I sees 'em; sometimes I'm right, and sometimes I'm wrong."

The second umpire says, "I calls 'em as I sees 'em, and I'm always right."

The third umpire says, "Until I calls 'em, they don't exist."

In the quantum world things are a bit more complicated: *There is no up or down* until someone *chooses* the line along which the two spins are to be measured. No one can say the particle points up—*until someone does the experiment.* Before that, the particle is in a superposition: as both up and down. It's that Uncertainty Principle at work, and it isn't like anything we experience in our ordinary life—so it takes imagination to get it—but we now know that uncertainty enhances our lives (or, at least, I think so). It also makes us participants in the greater world.

We now know that things in the quantum world aren't neat and decided in advance. When you measure or interact with quantum particles, you become an actor in the quantum drama.

Einstein goofed. He and his friends got caught in classical, macro-world thinking. They were sure those entangled particles must send a message to each other, with the up particle telling the other to become a down. If that were to happen, instantaneously, then the speed-of-light rule would be broken. That's what bothered EPR. They knew that couldn't happen. But they just weren't ready to accept quantum reality.

So is there any form of instant communication between entangled particles?

Recall that quantum particles exist as a smear of potential called a superposition. Electrons, for instance, in a superposition are neither particle nor wave, but they have the potential to be either. Measure them and *then* they make a choice. They become particles or waves, not both.

Bohm: Brilliant? Bonkers? Take Your Pick

Physicist David Bohm (1917–1992) was among the first to see the potential of entanglement. As a graduate student at the University of California at Berkeley he discovered that, in a plasma, electrons separated from atoms do not behave as individual particles but rather as part of a larger whole. Those electrons act in a highly organized way, as though an orchestra conductor were leading their behavior.

The universe is more than building blocks, said Bohm, who believed there's an underlying connectedness in everything. He was challenging the Newtonian way of thinking.

Bohm was also challenging a foundation of Einstein's physics: the idea of locality. Locality is the classic concept that everything has a specific position in time and space and that nothing can travel the spacetime barrier faster than the speed of light. Einstein's special relativity (like Newton's laws of motion) is based on locality.

According to Bohm, each unmeasured particle has nonlocality, meaning things can be interconnected at the quantum level, even if that connection isn't pinned down to a location. And—take note—*non*locality allows for instantaneous happenings anywhere in the universe. The speed of light is beside the point.

Bohm's work is controversial. He stood with Einstein in some of his interpretations of quantum theory. His ideas break from all the other fundamental theories of physics. Nonlocality suggests that our very idea of reality, which assumes that phenomena are separate from one another, might be wrong.

David Bohm (left), an assistant professor at Princeton in 1949, was working closely with Einstein at the nearby Institute for Advanced Study when he was called before the House Un-American Activities Committee in Congress. As a young man in the 1930s, Bohm had joined Communist and peace organizations. He was asked to name other members. He refused. (Under the Fifth Amendment of the Constitution, he had the right to stay silent.) This was the McCarthy era; Bohm was arrested. He was acquitted in 1951, but he had lost his job at Princeton. He accepted a physics chair in Brazil (and later, others in Israel and London). Today, he is known for a philosophy that brings spirituality to physics and focuses more on wholeness than on analysis of separate parts.

Nope. As far as we know, there is no message between the two entangled particles, only the fact of entanglement.

It took a lot of detective work for physicists to figure this one out.

What is new in this quantum story is that you get to choose the direction along which the spin will be measured. You can wait until the particle arrives at your spaceship; the particle will then be one light-year from its source and two

Shoot a stream of electrons at a barrier. They might bounce off, but they might show up on the other side of the barrier. Weird? Yes. But the electronics industry depends on that quantum weirdness—physicists call it quantum *tunneling*.

Entangled Particles Found!

In the 1930s, the technology to experiment with entangled electrons or photons (or even prove they exist) was not available. Physicist Wolfgang Pauli said that since we can't prove entanglement, let's move on to other things. But the EPR paper had other physicists thinking.

In the 1960s, Irish scientist John Bell (1928–1990), working in Geneva, Switzerland, devised an entanglement experiment (now very famous); it took a while before it could be carried out. Finally, in the 1970s and 1980s, some modest experiments by Bell and others showed proof of entanglement over very short distances.

Zoom ahead to Austria in 1997. Researchers at the University of Vienna and the Austrian Academy of Science fed 800 meters (875 yards) of optical fiber through the public sewer to connect labs on opposite sides of the Danube River; they teleported the properties of photons from one lab to the other.

In this 2003 quantum entanglement experiment, the photons of laser beams aren't teleporting themselves. They're teleporting information about their quantum state (their spin direction or polarization, for example). The aim is to explore the breakdown of causality—that is, the effect coming before the cause. How is that possible? Remember that entanglement is instantaneous, which means independent of both space and time.

In June 2003, the Austrian physicists took the experiment further. They sent a laser beam through a barium borate crystal, splitting photons into pairs of photons. Those entangled photons, with wavelengths of 810 nanometers (a nanometer is one-billionth of a meter), traveled through free space (rather than optical fiber) from a projecting telescope to two receiving telescopes: One was 150 meters (164 yards) away, and the other was across the Danube River 500 meters (547 yards) away. There was no direct line of sight between the telescopes.

What happened? When one photon was disturbed, the other instantaneously showed the effects. The photons remained entangled over a distance of 600 meters (656 yards). Spooky? Yes, but this experiment showed that entanglement is a real phenomenon.

Too bad Einstein and Bohr weren't around to hear of it—both men were dedicated to finding the truth, even when it proved them wrong. (Einstein died in 1955, Bohr in 1962.) What would Einstein say now?

light-years from its entangled twin. (But it doesn't matter how far it has traveled.)

Fast-forward to today. The EPR effect is very much on the scientific blackboard. Why? Entanglement theory suggests there are underlying connections in the universe that we never suspected. How extensive is this quantum-level connection between particles? And will it make a difference to us in our macro world?

We don't yet know. On a practical level, scientists are working to develop quantum computers that would use entangled particles (more on this in chapter 47). If they are successful, a single laptop will be more powerful than all of today's computers put together. That's just one possibility. Teleportation is another. Unbreakable secret codes are yet another.

Esoteric stuff, but fascinating, too.

Decades before experiments proved that quantum teleportation is possible, science-fiction writers took the concept to the next imaginative level. They wondered: What if we could teleport any object—even ourselves? In *The Fly*, a 1957 short story adapted into several movies (notably in 1958 [1], and in 1986), a fly happens to enter a scientist's experimental teleportation chamber at just the wrong moment. Terror ensues after their molecules are mingled. The creators of *Star Trek* (2) "invented" their famous transporter ("Beam me up!") in 1964 for the pilot episode. An occasional malfunction generated interesting plots, such as "evil twins" for Captain Kirk and the crew in "The Enemy Within." Many video games (3 and 4) make convenient use of teleportation to instantly send characters from one locale to another, but, unfortunately, the technology is only possible in the quantum world.

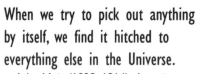

When we try to pick out anything by itself, we find it hitched to everything else in the Universe.
—John Muir (1838–1914), American naturalist, *My First Summer in the Sierra*

Super Stars

We shall not cease from exploration
And the end of all our exploring
Will be to arrive where we started
And know the place for the first time.
—T. S. Eliot (1888–1965), American-British poet, from *Four Quartets*, "Little Gidding"

We are lucky to live in an age in which we are still making discoveries. It is like the discovery of America—you only discover it once. The age in which we live is the age in which we are discovering the fundamental laws of nature.
—Richard P. Feynman (1918–1988), American physicist, from a 1964 speech

Science depends on organized skepticism, that is, on continual, methodical doubting. Few of us doubt our own conclusions, so science embraces its skeptical approach by rewarding those who doubt someone else's.
—Neil deGrasse Tyson and Donald Goldsmith, *Origins: Fourteen Billion Years of Cosmic Evolution*

Brian Schmidt runs cross-country track and plays the French horn at his Alaskan high school in the 1980s. He isn't sure what he wants to do with his life until a career counselor tells him, "You should do what you would do for free. That's the best career." Schmidt realizes that "The only thing I would actually do for free [is] science, and specifically astronomy." And so he goes off to college (the University of Arizona), and then graduate school (Harvard), and in 1993 earns a Ph.D. in astronomy. At Harvard, the professor who guides Schmidt's research, Robert Kirshner, is an expert on the giant exploding stars called supernovas.

Fritz Zwicky (see page 424) came up with the term *supernova* in 1931. It describes a blast that's rarer and more powerful than a *nova* explosion. Unlike a supernova, a nova outburst doesn't completely destroy the star, allowing some stars to erupt several times.

Finding a supernova means knowing the heavens well. It means scanning the skies for a bright spot that wasn't there last month. Those star bombs will be important in Schmidt's future. He will devise a computer program that will help automate the search for them.

Also important to his future is an Australian named Jenny he meets at Harvard. She is getting her doctorate in economics. After they marry, Schmidt takes a job that leads him to the world's first large, fully automated telescope, at Mount Stromlo near Australia's capital, Canberra.

By the late 1990s, Brian Schmidt is leader of the High-Z Supernova Search Team, an international group that includes Kirshner and more than 20 tenacious astronomers who come from five continents and spend their time exploring the heavens. Their computer screens are linked to

The **Z** in High-Z Supernova Search Team is a symbol for redshift (see page 331)—a shift toward the red end of a star's spectrum that means the light source is moving away from us. **HIGH-Z** means a large redshift, which indicates that the star is moving away *fast*. Hubble's law tells us a fast-receding star is far away. So, the High-Z team is looking for stars that are several billion light-years away, which exploded several billion years ago, and whose bright supernova light is just now reaching Earth.

giant telescopes perched above the clouds on mountaintops as well as to the Hubble telescope in Earth orbit. They have a hypothesis to prove.

Schmidt, Kirshner, and their peers believe that the mutual gravitational attraction of stars, planets, and other masses should act as a kind of universal brake and, as new galaxies form, slow down the very fast outward push that has been stretching spacetime since the Big Bang. In other words, mathematical equations predict that the pull of gravity will,

Eagle-eyed teenagers can search for supernovas like the pros through Hands-On Universe, an educational program developed by the Lawrence Hall of Science at Berkeley. They download fresh telescope images, again and again, of the same patch of sky and scour them for the sudden appearance of an exploded star. Harlan Devore, a teacher at Cape Fear High School in Fayetteville, North Carolina, processed this computer image of Supernova 2006al, which he co-discovered in February 2006.

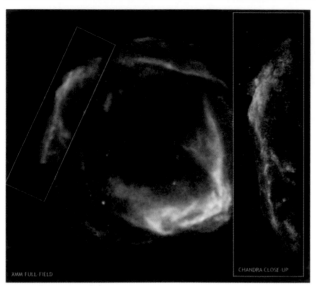

XMM FULL FIELD   CHANDRA CLOSE-UP

Note the gaping hole at the center of this supernova remnant, called RCW 86, within our galaxy. It's a sign that the debris ring has had plenty of time to expand—roughly 2,000 years. That makes the time frame of the explosion consistent with one of the earliest recorded mentions of a supernova, by Chinese astronomers in 185 C.E.

Our Sun will never be a Type Ia supernova because it is a *lone star* (no Texas pun intended)—it has no companion star to feed it mass or steal mass from it. It will never be a Type II supernova because it isn't massive enough. Stars many times more massive than the Sun end their lives as Type II supernovas, which involves a runaway nuclear reaction and an enormous detonation, leaving a neutron star or a black hole. Unlike the Type Ia explosions, Type II supernovas vary in intensity.

over time, slow the universe's expansion. But that prediction needs verifying by comparing data from supernovas, old and new.

At the Lawrence Berkeley National Laboratory in Berkeley, California, lean, intense Saul Perlmutter leads a group of astronomers known as the Supernova Cosmology Project (SCP). Like High-Z, they intend to document a slowing of the universe's expansion by studying supernovas.

SCP has a head start. At a meeting in Princeton in 1996, Perlmutter presented their preliminary findings based on an analysis of eight supernovas. He announced that, as predicted, the growth rate of the universe seems to be decelerating.

But SCP's initial results aren't conclusive. That's why Schmidt and his High-Z colleagues decide to go for the same goal as SCP. They, too, hope to pin down the rate of deceleration of the universe. They have been studying relatively nearby supernovas; the SCP astronomers have studied faraway ones. Now, High-Z will also study distant supernovas in order to make comparisons. They are the tortoises, the latecomers. The SCP astronomers, the hares, are way ahead in the search. Each group uses its own very sophisticated computer programs.

Only a few giant telescopes in the world are suitable for the kind of viewing they plan to do; astronomers everywhere vie for viewing time. The High-Z astronomers make a case that this is an important quest and that more than one group should be gathering information. They are allotted limited time.

Supernovas are not all alike; astronomers have classified them into types. The High-Z and SCP astronomers are only interested in **Type Ia supernovas**. These are products of medium-sized stars that were part of binary systems. That means they had companion stars. One of that pair, having used up most of its hydrogen, becomes compressed into a white dwarf. The other swells up and becomes a red giant.

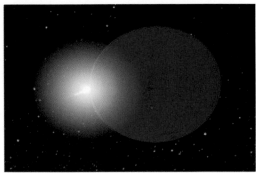

As the red giant expands, its outer regions extend close to the white dwarf. Not a good idea; it is invading the white dwarf's personal space and will get wolfed down. (The small, dense star—the white dwarf—steals mass from the less dense red giant by gravitational attraction.) When the white dwarf reaches a predictable mass (1.4 Suns, the Chandrasekhar limit), it will collapse and explode.

Type Ias are reliable; each has almost the same mass and luminosity when it blows up. According to Schmidt, "If you've seen one of them, you've more or less seen them all. They're all the same brightness. And so simply by looking at how bright these objects are, we can measure their distance: the fainter they [appear to us], the farther away they are." For an astronomer, that makes a Type Ia supernova special. It can be used as a kind of standard measure.

Like a capricious geyser, a neutron star named EXO 0748-676 (the white dot in this illustration) "goes nuclear" in a massive explosion whenever gas collected from its red giant companion reaches a high enough pressure and temperature for nuclear fusion.

Remember, luminosity is how bright the star actually is. Apparent brightness is how bright it looks to us— the farther away a star is, the fainter it appears.

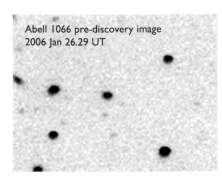

Abell 1066 pre-discovery image
2006 Jan 26.29 UT

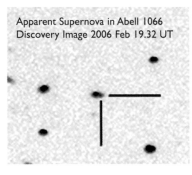

Apparent Supernova in Abell 1066
Discovery Image 2006 Feb 19.32 UT

Here's the before (far left) and after picture that revealed the sudden presence of Supernova 2006al (a dot turned into a messy smear) in Abell Galaxy Cluster 1066.

Both teams search the heavens looking for beacons of light that weren't there before. They study broad vistas with billions of stars, checking each day's data against previous photographs. The computer programs help. When they find a Type Ia, they must work quickly. Kirshner says, "New supernovae are like fresh fish. If you don't use them right away, they spoil." The search and follow-up measurements are intense.

Type Ia supernovas make great standard candles for measuring the early universe for two reasons: Besides being uniformly luminous, these exploding stars are super-bright; even the very distant ones are easy to spot from Earth.

At the Cerro Tololo Inter-American Observatory, perched on a flat mountaintop in Chile, high-soaring condors sometimes drop by and rap on the windows in the hopes of scoring a meal from visiting astronomers.

On a night in 1997, High-Z's Peter Challis is in Chile peering at a computer screen keyed to the colossal Cerro Tololo Inter-American telescope way up on a mountaintop where the viewing is excellent. Challis, a big guy in cargo shorts and tennis shoes, is under pressure to find Type Ia supernovas for the team to study. Alex Filippenko (from the University of California at Berkeley) is flying to Hawaii.

High-Z has been given precious viewing time at the Keck telescope there. Bruno Leibundgut (from the European Southern Observatory) has time scheduled on a giant telescope at another site in Chile. Filippenko and Leibundgut are expecting Challis to pinpoint supernovas for them to document.

Peter Challis and some colleagues are eating pizza when Challis shouts, "Bingo! We got one!" But one isn't enough.

Science Inspires Poetry

American poet Robert Frost (1874–1963) cornered Harlow Shapley at a college faculty party back in the 1910s. Shapley was a well-known astrophysicist at Harvard University.

Robert Frost

Harlow Shapley

"Professor Shapley, you know about astronomy," Frost reportedly said. "Tell me how the world is going to end." Shapley thought Frost was joking, but Frost was serious.

The poet asked again. "So," said Shapley, "I told him that either the Earth would be incinerated, or a permanent ice age would

gradually annihilate all life on Earth."

Some 40 years later, Shapley described the encounter to an audience. "Imagine my surprise when just a year or two later I ran across this poem." He was referring to "Fire and Ice," one of Frost's best-known poems, published in 1920.

Here's how it begins:

Some say the world will end in fire,
Some say in ice.
From what I've tasted of desire
I hold with those who favor fire.

Robert Frost was a better poet than an astrophysicist. Experts now forecast a cold future for the universe billions of years from now (but there are still many unknowns).

As for Shapley's role as a poet's muse? He believed his science inspired "Fire and Ice." I agree. But do those last lines have anything to do with astronomy?

They spend the night peering at their computer screens. Before morning comes, they find more. Robert Kirshner describes what happens when he turns the list over to other team members:

> *They will gather the light from Pete's distant discoveries, spread each one into a spectrum, and note extremely diligently things that are invisible to the unaided eye. The spectrum will reveal each supernova's contribution to the stock of heavy elements in a distant galaxy and form the basis for a scientific prophecy of future cosmic expansion.*

There is no time to lose when a supernova is spotted. Briefly, it might be brighter than a galaxy, with hot gases shooting into space. But that intensity lasts only a few days or sometimes a month. Then the dying star fades, eventually disappearing from telescopes. What's left is a small, super-dense, spinning cinder made mostly of neutrons (a neutron star) or a black hole.

Like a time machine, the telescope takes the astronomers back into the past. The farther away a star is, the further back in history you're looking. A supernova that is 5 billion light-years away exploded 5 billion years ago (before Earth was formed).

The astronomers document and measure. Checking the redshift of a Type Ia supernova 5 billion light-years away tells them how fast the universe was expanding 5 billion years ago. The astronomers can compare that with today's expansion rate found in nearby stars.

Adam Riess, from the University of California, helps analyze the High-Z data. **It isn't telling him what he expects.**

At Lawrence Berkeley National Laboratory, Saul Perlmutter's SCP group is doing much the same thing. Science journalist Tom Yulsman describes Perlmutter: "As thin as a marathon runner and fueled by an almost manic enthusiasm—call it bright energy—Perlmutter seems determined to keep pace with the runaway universe." The SCP astronomers study 40 supernovas; they, too, intend to pin down the rate at which galaxies are moving away from one another.

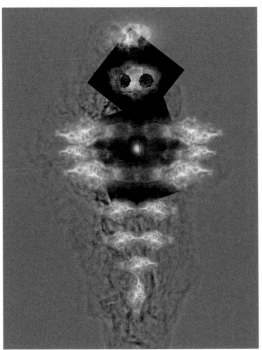

Astrophysicists Michael Bietenholz and Norbert Bartel of York University in Toronto, Canada, tapped into the talents of digital artist Balz Bietenholz, who helped turn their images of supernova remnants into this composition named *Novi*.

From featherweights (bottom of illustration) to super heavyweights (top), this picture graph shows how stars of seven different masses are born from protostars (red column at left), evolve (center), and die (right). Yellow, Sun-like stars are in the third row from the bottom. The top row, representing the heaviest stars in the universe, was added after the discovery of Supernova 2006gy, which provided the first observational evidence of a super-bright explosion that doesn't produce a black hole.

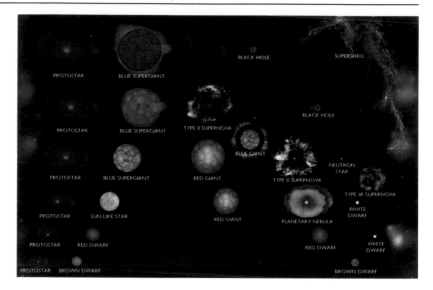

Astronomy brought Saul Perlmutter, Adam Riess, and Brian Schmidt 1 million dollars! They shared the 2006 Shaw Prize in Astronomy for their discovery that the expansion of the universe is accelerating. Sir Run Run Shaw, a Hong Kong philanthropist, established the award for outstanding contributions to astronomy, life science and medicine, and mathematics. Riess's wife commented, "Hey, this astronomy stuff is pretty good."

Both groups keep reminding themselves that gravity, left alone, has a natural tendency to pull all of the cosmic mass together into one giant lump. That should slow down the universe's expansion. They are looking for conclusive proof of that deceleration.

Nailing down the expansion rate will allow astronomers to figure out a close-to-exact age for the universe. Knowing if the expansion rate is speeding up or slowing down will help them make predictions about the future. What do Schmidt, Kirshner, Riess, Perlmutter, and their fellow astronomers see when they map supernovas?

Something they have a hard time believing: "Our observations show that the universe is expanding faster today than yesterday," says Adam Riess. His analysis shows the expansion rate is about 15 percent greater today than it was 7 billion years ago. Perlmutter's group finds the same thing.

Neither discovers the deceleration that both expected to find.

"Try as we might, we have not found any errors," says Alex Filippenko (who works with both groups). Michael S. Turner, at Fermilab, says, "If it's true . . . [it] means that most of the universe is influenced by . . . some weird form of energy whose force is repulsive."

Repulsive? Turner is talking about an antigravity force that repels rather than attracts. If the universal expansion is accelerating, there must be a force countering gravitation—a dark energy. The word *repulsive* also means "ugly." And to most astrophysicists, there is something ugly about the dark energy solution to their problem. Kirshner calls his book describing their search *The Extravagant Universe*—extravagant as in "gaudy," like a peacock. MIT's Joshua Winn says, "Nature 'should' be simpler than that."

It is now 1998, and Einstein is long gone, but the announcement that the expansion of the universe is accelerating (the *New York Times* calls it "stunning" news) brings our friend Albert back into the headlines. (He might have chuckled.)

Remember Einstein's cosmological constant, the antigravity force named lambda? This repulsive force pushed outward, balancing the gravitational pull inward, thus keeping the universe stable. At that time, Einstein thought the universe was static and eternal with no beginning and no end and no change over time. His cosmological constant made the universe fit that static picture.

Here's the first direct proof of dark matter, detected by the gravitational deflection of light passing near and through it. Picture two huge galaxy clusters, 3 billion light-years from Earth, whamming into each other at 16 million kilometers per *hour* (10 million miles per hour)! In this composite image of the cosmic fender bender, the (digitally colored) pink blobs are visible matter, while blue highlights where most of the mass is. In a collision, visible matter is slowed by drag forces like air resistance, but dark matter interacts only through gravity and so moves through without the drag. Most of the matter in the clusters (blue) is separate from the normal matter (pink), giving direct evidence that nearly all of the matter in the clusters is dark.

> ### Dark Energy?
>
> Three possibilities lead the way as explanations for dark energy:
>
> 1. Einstein's cosmological constant, which means that dark energy is a property of space itself.
>
> 2. An unidentified energy field called "quintessence." If it exists, it fills space like a fog and is similar to the energy that drove inflation (chapter 44).
>
> 3. Neither of the above. Perhaps dark energy doesn't exist and is an illusion created by incorrect theories.

Data from the WMAP telescope tells us that the universe contains 74 percent dark energy and 22 percent dark matter (below). The stuff we can see (things composed of atoms) makes up only about 4 percent, while neutrinos are less than 1 percent. A Hubble telescope survey called COSMOS generated this three-dimensional map of dark matter (bottom right). The areas farthest away, in the back of the blue blob, date from the beginnings of the universe. Then, the distribution of matter was fairly uniform; over time, the relentless pull of gravity has made it clumpier and clumpier.

In 1929, after Edwin Hubble showed that the universe is not static, a cosmological constant didn't seem to be needed. Talking to George Gamow, Einstein called lambda "the biggest blunder of my life." He attempted to forget it. But *lambda wouldn't go away.*

In a 1947 letter to Georges Lemaître, Einstein was still trying to deal with lambda: "I am unable to believe that such an ugly thing should be realized in nature," he wrote. Lemaître was not so sure lambda was a blunder nor that it was ugly. He thought there might be a repulsive force countering gravity. So did Arthur Eddington.

Fast-forward 51 years to 1998 and the astonishing discovery by two teams of astronomers that the **galaxies are moving farther and farther apart at a velocity that is getting faster and faster.** How can that be explained? Can there actually be something in the universe countering universal gravity? **A lambda force? A dark energy?**

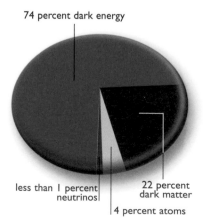

74 percent dark energy

less than 1 percent neutrinos

22 percent dark matter

4 percent atoms

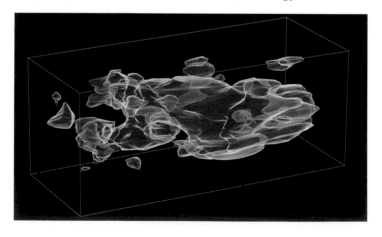

It seems there is.

Is lambda truly "constant"? Or has it varied over the life of the universe? What is it? No one is sure (lots of scientists are working on this), but MIT's precision cosmologist Max Tegmark says that dark energy might make up about 74 percent of the "stuff" of the universe. (Remember, energy and matter are two forms of the same thing.)

Meanwhile, the two supernova groups extend their work. By 2006, detailed cosmic maps of the CMB (see chapter 43) support their conclusions. Some of cosmology's fundamental questions are answered. We learn the age of the universe (about 13.7 billion years). We find that space, which takes interesting curves in the presence of matter, is flat on the biggest scale (which means overall) and is both homogenous and isotropic. **And we further confirm that the expansion of the universe is speeding up.** But we still don't know what dark energy is—or even if it actually exists.

Will understanding dark energy make a difference? Will it tell us something important?

Yes, the fate of the universe.

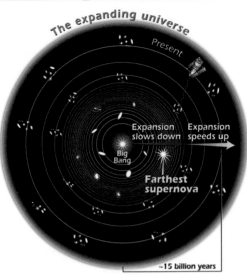

This diagram shows the changing speed of expansion since the Big Bang, slowing down initially (red ring) and then speeding up as some sort of dark force overtakes the pull of gravity. Expansion means the universe is growing colder. Will that continue? If so, our universe is heading for a big chill. If expansion slows, and gravitation pulls everything together, the universe will contract in a hot, big crunch. Don't worry; either way, we're talking billions of years from now.

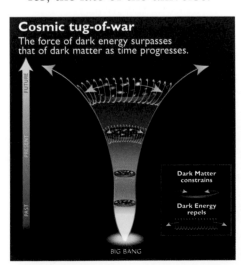

Cosmic tug-of-war
The force of dark energy surpasses that of dark matter as time progresses.

By peering billions of years back in time at extremely distant supernovas, the Hubble telescope found evidence for dark energy "pushing" on the expansion of the universe. The evidence hints that dark energy forces will eventually win over dark matter (diagram at left), but there are more questions than answers: Is dark energy stable (Einstein's cosmological constant)? Or does it fluctuate? Will it continue to push and speed up expansion, causing the "Big Rip"? Or will it reverse and become an attractive force, creating a "Big Crunch"?

Experts on the Dark Side

Are they speaking German or English? Fritz Zwicky (left) was educated in Zurich and later taught at Caltech. Otto Stern (right), a German experimental physicist, assisted Einstein at the University of Prague and later fled Europe. He helped confirm the wave property of matter and won a Nobel Prize.

Way back in the 1930s, the flamboyant Swiss-American astronomer Fritz Zwicky (1898–1974) figured out that something was going on in the heavens that didn't make sense. There didn't seem to be enough matter to make galaxies exist as they do. The gravitational force from the visible stars was not adequate to keep stars in circular orbits around a galaxy center. And yet stars were staying in orbit. Something unknown must be holding the galaxies together.

That wasn't the only problem for Zwicky. He knew that the galaxies interact through gravity. But when he observed galaxies and measured their movement, he discovered that the galaxies are moving much faster than the mass provided by the visible stars says they should move. He realized there must be strange stuff out there. He said galaxies are moving faster than their masses indicate because we can't see all their mass, that most of the mass in the universe must be dark. (*Dark* is the scientific term for "invisible.")

Let me repeat that: Zwicky said that **we can't see most of the matter in the universe.** Unknown matter must be hovering around, holding the galaxies together. But he couldn't figure out what that invisible stuff is. Perhaps, if Zwicky could see deeper and more clearly into space than the giant telescopes of his day allowed, he would find a clue.

Zwicky figured out that since gravity bends light rays, that refraction could turn the light from some stars into a kind of curved lens. A natural cosmic lens would magnify earthly lenses and allow us to see into the depths of space.

Zwicky was a maverick and not well liked by some of his peers, but he was onto something. Actually, he was onto several somethings. He said that galaxies exist in clusters and superclusters scattered in different directions. It was a much more interesting array than the boring, uniform picture held by most astronomers during the 1930s. Zwicky was also right about missing mass

What Is Dark Matter?

We're quite sure dark matter comes to us from the Big Bang and that it has nothing to do with ordinary atoms. Dark matter must escape detection because it isn't electromagnetic. It doesn't shine.

So what's the nature of that matter? We don't know. Some say it may be MACHOs (Massive Compact Halo Objects) like dwarf stars or black holes. Some say it is WIMPs (Weakly Interacting Massive Particles). Other particles have been predicted—but so far not found. This is embarrassing. There's more dark matter than atomic matter, and we can't identify it. We're trying. Watch for details on detection experiments at the Soudan Laboratory in Minnesota and at CERN in Switzerland.

Future astronomers have plenty of pieces of the dark matter puzzle to fill in.

That's Abell 2218 (the blue-ringed object) in the center of this diagram. This huge cluster of galaxies is one of the most impressive gravitational lenses we've found, acting like a natural telescope to the wonders beyond it. Light from galaxies behind Abell 2218 (gray arrows) gets bent by the cluster's formidable gravity. When the bent light reaches Earth (lower left corner), we see a mirage-type image of the more distant galaxies that's distorted into arcs and multiplied.

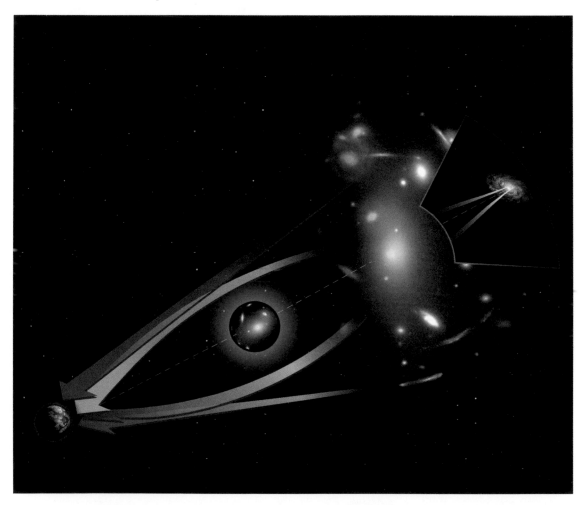

and about cosmic lenses.

Today we're using those natural starry lenses to look deep into spacetime. Gravitational lensing helps us measure the orbits and actions of distant stars and galaxies.

As for invisible matter, Zwicky was really ahead of his time on that. We now realize that the universe contains a whole lot of matter that can't be seen. We call it dark matter, and there is more of it, much more, than there is of matter we can see.

Vera Cooper Rubin (born in 1928) applied to Princeton University to study astronomy but was turned down because she was a woman. Undaunted, she became one of America's outstanding astronomers while raising four children and fighting for women's rights. Zwicky theorized that dark matter must exist; Rubin proved him right.

Vera Cooper Rubin was a kid when Zwicky was in his prime. From her bedroom window in Washington, D.C., she had a good view of the night stars and was hooked. At Vassar College, Rubin got a degree in astronomy. Then she went to Cornell University, where she was mighty lucky: Her teachers were Hans Bethe and Richard Feynman. After that, she earned a doctorate at Georgetown, where (lucky again) George Gamow was her advisor.

In the 1980s, using a telescope at the Kitt Peak Observatory in Arizona, Rubin decided to measure the velocities at which galaxies rotate. Working with astronomer Kent Ford, she saw something that startled them both. Stars and hydrogen gas moving at the outer edge of a spiral galaxy travel just as fast as those orbiting close to the starry center. What is out there on the fringe? There shouldn't be much gravitational pull at the outer edges of a rotating star system. And what keeps a galaxy from flying apart? There has to be some kind of matter—dark matter—that can't be seen.

Rubin remembered reading Fritz Zwicky's papers about what he called "missing mass." Her measurements of more than 200 galaxies proved that Zwicky's prediction was right.

Do you ever worry that you were born too late? That all the important discoveries have already been made?

Well, forget that worry. Cosmic explorers are needed. Twenty-first-century science is bursting with questions to be answered. Dark energy and dark matter are two of them. Here is a comment from Vera Rubin: "The joy and fun of understanding the universe we bequeath to our grandchildren—and their grandchildren. With over 90 percent of the matter in the universe to play with, even the sky will not be the limit."

A Surprising Information-Age Universe

Tomorrow we will have learned to understand and express all of physics in the language of information.
—John Archibald Wheeler, American physicist, "It from Bit" lecture (1989)

[I]nformation is as real and concrete as mass, energy, or temperature. You cannot see any of these properties directly, but you accept them as real. Information is just as real. It can be measured and manipulated.
—Charles Seife, American science writer, *Decoding the Universe*

[T]he universe computes, and because the universe is governed by the laws of quantum mechanics, it computes in an intrinsically quantum-mechanical fashion; its bits are quantum bits.
—Seth Lloyd, American quantum-mechanical engineer, *Programming the Universe*

Clocks were the most advanced technology of Isaac Newton's day. So, when Newton described the universe, the image he used was that of a cosmic clock.

This gorgeous 1410 clock (far right) on the Old Town Hall in Prague, Czech Republic, features a timepiece, a calendar, and 12 Christian apostles. The hand-cranked model of the Earth and Moon orbiting the Sun (near right) was built in 1712 for Ireland's 4th Earl of Orrery, Charles Boyle. Planetary mini-models, dubbed "orreries," were all the rage after Newton introduced his theory of gravitation.

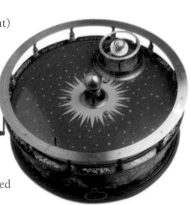

Then the steam engine arrived, and science began to view the universe as a vast, *energy*-powered engine. According to physicist Hans Christian von Baeyer:

> [I]n the brief span of twenty years [1840–1860], energy was invented, defined and established as a cornerstone, first of physics, then of all science. We don't know what energy is… but as a now robust scientific concept we can describe it in precise mathematical terms, and as a commodity we can measure, market, regulate and tax it.

Does technology lead us or vice versa? You won't find agreement on that. In a clock, the parts create the whole. So, in Newton's clock-driven universe, the goal was to understand those parts. In an energetic cosmos, the second law of thermodynamics (entropy) is in command, with fields as the vehicles of energy. And in a computer universe? Information rules.

Looking at the universe in digital terms is information-age thinking. Computer technology and space travel grew up side by side in the second half of the twentieth century. During the 1969 *Apollo 11* Moon landing mission, computers transmitted data at a mere 2,400 bytes per second compared to *3 million* bytes per second for these twenty-first-century computers (left) at the famous Mission Control Center at Johnson Space Center in Houston.

Those clock and engine metaphors led to astonishing leaps of science and technology. But today's technology is digital, and so—no surprise—the metaphor has changed. What's science's current image of the cosmos?

A computer.

Yes, some physicists are now describing the universe as a gigantic digital computer. It's the scientific metaphor of our times. America's eminent twentieth-century theoretical physicist John Archibald Wheeler writes, "The computer is built on yes–no logic. So, perhaps, is the universe."

MIT engineer Seth Lloyd says, "The original information processor is the universe itself. Every atom, every elementary particle registers information. Every collision between atoms, every dynamic change in the universe, no matter how small, processes that information in a systematic fashion."

The understanding of information—the science of information theory—is evolving. John Wheeler explains, "I think of my lifetime in physics as divided into three periods. In the first period . . . I was in the grip of the idea that Everything Is Particles. . . . I call my second period Everything Is Fields. . . . Now I am in the grip of a new vision, that Everything Is Information.

"Information may not be just what we learn about the world. It may be what makes the world," Wheeler adds.

In other words, information is not an abstraction. Like energy, "It can be **measured and manipulated**." For Wheeler and others, the universe is built on basic units of information. "It from bit" is a catchphrase of Wheeler's.

What's a *bit*? It's the smallest unit of information. A bit can be a particle or an antiparticle. It can be yes or no. True or false. Negative or positive. It can be a binary digit, either 0 or 1. Bit is actually short for "binary digit."

What's an *it*? An it is a system: an atom, a molecule, or the whole universe. "Its" make up the material world. When the bits in the universe are transformed into its, you get stars and galaxies and ants and elephants. **Everything in the universe is shaped by information's bits.** Or, think of it this way: Every *thing* (the "it") is only as real as the "bits" (the *information*) that describe it.

All the words, pictures, tunes, video clips, animated movies, and mouse-click commands on a computer boil down to an ocean of 0's and 1's (below), the two digits of binary programming code.

"Sure," say the skeptics. "Information is a useful idea, but it's not actual stuff." "Hold on," respond today's information theorists. "**Information is tangible.** It is real. It can be marketed and taxed."

Seth Lloyd is a spokesperson for the information-is-real camp. "Information is the ultimate 'substance' from which all things are made," he says. Run with this idea, and you have a digital universe that can be programmed.

Claude Shannon (1916–2001), born in Gaylord, Michigan, is worshiped as the founding father of information theory. Yes, I used the word *worshiped*. That may be going a bit far, but this man is a hero to a lot of followers. His mother was a high school principal; his father was a judge; and his grandfather, a Michigan farmer, invented farm equipment and a washing machine. Claude invented things, too, especially things that were fun: toys with clowns that juggled steel balls, a chess-playing machine, a specially balanced unicycle, a computer that calculated in Roman numerals, and lots more.

Math seemed like fun to him, and so he trained as a mathematician and an engineer at MIT and then went to work for Bell Labs in New Jersey. It was a freewheeling environment (pun intended). Shannon was sometimes seen juggling balls as he rode his unicycle down the halls of the laboratory. His employer didn't mind; Bell Labs was interested in his brain. Shannon was asked to figure out how many telephone conversations could be carried by one phone line, without one

Information is a substance; information is a metaphor. Which is it? Remember the Bohr-Einstein argument about the reality of photons? And the disbelief when Michael Faraday described energy-carrying fields as something substantial? We now know photons and energy are real. What about information?

Claude Shannon (below) demonstrates his mouse maze, one of the first experiments in artificial intelligence. Theseus the mouse is a magnetic machine controlled by relay circuits. "Born" in 1950, it searched the maze for a target and then used that experience to find the prize from any point within the maze.

Chapter 37 of *Newton at the Center* explains the second law of thermodynamics in more detail, but the short version is: Energy spreads out if it can, and the spread is always from hot to cold. Entropy is the measurement of the spread.

call interfering with another.

Engineers are used to analyzing things like bridges, where, if you know the mass and measurements of the structure, you can determine how many vehicles it will support. But measuring the size and weight of a telephone line won't tell you the number of conversations it can carry. To figure that out, Shannon used what he knew:

- the way telephone switching circuits turn on and off;
- Boolean logic (based on a binary system of 0's and 1's);
- and the second law of thermodynamics, which deals with the entropy or the randomness in a system.

Mathematician George Boole (1815–1864), profiled in this 1847 pencil sketch, developed a system of binary logic (0's and 1's only) now named after him. Boolean logic is the basis for today's computer-programming languages. Using "and" in a data search ("tigers *and* lions") means both items must be present to return a hit, shown here as an "out." Likewise, on the "AND" chart (bottom line), you must input two 1's to get a 1 (a hit). "Or," as in "tigers *or* lions," brings up data with either item. On the "OR" chart, either one or two 1s will return a hit (bottom three lines).

AND				OR		
IN	IN	OUT		IN	IN	OUT
0	0	0		0	0	0
0	1	0		0	1	1
1	0	0		1	0	1
1	1	1		1	1	1

Shannon saw a connection between telephone switches and Boolean logic. He linked entropy to a shortage of information in a message and discovered that many messages can be significantly shortened without losing their meaning.

Shannon came up with a way of analyzing a message or a system that allows it to be measured and dealt with mathematically, much as mass or energy is measured. His big idea was to translate the message (the bits) into binary code and then calculate with the 0's and 1's. Binary wasn't an arbitrary choice; Shannon, building on George Boole's mathematics, proved that binary's digital bits are the most energy-efficient way (by far) to process information. He paved a road that computer designers were soon walking.

Shannon laid the basis for *classical* information theory. Note I said "classical." Bits and its are classical entities. Shannon's formulas helped initiate the electronic communications age with its digital computers, iPods, and BlackBerries. But we live in a quantum world. Classical science explains only its surface.

"The world cannot be a giant machine, ruled by any pre-

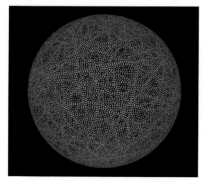

established continuum physical law," says Wheeler.

What's a continuum? For classical physicists, that much-used word means something without gaps, like the flow of water. We now know that nothing is continuous. Matter and energy are made of tiny lumps (Planck's quantum particles) that lead an uncertain life in a superposition—as both wave and particle—until they are measured and become one or the other.

Each of those quantum lumps has an identity and carries information about itself. That information can be added, subtracted, combined, and measured. **Classical quantum mechanics answers questions with a yes or a no, with hot or cold, but not both at the same time.**

John Wheeler and others have taken the next step, building on quantum mechanics, but going on to *quantum information theory*. Understanding it will take us far beyond the digital computers dominating the early twenty-first century.

With quantum information theory, you can have both yes and no at the same time. You can deal with Schrödinger's alive-and-dead cat (chapter 19). You can manage two or more calculations simultaneously. But, to do that, you need something in place of classical bits.

Physicists William Wootters and Benjamin Schumacher got together at Kenyon College in Ohio in 1992 (where Schumacher was a professor) and wished there were a quantum measure of information. Jokingly they called that measure a *qubit* (KYOO-bit), for "**quantum bit.**" "We laughed," said Schumacher, "but the more I thought about it, the more

Eric Heller of Harvard University created this computer model of a wave-particle duality. Those crazy paths along the surface of a sphere simulate a particle creating "wave trains" that collide and produce quantum chaos.

To a scientist, **CONTINUOUS** means "solid." The table I'm writing on seems solid, but it is made of atoms, and there's movement and emptiness inside those atoms. On a quantum level, nothing is solid or continuous.

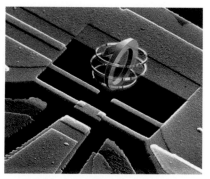

Qubits ("quantum *bits*") can be both 0 and 1 at the same time, existing in a superposition, which allows quantum computers to carry out two operations simultaneously. The image at left is a micrograph of highly magnified quantronium, a single quantum circuit, with a spinning symbol added.

Charles Babbage (1791–1871) never finished any of his early computing machines. Experts at the Science Museum in England wondered how his designs would work, so they constructed this Difference Engine No. 2 (top) from Babbage's 1847 plans. Today's quantum computers (bottom) take the same basic system of 0's and 1's—translated into on-off switches or open-closed gates in a circuit—to a new, powerful level.

I thought it was a good idea."

Today, quantum information theory uses Schumacher's qubits, which can be both 0 and 1 *at the same time*. Like a quantum wave/particle, a qubit exists in a superposition until it is measured.

And just as a photon or an electron can go through two slits at the same time, a qubit can perform two computations simultaneously. Seth Lloyd is a quantum-mechanical engineer. So he thinks of the universe as a quantum computer, and that's quite different from seeing it as a digital computer. (Digital technology follows the rules of classical science; quantum technology follows quantum science, which includes superposition and entanglement.)

Imagine a quantum computer: A quantum bit is telling that computer what to do. It instructs that 0 means "Do this!" (add 2 + 2) and 1 means "Do that!" (add 3 + 1). Now suppose that the quantum bit is in a superposition of 0 and 1. Then the bit tells the computer to "Do this" and "Do that" simultaneously. In other words, if a computer is a quantum computer, it will oblige by going into a quantum superposition. In one part of the superposition, the quantum computer is adding 2 + 2; in the other part, it is adding 3 + 1.

But quantum computers are not limited to two qubits working on simultaneous computations. There can be three or more. With each added qubit, there's an exponential leap in power. Lloyd says, "Even a small number of qubits allows an extraordinarily rich texture of interfering waves as they compute. A quantum computer given 10 input qubits can do 1,024 (or 2^{10}) things at once. A quantum computer given 20 qubits can do 1,048,576 (or 2^{20}) things at once. One with

300 qubits of input can do 2^{300} things at once, which is more things than there are elementary particles in the universe."

A bit represents a single binary digit—a yes or no. A qubit can potentially carry an unlimited amount of classical information.

But, like a quantum particle when it is observed, after you use a qubit in a calculation, it is no longer in an uncertain state. Observing a qubit changes it. It becomes a classic bit—a fundamental quantum of human knowledge: an up or down, a right or left, a 1 or 0.

That observation thing is at the heart of quantum weirdness. The observer becomes a part of the action. In the past, we were told that the greater universe had no need for us humans. We were insignificant. But quantum theory suggests that might not be so. Here's Wheeler again: "The physical world is in some deep sense tied to the human being."

In Newton's universe we had no role to play. We could observe but not determine scientific action. Galileo and Einstein put us (reluctantly) in the cosmic picture. With relativity, your point of view (from the ship's hold or the shore) determines what you see. As for quantum theories, they are observer dependent. A photon (or electron) doesn't

Information Puts Us in the Picture

We started our scientific venture standing with Aristotle and Ptolemy on a giant Earth moored at the center of the world. Like royalty living in isolation in a huge castle, we were convinced we held the universe's best real estate, although we didn't have a clue as to what else was out there.

Then along came Copernicus, Galileo, and Newton, and, in a surprising twist, as our vision expanded and our minds gained power, our place in the cosmos shrank.

Now, several hundred years after those giants walked this planet, science is telling us humans that we may actually be players in the universal drama. We may have a *participatory* role—and all because of a new science called information theory.

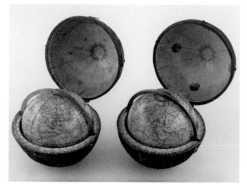

We earthlings are no longer isolated in a world of our own, as these nineteenth-century pocket globes (ca. 1830) seem to suggest. We're players in the greater universe.

How will quantum computing change your world? Imagine all the information in your library stored inside a computer small enough to hang from an earring. This quantum computer will be able to answer almost any question you can ask. But will you know the questions to ask? That's what will be important. Thinkers will be in demand.

These goggled physicists at the Australian National University are manipulating the spin state of a quantum information processor optically—with lasers. The team's claim to fame? Demonstrating "the first solid-state 2-qubit logic gate."

take on a particle or wave identity until something or someone engages with or measures it.

In science's evolving vision, we humans are *both* observers and participants. Some of today's physicists are now telling us that **ours is a "participatory universe."**

Murray Gell-Mann is a skeptic. "In quantum mechanics there's been a huge amount of mystical nonsense written about the role of the observer." But then he hedges his bets: "Nevertheless, there is some role for the observer in quantum mechanics, because quantum mechanics gives probabilities. And you want to talk about something that is betting on those probabilities. That's the real role of the observer."

One thing is clear: A universe with uncertainty at its core is not like a big machine. Reducing it to its parts does not provide all the answers. Considering the whole is more productive.

Today's physics has given us a dazzling array to consider. We not only have uncertainty to put in our mental backpacks, we also have black holes, dark energy, dark matter, entanglement, information theory, quantum gravity, string theory, teleportation, virtual particles, and wormholes—and that's just for starters. There's nothing mechanical or clock-like about any of them.

We may be inhabitants of a smallish planet off in a

Our cells are information-preserving engines, and they perform beautifully. Our genetic information remains virtually undisturbed after generations and generations of duplications.
—Charles Seife, *Decoding the Universe*

backwater of a galaxy not much different from billions of other galaxies, but we have something special going for us. Our information-processing DNA lets us think and compute. Most of the information processing in the universe—one atom whamming into another—is unthinking. We have the ability to analyze. Human thinkers, working collectively, wrought wonders in the twentieth century. Is there more to come?

John Wheeler seems to think science has been taking baby steps and that the next discoveries might truly change the way we live in our universe. Here he is asking for a leap in human thought: "How can physics live up to its true greatness except by a new revolution in outlook which dwarfs all its past revolutions? And when it comes, will we not say to each other, 'Oh, how beautiful and simple it all is! How could we ever have missed it so long!'"

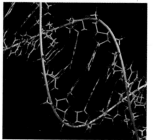

Like computers, DNA molecules use a code for processing information. The DNA code is not binary--it has four digits, representing chemical bases (here colored red, yellow, blue and green). The pattern of the bases determines which protein will get made. DNA usually exists as a pair of molecules twisted like a vine (a double helix) here shown as yellow-orange threads.

A Poet in Search of a Loom

In a 1999 essay titled "Pulling Diamonds from the Clay," physics Nobel laureate Murray Gell-Mann (who discovered quarks) considered the problem of finding meaning in a sea of information. For inspiration, he turned to these lines from American poet Edna St. Vincent Millay:

Upon this gifted age, in this dark hour,
Falls from the sky a meteoric shower
Of facts... they lie unquestioned, uncombined.
Wisdom enough to leach us of our ill
Is daily spun: but there exists no loom
To weave it into fabric; undefiled
Proceeds pure science, and has her say.

This picture looks like a meteoric shower; it's actually a simulation of a molecular solid (top) transitioning into a quantum liquid (bottom). It illustrates what we think happens to hydrogen under very high pressure. The data processing of this simulation set a speed record in 2006 on the Blue Gene/L supercomputer: 207.3 trillion floating-point operations per second (nicknamed "teraflops").

Is Anyone Out There?

Space, the final frontier. These are the voyages of the starship *Enterprise*. Her five-year mission: to explore strange new worlds, to seek out new life and new civilizations, to boldly go where no man has gone before.
—Gene Roddenberry (1921–1991), American television producer, *Star Trek*

A thousand years ago, everybody knew…the earth was the center of the universe. Five hundred years ago, they knew [the Earth] was flat. Fifteen minutes ago, you knew we humans were alone on it. Imagine what you'll know tomorrow.
—Tommy Lee Jones (playing Government Agent Kay) to Will Smith (Agent Jay) after meeting aliens in the Barry Sonnenfeld movie *Men in Black*

If aliens exist in our universe, they may understand and "package" aspects of reality very differently from the way our own minds do, but if we ever establish contact we're assured one common interest. They will be made of similar atoms and governed by the same physical laws.
—Martin Rees, from his essay "Cosmological Challenges: Are We Alone, and Where?"

Do you believe in aliens? Nuclear physicist Enrico Fermi was skeptical. He asked, "If extraterrestrials are commonplace, where are they?"

The search for intelligent life on another planet has, so far, turned up nothing. A few serious scientists think that conditions on Earth are unique and that we are alone in the universe. But others say we've only looked in our celestial neighborhood, we haven't looked very hard, nor for very long, and so we can't really know if others are out there.

According to Nobel Prize–winner Christian de Duve, a Belgian biochemist, "Life is almost bound to arise…

Enrico Fermi's casual question, "Where are they?", triggered a torrent of speculation and imagined aliens—that's ET below, a friendly variety. The question is now called the Fermi paradox.

If aliens exist on distant planets, what do these creatures look like? That's a question for fans of both science fiction (imaginative fun partly based on science) and science (facts verified by observation and experimentation). Scientists, not writers, custom-designed these bipedal aliens (left) as animals that would be ideally suited to the environment of an imaginary planet called Aurelia. The so-called "gulphogs" are taking shelter from a solar flare of their sun, a red dwarf star that never sets. These and other according-to-science aliens were featured in a National Geographic special called *Extraterrestrial* (2005).

wherever physical conditions are similar to those that prevailed on our planet some four billion years ago."

Astrophysicist Neil deGrasse Tyson, who is director of New York City's Hayden Planetarium, says, "To declare that Earth must be the only planet in the universe with life would be inexcusably egocentric of us."

Martin Rees, president of the Royal Society in London, says, "The prime exploratory challenge of the next fifty years is . . . to seek firm evidence for, or against, the existence of extraterrestrial intelligence."

Even Isaac Newton believed in the possibility of alien life. In his private papers, he wrote that, just as the Earth was populated, "so may the heavens above be replenished with beings whose nature we do not understand."

Many of today's cosmologists, astrophysicists, astrobiologists, and others are now convinced that some kind of life will be found elsewhere in the universe. Will it be intelligent life, and will we actually communicate? We'll never know unless we look. And we're doing just that. The search for extraterrestrial life, which a few decades ago was mostly the domain of science fiction, has now gone mainstream.

But, despite what science fiction and

The Space Age began on October 4, 1957, when the Soviet Union launched the first satellite, *Sputnik I* (below), into orbit. It circled Earth for 92 days. That same year, the first really large radio telescope was built at Jodrell Bank Experimental Station in Great Britain.

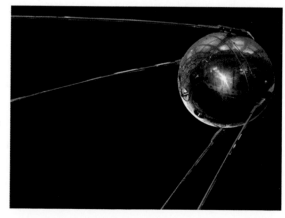

Light takes 1.3 seconds to travel from Earth to the Moon. A spaceship needs about 14 hours to make the same journey.

the movies say, we aren't going to find intelligent life beyond our solar system by climbing into a rocket ship and voyaging into space, at least not with our current technology. The Milky Way galaxy contains about 100 billion stars. Those stars are typically some 50 trillion kilometers (30 trillion miles) apart. Light needs about five earth years to travel that distance. But *only light can go at the speed of light*. And, right now, we don't have the know-how to send a vehicle at anything close to the speed of light.

But we do have the technology to make **interstellar communication** a possibility—if there are other civilizations with the same idea and they are transmitting or listening to signals. About 10 percent of the stars in our galaxy are Sun-like; we now know that many of them have orbiting planets. One thousand of those stars are within 100 light-years of us. Their planetary systems are close enough to make contact.

Where do we start?

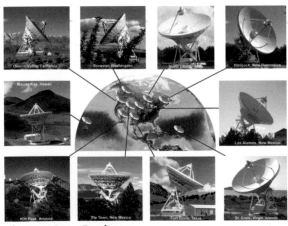

The Very Long Baseline Array (VLBA) is a network of 10 radio telescopes controlled by the National Radio Astronomy Observatory in New Mexico. Astronomers are tuning into what the universe has to tell them. What if life is found elsewhere? Will that change our relationship to the universe?

• One way to search for intelligent life is by sending a radio-wave message from Earth and waiting for an answer. That's a long shot. Our signal broadcasting technology is less than a century old, and we can't yet broadcast signals strong enough to travel vast distances.

• Another strategy is to listen for a cosmic "hello" coming from space. We're assuming there are alien civilizations more advanced than ours capable of beaming very strong radio signals.

• Some astrophysicists are suggesting that we listen for the everyday signals of civilization that get leaked into space—like radar, TV, and FM radio. A new radio telescope array in Australia, the Mileura Widefield Array, is designed to detect hydrogen molecules from the early universe but could also look for suspicious broadcasts from nearby stars.

Roman Atoms

The Roman poet Titus Lucretius Carus, known as Lucretius, wrote of "atoms" and "other worlds" in the first century B.C.E. These lines come from Book II of his six-volume work *De Rerum Naturae* (*On the Nature of Things*). In it Lucretius puts the theories of Democritus and Epicurus on the origins of the universe into exquisite Latin verse. Lucretius, who was also a philosopher, believed that superstition and ignorance were the cause of most evil.

*And so I say again, again you must confess
That somewhere in the universe
Are other meetings of the atom stuff resembling
   this of ours,
And these the aether holds in greedy grip.
For when the atom stuff is there,
And space in which the atom stuff may move,
And neither thing nor cause to bring delay,*

*The process of creation must go on; things
   must be made. . .
Why then confess you must
That other worlds exist in other regions of
   the sky,
And different tribes of men, kinds of wild beasts.
This further argument occurs:
Nothing in nature is produced alone;
Nothing is born unique, or grows unique, alone.*

Cornell astronomer Frank Drake was working with radio telescopes in Green Bank, West Virginia, in 1958 when he asked himself how we on Earth could make contact with others elsewhere in the universe—if others exist. The following year, Cornell astrophysicists Philip Morrison and Giuseppe Cocconi, working independently, answered that question in the respected British journal *Nature*. They said the most efficient way to communicate through space is by using the microwave frequencies of radio waves. They even suggested starting with the frequency of neutral hydrogen atoms (1,420 megahertz). "The probability of success is difficult to estimate," they wrote, "but if we never search the chance of success is zero."

In 1960, Drake tuned to that radio frequency and spent two months intensely listening for organized signals coming from two nearby Sun-like stars: Tau Ceti and Epsilon Eridani. He called his effort "Project Ozma," after the princess in L. Frank Baum's *Wizard of Oz* books. Drake

Frank Drake stands firm in his belief that aliens exist. Behind him is the National Radio Astronomy Observatory at Green Bank, West Virginia, where he carried out early astrobiology research in 1960.

Drake's Equation

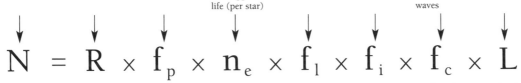

| Number of communicating civilizations in our galaxy | Rate of suitable stars formed per year in our galaxy | Percentage of suitable stars that have planets | Average number of planets (or moons) that can support life (per star) | Percentage of hospitable planets where life exists | Percentage of populated planets with *intelligent* life | Percentage of intelligent life than can communicate with radio waves | Average lifetime of communicating civilizations |

$$N \;=\; R \;\times\; f_p \;\times\; n_e \;\times\; f_l \;\times\; f_i \;\times\; f_c \;\times\; L$$

| Drake's 1961 estimate was 10,000. The lowest is 1, meaning us—and only us—in the galaxy. | Drake chose 10; current estimates are closer to 1 with some up to 6, if you include stars unlike the Sun. | Drake chose 50%; newly found exoplanets puts this as high as 90%, but it could be as low as 10%. | Drake chose 2. Our Sun has at least 1 (Earth), but Mars and a couple moons might qualify. | Drake chose 100%, meaning that if life can form, it will. Others are less optimistic. | We have no idea. Also, how do you define "intelligent" life? There's a very wide debate. | We have little clue. We're the only species that can do it out of many millions on Earth. | Drake chose 10,000 years, but who knows? Our "radio days" number less than a century —since 1937. |

The Drake equation estimates the number (*N*) of civilizations in the Milky Way galaxy that can send radio waves into space. The challenge is that all the values on the right side are either inexact (like *R*) or just plain unknown. The last figure, *L*, is the haziest but may be the most important: How long do advanced civilizations last? We have only ourselves as a guide—with our few thousand years of art, religion, warfare, science, and historical records. How much longer will we stick around? The longer we have to develop our capabilities, the better our chance of connecting with others.

Sun

A 2005 survey of 30 million stars gave us this illustrated map of the Milky Way galaxy, as viewed from above.

didn't hear any conversations from outer space—but Project Ozma got him started.

Soon after that, Drake invited all the scientists and engineers he knew who were serious about finding life elsewhere to a meeting at Green Bank. All 12 came. To help focus their search, Drake devised a formula, now known as the Drake equation (see above).

The goal of the equation is to find N, which stands for the number of civilizations in the Milky Way galaxy capable of communicating with us. The figure R is the approximate number of stars born in our galaxy each year. If you know R you can begin to narrow the estimate for N by asking informed questions: How many stars are stable? How many

The **HABITABLE ZONE** is a region in which a planet is warmed enough by its sun—and inner radioactivity—to let life exist and the right distance away from that sun so that water doesn't all boil away or freeze. Our solar system is within the galactic habitable zone (GHZ)—just the right distance from the Milky Way's dense, hot center.

Nature is strictly governed by impersonal mathematical laws.... If we ever discover intelligent creatures on some distant planet and translate their scientific works, we will find that we and they have discovered the same laws.
—Steven Weinberg, American physicist

have orbiting planets in what is known as a habitable zone? And how many have the right chemistry to be life-friendly?

The Drake equation supposes that life on Earth came about through ordinary processes and so life is likely to appear elsewhere. "No freak events are required," wrote Drake in the 1980s. "In the Universe, anything that can happen will happen.... The existence of life, and intelligent, technology-exploiting life, in large quantities, should not be a surprise."

Back in 1961, Drake and his colleagues at the Green Bank meeting conjectured that there should be a thinking civilization within 1,000 light-years of Earth and an estimated 10,000 such civilizations in the Milky Way galaxy. Lots of people thought this was pie-in-the-sky thinking, but a charismatic Harvard professor named Carl Sagan was one who didn't. By 1968, Sagan was at Cornell following dual careers. He was a respected astrophysicist as well as a writer of popular books and TV and movie scripts. Sagan took the ideas and optimism behind the search for alien life and brought it to a mass audience.

Sagan, Drake, and others founded the SETI (Search for Extraterrestrial Intelligence) Institute in 1984. Observations began on October 12, 1992. That October date was picked carefully: It was the 500th anniversary of

Astronomer Carl Sagan (1934–1996) extended Drake's formula to one that includes the whole universe. In 1966, Sagan (above) and Iosif Shklovskii (a Soviet astronomer) published a pioneering book, *Intelligent Life in the Universe*. It was a rare example of Soviet-American collaboration during the Cold War, and it led to joint space exploration programs.

Love and War

The symbol *L* in the Drake equation stands for longevity—how long a technological society lasts before it obliterates itself. The Roman poet Ovid commented on the human tendency to self-destruct using "clever" inventions. Here's a bit of his long love poem, *Amores*:

Clever human nature,
Victim of your inventions,
Disastrously creative,
Why cordon cities with towered walls?
Why arm for war?

Using the Drake equation means working with probability theory: When you can't pinpoint most specifics in advance, you use statistics to help you understand what you do have. But, because there are many unknowns, the equation gives an incredibly wide range of results: from one (Earth) to billions. Some say that's useless, while others see it as a way to organize and frame the search.

Most radio telescopes are devoted to astronomical pursuits other than SETI. The Southern SETI project based in Argentina (right) and the Allen Telescope Array (ATA), illustrated at left, are two exceptions. Southern SETI has a clear view of the center of the galaxy, where stars and their presumed planets are more plentiful. When ATA is complete, some 350 radio dishes will collect radio energy in the hopes of finding a signal that just might indicate intelligent life.

what was probably the greatest *Earth-changing* feat of exploration ever. Then, and now, those involved with SETI, NASA, and other space exploration efforts know that **if intelligent life is found elsewhere, it will be the greatest act of human exploration ever.**

Today, SETI conducts a round-the-clock cosmic search using the Allen Telescope Array located near Mount Lassen in northern California. That array (funded by Microsoft co-founder Paul G. Allen) is made up of relatively small radio dish receivers 6.1 meters (23 feet) in diameter tethered on a mountain ranch about 466 kilometers (290 miles) northeast of San Francisco. Plans call for 350 receivers. Each dish can collect EM waves in several radio bands at the same time. The most tuned-into span of frequencies is known as the "water hole" because it is a relatively noise-free region in the radio spectrum. It also includes the emission frequencies of hydrogen (H) and of the hydroxyl molecule (OH), which together form water (H_2O), a necessary component for life as we know it. Just as animals congregate at local water holes, we're guessing that intelligent civilizations will congregate around these optimal radio frequencies.

The Allen Telescope Array is expected to check out 100,000 stars by 2015. Data from its radio telescopes and others is crunched at the SETI Institute with the aid of computer users around the globe. The search is for any unnatural patterns that might indicate *intentional* cosmic radio signals.

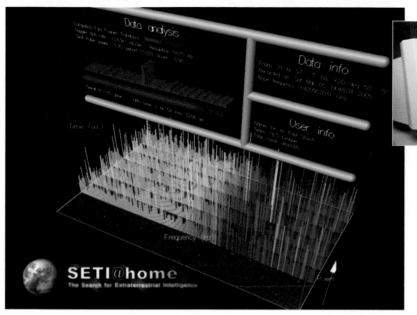

SETI@home
The Search for Extraterrestrial Intelligence

The popular SETI@home screen saver allows anyone with a computer to join in the search for ETs. When otherwise idle, millions of machines crunch an astronomical amount of data collected by radio telescopes. The collective hope is to find an unnatural pattern—a persistent signal at the frequency of hydrogen or a series of prime numbers, for instance.

The hunt for life-supporting planets is often called the **"Goldilocks" search** (after the three bowls of porridge part of "Goldilocks and the Three Bears"). The right planets can't be too hot or too cold. (It's that habitable zone.) And, if the life-forms are anything like us, a suitable planet should be orbiting a star like our Sun that isn't too big or too little.

Besides that, elements such as carbon, hydrogen, oxygen, and nitrogen should be present (known collectively as CHON). And water is needed as a transporting medium for cell growth. So the search is for planets with liquid water and those key CHON elements.

Meanwhile, biologists and astrobiologists are using new technologies to analyze Earth's life-forms to a degree never before attempted. All earth life that we know about speaks a genetic language based on DNA. "It is as if, instead of the thousands of different human languages, everything were written in the same alphabet using words with the same meaning," say physicists Joel R. Primack and Nancy Ellen Abrams. Could there be a different language for life elsewhere? Might it not be based on DNA? Are the conditions that support life on Earth essential throughout the universe? Does alien life have to resemble us? No one knows.

Keep in mind, if a civilization 300 light-years away is studying Earth, they're seeing us as we were 300 years ago, in the early 1700s—without airplanes, without automobiles, without telephones, without electric lights, and without radio technology through which to talk back. There were no radio telescopes before 1937.

Kepler (illustrated right) is a satellite telescope (due for launch in 2008) whose sole mission is to stare at a patch of stars for signs of small, rocky planets. It will look for periodic changes in brightness that happen when a planet transits (moves in front of) a star.

If we find evidence of any kind of life, past or present, on a planet or moon elsewhere in our solar system, it will suggest that basic life-forms may be widespread in the cosmos. And, in December 2006, scientists studying images taken by NASA's Mars Global Surveyor announced that they found convincing signs of recent mud flows on Mars. Mud flows? They mean liquid water and the

Questions on the Astrophysicist's Blackboard

Did life arise on Earth, or was it brought here from another planet?

Is ours the only kind of life that has appeared on Earth, or did earlier forms develop and then get wiped out?

Intelligent life has needed almost all of Earth's 4.5-billion-year history to evolve. Despite mass extinctions, major asteroid strikes, and big climate swings, our planet has remained essentially stable during that whole period. How important is that stability?

Do technologically advanced civilizations destroy themselves?

Geological activity on Earth—like volcanic eruptions and shifting continental plates—helps to recycle carbon and other life-essential elements. Are there other ways to provide for those

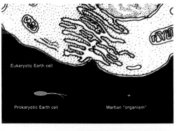

elements? Or might life arise from different elements elsewhere?

Our Moon stabilizes Earth's rotation and climate. How many planets have big moons as we do? Could life on Earth have evolved without that moon?

Regions close to the center of the Milky Way galaxy (where at least one huge black hole exists, along with energetic supernovas) are filled with dangerous radiation that probably would make life impossible. Regions far from the center have few supernovas and thus little stardust; stardust contains the heavy elements that are essential for rocky planets and life. So, what is the galactic habitable zone (GHZ)?

Suppose we are unique; could we possibly "seed" other planets and spread life as we know it? We know that our planet will not live forever. Can we find another home?

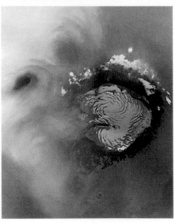

Did Mars (bottom) harbor life in a wetter, warmer past? Could fossil traces of that life travel to Earth by way of meteors? A Mars rock named ALH84001, found in 1984 in Antarctica, harbors a mysterious microscopic form. Is it life? Not as we know it. The suspected microorganism is far smaller than any Earth cells—a tiny dot (lower right corner) compared to familiar cells (lower left, top illustration).

possibility of microbes. Finding microbial life on Mars would be huge news—you can be sure you'll hear of it.

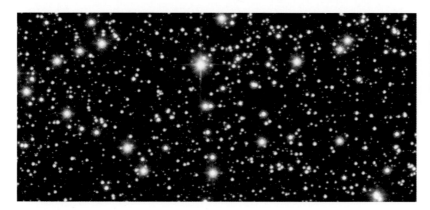

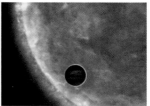

How do you spot a tiny, dark dot of a planet among a dazzling display of stars (left)? Indirectly, for the most part. The illustration above shows one technique: the transit of a planet in front of its star. The crossing causes the star's apparent brightness to drop by a small but detectable percent. Astronomers use powerful space telescopes to search for and measure these brief but periodic changes. The first exoplanets found during transits were, not surprisingly, very large planets extremely close to their stars. Nicknamed "hot Jupiters," they complete an orbit within days or even hours.

Just a few years ago, the only planets we knew were those in our solar system. Since then, for believers, there's been lots of good news: Thanks to sophisticated planet-finder telescopes, we have now discovered several hundred planets orbiting stars other than our Sun. They're called **extrasolar planets** (dubbed exoplanets). We can reasonably assume that the universe is filled with exoplanets. What we've found so far is a great variety of planets. The diversity was unexpected.

The first exoplanet was discovered in 1992 orbiting a pulsar. As you'll recall, pulsars are rotating neutron stars that send out intense radio waves—not a likely place to look for life-forms. But in 1995, Didier Queloz and Michel Mayor of the Geneva Observatory in Switzerland announced that they had found an exoplanet circling a Sun-like star. It turned out to be a huge, gassy, hot, Jupiter-like planet and so near its parent star that it completes an orbit every 4.2 earth days. A "hot Jupiter" is not a good candidate for life, but having a Jupiter-like planet in the neighborhood (as we do) is a good thing, since large planets help to protect smaller ones from comet and asteroid devastation. The discovery surprised Queloz, who said, "You realize that our solar system is just one example of the many ways that nature is building planets."

Once one was found, other exoplanets soon turned up. Most have elongated, elliptical orbits, in contrast to the close-to-circular planetary orbits in our solar system. Planets

Picture a watery, low-gravity world with a thick, carbon-rich atmosphere. What alien plants would flourish there? A tall forest of pagoda-shaped trees (background)? Hydrogen-filled balloon plants rising to soak in more sunlight? These are two more scientifically designed life-forms from the show *Extraterrestrial.*

Some "World" History

When you talk about the history of the world, we earthlings are newborns in the parade of time. The universe is thought to be 13.7 billion years old. Our solar system was formed about 4.5 billion years ago. Earth's birthday was some 4.65 billion years ago.

And people? While some of our hominid ancestors were walking on two feet at least 4 million years ago, our human ancestor—*Homo sapiens sapiens* (remember, that's us)—might only be about 300,000 years old (archaeological finds keep changing that date).

In universe time, we people arrived on Earth a hiccup ago.

with elliptical orbits have extremes of hot and cold, not a likely environment for advanced life. Nor is a gassy planet. What we're looking for are rocky planets in a habitable Goldilocks zone.

Frank Drake, who believes we will find them, says:

> *With the recent results suggesting the possible earlier existence of simple life on Mars, and the chance that undersea life may be found on Europa [a Jupiter moon], we've witnessed a shift in our perceptions about the prevalence of biology. I am sure that life is not a rare occurrence at all, but is as natural throughout the universe as the formation of planets and stars.*

Celestial Goggles

When it comes to "seeing" the universe, modern telescopes have given us a new set of glasses. As long as we thought we were:
- top dogs
- the only important planet
- the focus of the universe
- at the center of everything,

then it wasn't terribly important to look beyond ourselves. We were like the ancient Greek pretty boy Narcissus—in love with our own image. Our perspective was limited by the space we occupied.

As we began to extend our vision, the questions changed and so did the answers. We realized that our poets and science-fiction writers—like H. G. Wells and Jules Verne and Isaac Asimov—were onto something. The unexpected is sometimes to be expected.

Fishing for Planets

Like fishermen sitting in the middle of a school of trout, astrophysicists are finding that our home planet floats in a celestial sea filled with solar systems. They've discovered that many of those solar systems include noteworthy planets. Do some of them sustain life?

"Rocky planets [like Earth] are almost certainly common," says Geoff Marcy of the University of California at Berkeley.

A red dwarf star named Gliese 581 seems to have at least three planets in orbit, including one that is about one and one-half times larger in radius than Earth, five times heavier, and perhaps composed of rock and water.

The red dwarf Gliese 581 (top right of the illustration) has at least three planets that are 5, 8, and 15 times the mass of Earth.

Astronomers discovered this so-called "super Earth" by the wobble it creates in Gliese 581's orbit. The planet is extremely close to its star, orbiting every 13 days, but red dwarfs are much less luminous than the Sun, and so the planet is within a habitable zone.

"On the treasure map of the universe, one would be tempted to mark this planet with an X," says Xavier Delfosse, a French member of the multinational team that made the discovery.

Epsilon Eridani is a young star system that, at just 10.5 light-years away, harbors the closest known exoplanet to us (left illustration). The gas giant is unlikely to support life but perhaps it has hospitable moons. On the European Space Agency drawing board is a posse of six infrared space telescopes called Darwin (right). They'll hunt for water and other vital life-support molecules in the atmospheres of exoplanets.

Can we go there and check it out? Geoff Marcy, speaking of the star Tau Ceti, said, "Ultimately we need to go with robotic spacecraft and a small digital camera." Tau Ceti is only 12 light years away. If we could travel at 99 percent of the speed of light, we could make it in your lifetime. But at current rocket speed, the journey would take about 500,000 years. Gliese 581, at 20.5 light-years away, would take even longer. So, for now, our exploration of alien worlds will have to be strictly long-distance.

Albert Einstein asked a profound question: "Did God have any choice in the creation of the world?" In other words, do things have to be as we know them, perhaps for deep mathematical reasons? Or are other rules possible? Could there be other kinds of universes? As yet, we can't answer those questions.

This Is the Last Chapter, but It's Not the End

For myself, I like a universe that includes much that is unknown and, at the same time, much that is knowable. A universe in which everything is known would be static and dull....A universe that is unknowable is no fit place for a thinking being. The ideal universe for us is one very much like the universe we inhabit. And I would guess that this is not really much of a coincidence.
—Carl Sagan (1934–1996), American astronomer, "Can We Know the Universe? Reflections on a Grain of Salt" essay

The good mate said: "Now must we pray,
For lo! the very stars are gone.
Brave Admiral, speak, what shall I say?"
Why, say, "Sail on! sail on! and on!"
—Joaquin Miller (1837–1913), American poet,
from "Columbus"

A meteorologist launches a radiosonde, a device designed to measure and transmit weather data at varying altitudes.

Will it rain on the parade you're planning for two weeks from today? Sorry, no one can predict that—and be sure. We can tell you with astonishing detail what happened billions of years ago in the universe, and we can tell you what is going on in the nuclei inside the atoms of your body. But will it rain in two weeks? Feed anything you want into the computer, like weather data from around the world. It doesn't matter. It could rain on your parade. Or maybe the sun will shine. No one can tell you for certain which it will be.

You see, sometimes repeated actions—scientists call them iterations—can lead in surprising directions. One butterfly's breeze? One car's exhaust? Neither amount to much. But iterating that breeze or that exhaust can stir up a hurricane or threaten a planet. Is this an exaggeration? Maybe not. Small things that get iterated can lead to big happenings—and then sensible predictions get blown away.

Can the whisper of wind from a butterfly's flutter-by lead to a hurricane? (That's Hurricane Rita above, color-coded by rainfall.) We can't predict that chaotic chain of events, but quantum experiments have given us new tools for dealing with the ways of the universe. This computer graphic (left) maps the dance of cosmic particles called muons, with a half-life of two-millionths of a second, through a detector deep in a salt mine near Cleveland, Ohio.

The classical scientists believed that understanding nature would bring the power of exact prediction. But the new sciences, dealing with broad information, lost that surety; the new reality turned out to be much more complex than we had imagined (and much more interesting). Even our

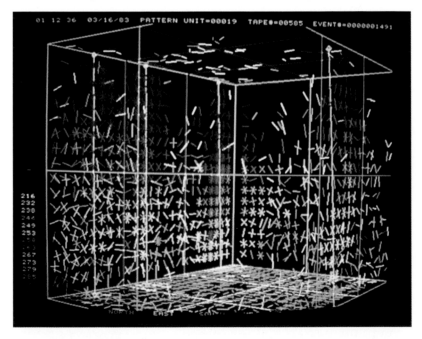

fanciest computers and our most brilliant minds can't seem to figure out some things in advance—like the way the cream in your coffee cup will pattern its swirl, or why one person and not another has a heart attack, or exactly when an epidemic will run its course, or what the stock market will do next month. According to biologist Stuart Kauffman,

The Flea Effect

So nat'ralists observe,
a Flea
Has smaller Fleas
that on him prey,
And these have
smaller still to
bite 'em,
And so proceed
ad infinitum.

—Jonathan Swift (1667–1745), English satirist, from "Poetry: A Rhapsody"

Quantum mechanics, as we know it, doesn't quite work with gravitation. That's messing up our theories. We think there might be a missing particle; we're calling it a Higgs boson. If it exists and if we can find it—well, it will answer questions and set the direction for twenty-first-century physics. Watch for details on the search.

in *At Home in the Universe*, "Failure to predict does not mean failure to understand or to explain."

British playwright Tom Stoppard had this to say (in a play titled *Arcadia*):

> *Relativity and quantum theory looked as if they were going to clean out the whole problem between them. A theory of everything. But they only explained the very big and the very small. The universe, the elementary particles. The ordinary-sized stuff which is our lives, the things people write poetry about—clouds—daffodils—waterfalls—and what happens in a cup of coffee when the cream goes in—these things are full of mystery, as mysterious to us as the heavens were to the Greeks.*

Had we been so busy congratulating ourselves on the scientific achievements of modern times, and how we were understanding the very big and the very small, that we'd missed something mighty important: the everyday world and its unpredictable complexities?

Remember what Plato said about Thales? "When stargazing and looking upward, [he] fell into a well" and was

Some astronomers call this "God's eye," but when they are being scientific they explain that it is the Helix Nebula and our best example of the end-of-life process for a star about the same size as our Sun. The nebula, which spans 2.5 light-years, is throwing off its outer gaseous wrapping. The central core is destined to become a dense white dwarf. The expelled gases create a fluorescent planetary nebula, which, after thousands of years, will fade away. Is this our future?

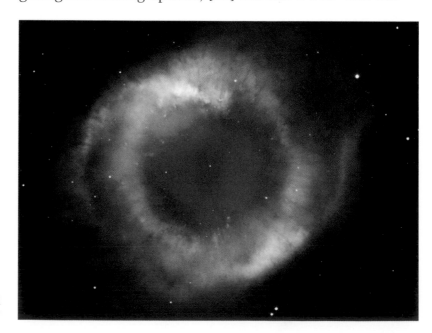

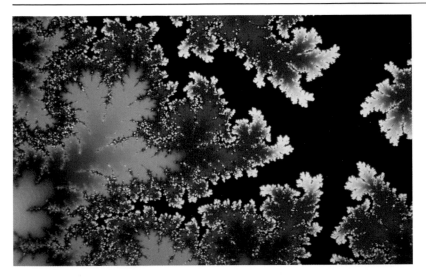

The mathematical equations of chaos theory produce beautiful geometric iterations called fractals. *Iteration* means "repetition." The tiny zigs and zags are echoed in ever-larger patterns. In his book *Chaos*, James Gleick explains, "It is a geometry of the pitted, pocked, and broken up, the twisted, tangled, and intertwined. . . . The pits and tangles are more than blemishes distorting the classic shapes of Euclidian geometry. They are often the keys to the essence of a thing." Research "Mandelbrot set" to learn more.

teased by a maidservant because he was "so eager to know what was happening in the heavens, that he did not notice what was in front of him, nay, at his very feet."

Once a few thinkers started crunching numbers on their computers—numbers that reflected the way the small-scale world actually seems to work—others jumped in and began studying what some call **chaos theory** and others describe as **complexity**. These new sciences find order and regularity in the seemingly arbitrary and disorderly. Pioneers studying the complexity of the universe now believe that nature tends to self-organize. Why? How? They're working to answer those questions.

"The understanding of nature's complexity awaited a suspicion that the complexity was not just random, not just accident," says James Gleick in a book titled *Chaos*. Today's scientists, using computer simulations (instead of microscopes and cyclotrons), have been finding order-without-absolute-predictability in seeming disorder. It is among the ideas bringing new excitement to the world of science. And that's not all.

Lewis F. Richardson (1881–1953) was an English scientist who studied wind and weather and the problems of mapping a raggedy coastline back in the 1920s. He wrote:

> Big whorls have little
>   whorls
> That feed on their velocity,
> And little whorls have
>   lesser whorls
> And so on to viscosity.

Physicists and mathematicians want to discover regularities. People say, what use is disorder. But people have to know about disorder if they are going to deal with it. The auto mechanic who doesn't know about sludge in valves is not a good mechanic.
—James Yorke, mathematician/physicist who named chaos theory

Read On

For the general reader, who wants more of the story behind today's science, I especially recommend *Short History of Nearly Everything* by Bill Bryson (Broadway Books, 2004) and *Einstein: His Life and Universe* by Walter Isaacson (Simon & Schuster, 2007). To glimpse the beauties of chemistry, read *The Same and Not the Same* by Roald Hoffmann, a Nobel laureate who is also a poet (Columbia University Press, 1995).

The National Science Teachers Association (NSTA) has rated these books for middle school level and up as "outstanding." You'll find descriptions and reviews, along with more great science books, at this address: http://www.nsta.org/ostbc.

Biographies

Albert Einstein: A Biography, by Alice Calaprice and Trevor Lipscombe (Greenwood Press, 2005).

Annus, Mirabilis: 1905, Albert Einstein, and the Theory of Relativity by John Gribbin and Mary Gribbin (Penguin Books, 2005).

Atomic Universe: The Quest to Discover Radioactivity, by Kate Boehm Jerome (National Geographic Science Quest series, 2006).

Dear Professor Einstein: Albert Einstein's Letters to and from Children, edited by Alice Calaprice (Prometheus Books, 2002).

Einstein: The Passions of a Scientist, by Barry Parker (Prometheus Books, 2003). Also available in Spanish.

The End of the Certain World: The Life and Science of Max Born, by Nancy Thorndike Greenspan (Perseus, 2005).

Enrico Fermi: Trailblazer in Nuclear Physics, by Erica Stux (Enslow Nobel Prize– Winning Scientists series, 2004).

Ernest Rutherford: And the Explosion of Atoms, by J. L. Heilbron (Oxford University Press, 2003).

Genius: A Photobiography of Albert Einstein, by Marfé Ferguson Delano (National Geographic Children's Books, 2005).

The Great Minds of Science series (Enslow Publishers), especially *Niels Bohr* (2003) and *Ernest Rutherford* (2005), both by Naomi Pasachoff, and *Stephen Hawking* (2005), by John Bankston.

Inventing the Future: A Photobiography of Thomas Alva Edison, by Marfé Ferguson Delano (National Geographic Children's Books, 2002).

Marie Curie: The Woman Who Changed the Course of Science, by Philip Steele (National Geographic Children's Books, 2006).

Something Out of Nothing: Marie Curie and Radium, by Carla Killough McClafferty (Farrar, Straus and Giroux, 2006).

Thomas Edison: A Brilliant Inventor, by the editors of *TIME for Kids* with Lisa deMauro (HarperCollins, 2005).

Science Books

The Elements, by Ron Miller (Twenty-First Century Books/Lerner Publishing Group, 2004).

Light Years and Time Travel: An Exploration of Mankind's Enduring Fascination with Light, by Brian Clegg (John Wiley and Sons, 2002).

One Two Three...Infinity: Facts and Speculations of Science, by George Gamow (Dover, 1988). Originally published in 1947.

Relativity and Quantum Mechanics: Principles of Modern Physics (2001) and *Waves: Principles of Light, Electricity, and Magnetism* (2001), by Paul Fleisher (Lerner Publications).

Symmetry and the Beautiful Universe, by Leon M. Lederman and Christopher T. Hill (Prometheus Books, 2004).

Picture Credits

Grateful acknowledgment is made to the copyright holders credited below. The publisher will be happy to correct any errors or unintentional omissions in the next printing. If an image is not sufficiently identified on the page where it appears, additional information is provided following the picture credit.

Abbreviations for Picture Credits
Picture Agencies and Collections
AR: Art Resource, New York
AIP: American Institute of Physics
BAL: Bridgeman Art Library, London, Paris, New York, and Berlin
Caltech: California Institute of Technology
PR: Photo Researchers, Inc., New York
SPL: Science Photo Library, London
COR: Corbis Corporation, New York, Chicago, and Seattle
GC: Granger Collection, New York
SSPL: Science Museum /Science & Society Picture Library, London
NASA: National Aeronautics and Space Administration
  JPL: Jet Propulsion Lab
  GSFC: Goddard Space Flight Center
  MFSC: Marshall Space Flight Center

Maps
All base maps (unless otherwise noted) were provided by Planetary Visions Limited and are used by permission. Satellite Image Copyright © 1996–2005 Planetary Visions.
PLV: Planetary Visions Limited
SR: Sabine Russ, map conception and research
MA: Marleen Adlerblum, map overlays and design

Illustrator
MA: Marleen Adlerblum (line drawings)
All timelines were drawn by Marleen Adlerblum. All timeline photographs are public domain, unless otherwise listed.

Frontmatter
ii: Brookhaven National Laboratory/PR; ix: (Composite) NASA/JPL-Caltech/Max-Planck Institute/ P. Appleton (SSC/Caltech)

Chapter 1
Frontispiece: Caltech; 2: (top left and right) Hebrew University of Jerusalem Albert Einstein Archives, courtesy of AIP; (bottom) © Bettmann/COR; 3: PLV; 4: (top and bottom) akg-images; 5: GC; 6: (left and right) MA; 7: NASA/MSFC/NSSTC/Hathaway 2007/04; 8: Erich Lessing/AR; 9: Tesla Memorial Society of New York; (bottom) Clive Goddard, CSL Cartoon Stock

Chapter 2
10: Erich Lessing/AR; 11: (left) Edward Owen/AR; (right) Scenographia: Systematis Copernicani Astrological Chart, c. 1543, devised by Nicolaus Copernicus (1473–1543) from "The Celestial Atlas, or the Harmony of the Universe" c. 1660, British Library, London/BAL; 12: (top) Smithsonian Institution; (bottom) SSPL; 13: (top) GC; (bottom) Keith Kent/PR; 14: TIMELINE; 15: *Dr. William Gilberd (1540–1603) Showing his Experiment on Electricity to Queen Elizabeth I and her Court*, detail of Gilberd, 19th century oil on canvas, Arthur Ackland Hunt, private collection, The Royal Institution, London/BAL; 16: NASA/PR; 17: (top) SSPL; (bottom) GC

Chapter 3
18: The Ampere and Electricity Museum, Lyon, France; 19: (top) courtesy of John Jenkins, www.sparkmuseum.com; (bottom) Magnetic experiment by Michael Faraday (1791–1867)/Bibliotheque Nationale, Paris, France, Giraudon/BAL; 20: Courtesy of Prof. David S. Ritchie, James Clerk Maxwell Foundation, Edinburgh; 21: Eric Heller/PR; 22: (top) GC; (bottom) MA; 23: MA; 24: (top) AR; (bottom) GC; 25: PR; 26: (top) GC; (bottom) AR; 27: (top left and middle) GC; (top right) courtesy of ACDC, Atlantic Records; (bottom) GC

Chapter 4
28: Cosmography or Science of the World (engraving) (b/w photo), French School, (17th century)/Bibliotheque Nationale, Paris, France, Giraudon/BAL; 30: AIP/Emilio Segrè Visual Archives; 31: (top) AIP/Emilio Segrè Visual Archives; (bottom) MA; 32: Jan Jerszynski; 33: Collection of the Oakland Museum of California, Gift of Anonymous Donor; 35: AIP/Emilio Segrè Visual Archives

Chapter 5
37: SSPL; 38: *Dawn near Reading*, 1870 (w/c on paper), English School, (19th century), Ironbridge Gorge Museum, Telford, Shropshire, UK/BAL; 39: (top) W.H. Hales, AIP Emilio Segrè Visual Archives; (bottom) University of Cambridge, Cavendish Lab; 40: TIMELINE; 41: *Professor Sir William Crookes (1832–1919)* from "Illustrated London News", May 1914 (colour litho), English School, (20th century), Bibliotheque Nationale, Paris, France, Archives Charmet/BAL; 42: *The Phenomenon of Electrical Luminosity*, from "l'Univers et l'Humanite" by Hans Kraemer, c. 1900 (colour litho), French School, (20th century), Private Collection, Archives Charmet/BAL; 43: (top) Charles D. Winters/PR; (bottom) C.R.T. Wilson/PR; 44: (top) SPL/PR; (middle) H. Turvey/PR; (bottom) MA; 46: Phanie/PR; 47: S. Horrell/PR; 48: Science Source/PR

Chapter 6
50: Scimat/PR; 51: Sidney Harris; 52: MA; 53: (left) LBNL/Science Source/PR; (right) Steve Allen/PR

Chapter 7
54: Muzeum Marii Sklodowskiej Curie, Poland; 55: GC; 56: (top) *The Eiffel Tower*, 1889 (panel), Georges Seurat (1859–1891), Fine Arts Museums of San Francisco, CA/BAL; (bottom) SPL/PR; 57: (top) SPL/PR; (bottom) TIMELINE

Chapter 8
58: *The Courtyard of the Old Sorbonne*, 1886 (oil on canvas), Emmanuel Lansyer (1835–1893), Musee de la Ville de Paris, Musee Carnavalet, Paris/Giraudon/BAL; 59: Astrid & Hanns-Frieder Michler/PR; 60: (top) Pierre (1859–1934) and Marie Curie (1867–1934) in their laboratory (b/w photo), Gribayedoff © Collection Kharbine-Tapabor, Paris/BAL; (bottom) MA; 61: Science Museum/SSPL; 62: Bettman/COR; 63: Science Museum/SSPL; 64: C. Powell, P. Fowler & D. Perkins/PR; 65: PR; 67: Photo courtesy of Jay Pasachoff; 68: (top) SPL/PR; (bottom) Musee Curie; 69: (top left) Tomasz Barszczk/Super-Kamiokande Collaboration/PR; (top right) SPL/PR; (bottom) SPL/PR

Chapter 9
70: AR; 71: H. David Seawell/COR; 72: Alfred Pasieka/PR; 73: (left) Krafft/PR; (right) Celestial Image Co./PR; 74: Department of Physics, University of Adelaide, Australia; 76: *Man with Violin*, 1911 (oil on canvas), Pablo Picasso (1881–1973), Private Collection, © DACS/BAL; 77: (left) NOAO/PR; (right) NASA

Chapter 10
78: GC; 79: (top) courtesy of Einstein-Haus, Bern; (bottom) courtesy of the Schweizerisches Literaturarchiv (SLA), Bern; 80: (top) GC; (bottom) courtesy of the US Patent and Trademark Office; 81: (top) SPL/PR; (bottom) SSPL; 83: MA; 84: Sidney Harris; 85: Erich Lessing/AR

Chapter 11
87: GC; 88: (left) MA; (right) Erich Schrempp/PR; 89: Gemini Observatory/AURA/NSF; 90: (top) Jerome Wexler/Science Photo Library; (bottom) Jean-Francois Colonna; 91: *Haystacks at Sunset, Frosty Weather*, 1891, Claude Monet (1840–1926), Private Collection/BAL; 92: Novosti Photo Library/PR; 94: SPL/PR; 97: (left) Richard R. Hansen/PR; (right) Jerome Wexler/PR

Chapter 12
98: Andrew J. Russell/ Prints and Photograph Division of the Library of Congress; 99: (top) James Cavallini/PR; 100: (top) *Democritus* (c. 460–c. 370 BC) 1692 (oil on canvas), Antoine Coypel (1661–1722), Louvre, Paris, France/BAL; (bottom) *Portrait of Ernst Mach (1838–1916)* (b/w photo), French photographer, Archives Larousse/BAL; 101: (top left and right) GFSC/NASA; (bottom) Average Pagan Landscape, 1937 (oil on panel), Salvador Salvador Dali (1904–1989), Gala-Salvador Dali Foundation, Figueres, Spain © DACS/BAL; 102: MA; 103: SPL/PR

Chapter 13
104: Advertisement for "Caley's Christmas Pudding Crackers" (litho), Sir Alfred Munnings (1878–1959) © Norwich Castle Museum and Art Gallery/BAL; 105: A Pictorial Recipe for your Plum Pudding, Eliot Hodgkin, Illustrated London News Christmas Number, 1961, The Illustrated London News Picture Library, London/BAL; 106: C. Powell, P. Fowler & D. Perkins/PR; 107: Sidney Harris; 108: MA; 109: (top) Bettmann/COR; (bottom) MA; 110: (top) *The Alchymist*, 1771 (oil on canvas), Joseph Wright of Derby, Derby Museum and Art Gallery, UK/BAL; (bottom) Tommaso Guicciardini/INFN/PR; 111: Angelo Cavalli/zefa/COR; 112: (top) Niels Bohr Archive, Copenhagen; (bottom) David Copperfield and

Little Em'ly, illustration for "Character Sketches from Dickens" compiled by B. W. Matz, 1924 (colour litho), Harold Copping (1863–1932) Private Collection/BAL; 113: courtesy of Keith Papworth, University of Cambridge

Chapter 14
115: MA; 116: David Parker/PR; 117: MA; 118: (top) CERN/PR; (bottom) Alfred Pasieka/PR; 119: (top) GC; (bottom) SPL/PR; 120: Freeman D. Miller/PR; 121: (top) Dr. Jeremy Burgess/PR; (bottom) A. Syred/PR; 122: Scott Camazine/PR; 123: MA; 124: The Museum of Modern Art/Scala/AR; 125: JPL/NASA; 126: European Space Agency/PR; 127: MA

Chapter 15
129: (top) Courtesy of John D. Jenkins, American Museum of Radio and Electricity, Bellingham, Washington; (bottom) GC; 130: (top) SSPL; (bottom) TIMELINE; 131: (top and bottom) SPL/PR; 132: (top) I. Curie & F. Joliot/PR; (bottom) GC; 133: (top) Berkeley-Lab; (bottom) MA; 134: MA; 135: James King-Holmes/PR

Chapter 16
137: Niels Bohr (1885–1962) reproduction of a 1939 mosaic in the Royal Theatre (Staerekassen) Copenhagen (colour litho), Ejnar Nielsen (1872–1956) (after), Private Collection, © DACS/Archives Charmet/ BAL; 138: (top) Bettmann/COR; (bottom) Copenhagen by Gaslight, W. Behrens (20th century), Private Collection, Photo © Bonhams, London, UK/BAL; 139: (top) Prints and Photographs Division of the Library of Congress; (bottom) Published by Cambridge University Press; 140: Prints and Photographs Division of the Library of Congress; 142: (left) Portrait bust of Zeno of Citium (334–262 BC), 3rd century BC) (marble), Museo Archeologico Nazionale, Naples, Italy/BAL; (right) Photothèque R. Magritte-ADAGP/AR; 143: Niels Bohr Archives, Copenhagen

Chapter 17
145: (top) SPL/PR; (bottom) GC; 146: Dr. Erwin Mueller/PR; 147: GC; 148: Michael W. Davidson/PR; 149: SPL/PR

Chapter 18
151: Eric Heller/PR; 152: Max-Planck-Institut/AIP/Emilio Segrè Visual Archives; 153: published by Random House Publishing Company; 154: G.T. Jones, Birmingham University/Fermi National Accelerator; 155: (top) MA; (bottom) Lawrence Berkeley National Laboratory/PR; 156: Scala/AR; 157: (top) Detlev van Ravenswaay/PR; (bottom) MA

Chapter 19
158: Prints and Photographs Division of the Library of Congress; 159: Edward Kinsman/PR; 160: (top) SPL/PR; (bottom) AIP/PR; 161: Don Quixote and Sancho Panza (oil), Honore Daumier (1808–1879), courtesy of private collector © Agnew's, London, UK/BAL; 162: (left) courtesy of Scientific American Magazine; (right) Tom Swanson; 163: National Institute of Standards and Technology; 164: (top) MA; (bottom) Lawrence Berkeley National Laboratory Image Library; 165: AIP; 166: IBM Almaden Research Center Visualization Lab; 167: Michael Gilbert; 168: (top left, middle and right) Clive Freeman/Biosym Technologies/PR; (bottom) SPL/PR; 169: (top) Wolkow Lab, University of Alberta; (bottom) Vintage Views, IBM

Chapter 20
170: Fermi National Accelerator Laboratory, US Department of Energy; 171: (left) Courtesy of The Cavendish Lab; (right) AIP/Emilio Segrè Visual Archives; 172: Ernest Orlando, Lawrence Berkeley National Laboratory; 173: Dave Judd and Ronn MacKenzie/LBNL; 174: collection of LBNL; 175: courtesy of Phil Broad, Movie Sets & Vehicles, Cloudster.com; 176: AIP/PR; 177: Jordan Goodman, Particle Astrophysics Group, University of Maryland

Chapter 21
179: ArSciMed/PR; 180: MA; 181: NASA/PR; 182: Tom Hollyman/PR; 183: (top) Special Collections, The Valley Library, Oregon State University; (bottom left) A. Barrington Brown/PR; (bottom right) Omikron/PR; 184: Eye of Science/PR; 185: TIMELINE; 186: Arnold Fisher/PR; 187: Illustrations by MA; 188: Spencer Grant/PR; 189: (top) GC; (bottom) Alfred Pasieka/PR

Chapter 22
191: Sidney Harris

Chapter 23
193: (top) Hulton-Deutsch Collection/COR; (bottom) courtesy of Dr. Robert D. Brooks; 194: (top) Ullstein Bild/GC; (bottom) GC; 195: (top) Arthur Szyk, Prints and Photographs Division of the Library of Congress; (bottom) Austrian Archives/COR; 196: (top) Hulton-Deutsch Collection/COR; (bottom) GC; 197: (top) Frank Frazetta; (bottom) GC; 198: (top) Star of David cloth patch, printed "Jude", as compulsorily worn by Jews in Nazi Europe, Nationalmuseet, Copenhagen/BAL; (bottom) Prints and Photographs Division of the Library of Congress; 199: (top right) Oath of Allegiance, 1925 (oil on canvas), Fedor Pavlovic Resetnikov (1906–1983), State Russian Museum, St. Petersberg, Russia © DACS/BAL; (middle) CineMasterpieces; (bottom right) AIP/Emilio Segrè Visual Archives, Wigner Collection; 200: (top) Museo Nacional Centro de Arte Reina

Sofia, Madrid/AR; (bottom) Criterion Collection-1938; 201: (bottom) GC; 202: (top) Bettmann/COR; (middle) GC; (bottom) Ullstein Bild/GC; 203: (top) Ullstein Bild/GC; (bottom) GC

Chapter 24
204: courtesy of Mandeville Special Collections Library, University of California at San Diego; 205: View of Pest and Buda, 1870s (w/c on cardboard), Hungarian School (19th century), The Nicolas M. Salgo Collection, USA/BAL; 206: GC; 207: GC; 208: GC; 209: (top) courtesy of the Max Planck Society, Munich; (bottom) University of California/AIP/PR; 210: (top) AIP/Emilio Segrè Visual Archives; 212: C.T.R. Wilson/PR; 213: SPL/PR; 214: View of the Graben, Vienna, c. 1860–1880 (b/w photo), Austrian Photographer, (19th century), Private Collection, Archives Charmet/BAL; 215: SPL/PR; 216: Adam Hart-Davis/PR; 217: (top) MA; (inset) Nancy Rose; 218: (top) Dr. Kari Lounatmaa/PR; (bottom) AIP/Emilio Segrè Visual Archives

Chapter 25
220: GC; 221: Bettmann/COR; 223: (top) © Sergey Konenkov/Sygma/COR; (bottom) DOE/Science Source/PR; 225: (top) Image © The Albert Einstein Archives, The Jewish National & University Library, The Hebrew University of Jerusalem, Israel; (bottom) SPL/PR; 226: (top) Prints and Photographs Division of the Library of Congress; (bottom) GC; 228: M-Sat Ltd/PR; 230: Science Museum/SSPL; 231: (top) Kenneth Eward/BioGrafx/PR; (bottom) Jacana/PR; 233: US Navy, Office of Public Relations, Washington; 235: SPL/PR

Chapter 26
237: Prints and Photographs Division of the Library of Congress; 238: NASA; 239: (left) COR; (right) Los Alamos National Laboratory/PR; 240: COR; 241: National Archives/PR; 242: Bettmann/COR; 243: (top) COR; (bottom) AIP; 245: Prints and Photographs Division of the Library of Congress; 246: AIP/Emilio Segrè Visual Archives; 247: Prints and Photographs Division of the Library of Congress; 248: GC; 249: (top) Smithsonian Institution, National Museum of American History, AIP/Emilio Segrè Visual Archives; (bottom) photograph by Pach Brothers/AIP/Emilio Segrè Visual Archives; 250: COR; 251: The Cosmic Form of the God Vishnu, c. 1800 (gouache on card), Indian School, (19th century), © Oriental Museum, Durham University, UK/BAL; 252: (left) GC; (right) Scott Camazine/PR; 253: (top) Bettmann/COR; (bottom) Bettmann/COR

Chapter 27
255: (bottom) Physics Today Collection/AIP/PR; 256: (top left and right) AIP/Emilio Segrè Visual Archives; (bottom) published by W.W. Norton (USA); 257: (top) SPL/PR; 259: (top left) David Parker/PR; (top middle) Volker Steger/PR; (top right) Brookhaven National Laboratory/PR; (bottom left) DOE/PR; (bottom middle) Fermilab/PR; (bottom right) David Parker/PR

Chapter 28
261: Larry Landolfi/PR; 262: Alinari/AR; 263: Tourists Taking Scenic Photographs from the Rear Observation Platform of a Union Pacific Train, 1910 (b/w photo), American Photographer, (20th century), Private Collection, Peter Newark American Pictures/BAL; 264: Detlev van Ravenswaay/PR

Chapter 29
267: Banque d'Imagese, ADAGP/AR; 269: Apollo 8/NASA; 270: Digital Image © The Museum of Modern Art/Licensed by SCALA/AR; 272: Sidney Harris; 275: David J. Grossman

Chapter 30
277: Caltech; 278: Agence Vandystadt/PR; 279: Antony Searle/The Australian National University, Canberra; 280: MA; 281: (top) Tony Craddock/PR; (bottom) Daniel Sambraus/PR; 282: AR; 283: MA

Chapter 31
285: Einstein Juggling with Time, 2000 (oil and tempera on panel), Frances Broomfield (contemporary artist), Private Collection, © Frances Broomfield/Portal Gallery, London/BAL; 287: Michael Pead; 288: (top) Portrait of Franz Kafka (1885–1924) c. 1908 (b/w photo) by Czech School, © Private Collection/Archives Charnet/BAL; (bottom) Illustration for the "Metamorphosis" by Franz Kafka, 1946 (litho) by Hans Fronius (1903–1988), © Bibliotheque Nationale, Paris/BAL

Chapter 32
290: Erich Lessing/AR; 290: MA; 291: MA; 293: image credit/NASA, illustration/MA (based on a flipbook by Adler Planetarium); 294: (top and bottom) MA; 295: Philippe Psaila/PR

Chapter 33
296: Sidney Harris; 297: Phil Rose; 298: (top) JPL/NASA; (bottom) MA; 299: (left) Coneyl Jay/PR; (right) Mark Clarke/PR; 300: (top) Spencer Grant/PR; (bottom) CERN/PR; 301: Alexander Tsiaras/PR

Chapter 34
302: The Soundness of Newton's Laws, illustration from Inventions (litho) by

William Heath Robinson © Private Collection/BAL; 303: Eric Schrempp/PR; 304: illustrations by G. Lasellaz, *How a Cat Falls, from Les Dernieres Merveilles de la Science* by Daniel Bellet (chromolitho); 305–306: MA (adaptations from *How a Cat Falls*); 307: MA; 309: (top) Carl Goodman/PR; (bottom left) NASA/PR; (bottom right) Chris Butler/PR; 311: NASA

Chapter 35
313: (top) Laguna Design/PR; (bottom) Christian Darkin/PR; 314: Erich Lessing/AR; 315: NASA/MSFC; 316: Frank Zullo/PR; 318: (top) GC; (bottom) US Army Signal Corp; 319: Transocean, Berlin/Dibner Library of the History of Science and Technology; 320: (left and right) F.W.Dyson, A.S. Eddington, and C. Davidson, "A Determination of the Deflection of Light by the Sun's Gravitational Field, from Observations Made at the Total Eclipse of May 29, 1919" *Philosophical Transactions of the Royal Society of London. Series A, containing Papers of a Mathematical or Physical Character* (1920); 321: *New York Times*, November 10, 1919

Chapter 36
323: Plate XXXI from the "Original Theory of the Universe" by Thomas Wright (1711–1786), 1750 (engraving) (b/w photo), English School, (18th century), Private Collection/BAL; 324: (top) Illustration from "From the Earth to the Moon" by Jules Verne (1828–1905) Paris, Hetzel, published in 1865 (engraving) (b/w photo) by Emile Antoine Bayard (1837–1891) © Bibliotheque Nationale, Paris/Lauros/Giraudon/BAL; (bottom) Julian Baum/PR

Chapter 37
326: Caltech; 327: Hale Observatories, AIP/Emilio Segrè Visual Archives; 328: Harvard College Observatory, AIP/Emilio Segrè Visual Archives; 329: (top) STSI/NASA; (bottom) SPL/PR; 330: Victor de Schwanberg/PR; 331: David Parker/PR; 332: (left, middle, right) JPL-Caltech/STScI/NASA; 333: (left, right) NASA/JPL-Caltech

Chapter 38
334: Gale Gant/Mt. Wilson Observatory Association; 335: (top) Shigemi Numazawa/Atlas Photo Bank/PR; (bottom) Dr. Jean Lorre/PR; 336: (top) NRAO/AUI; (bottom) NASA/ESA/R. Bouwens and G.Illingworth (University of California, Santa Cruz); 337: (top) AIP/Emilio Segrè Visual Archives; (bottom) Gerard Lodriguss/PR; 338: NASA/ESA/the Hubble Heritage Team/STScI/AURA/A. Riess (STScI); 339: Caltech Archives

Chapter 39
340: SPL/PR; 342: (top) NASA/H.E. Bond and E. Nelan (STSI)/M. Barstow and M. Burleigh (University of Leicester, UK)/J.B. Holberg (University of Arizona); (bottom) Charles D. Winters/PR; 343: NASA/ESA/the Hubble Heritage Team (STScI/AURA)/ ESA/Hubble Collaboration

Chapter 40
346: Julian Baum/PR; 347: F. Walter (State University of New York at Stony Brook)/NASA; 348: NASA/MSFC; 349: J. Hughes (Rutgers) et al./CXC/NASA; 350: NASA/CXC/SAO; 351: Data from the Digitized Sky Survey/Image processing by Davide De Martin; 353: Ullstein Bild/GC; 354: NASA/MSFC; (inset) Chandra X-ray Center (CXC); 355: NASA/the Hubble Heritage Team (STScI/AURA)

Chapter 41
357: (top) NASA/JPL-Caltech/Tim Pyle (SSC); (bottom) Tony Craddock/PR; 358: AIP/Emilio Segrè Visual Archives; 359: NASA/CXC/SAO/H. Marshall et al.; 360: (all photos) NASA/Apollo Image Gallery; 361: (top left) Axel Mellinger/NASA; (top right) S. Digel and S. Snowden (GSFC)/ROSAT Project/MPE/NASA; (bottom left) DIRBE Team/COBE/NASA; (bottom right) NASA/Goddard Space Flight Center; 362: ESA/NASA and Felix Mirabel (the French Atomic Energy Commission & the Institute for Astronomy and Space Physics/Conicet of Argentina); 363: (top) AIP/Emilio Segrè Visual Archives/Physics Today Collection; (bottom) NASA/CXC/M. Weiss; 364: courtesy of V.I. Goldanskii; 365: (top) H. Bond (STScI) and B. Balick (University of Washington)/NASA; (bottom) NASA/MSFC; 366: (top) ESA/PR; (bottom) Hubble Space Telescope/NASA; 367: (top) Daniel Wang (University of Massachusetts); (bottom) ; 368: (top) NASA/CXC/Caltech/D. Fox et al.; illustration by NASA/D.Berry; (bottom and inset) Spectrum/NASA/E/PO/Sonoma State University Aurore Simonnet; 369: (top) Jon Lomberg/PR; (bottom) NASA/JPL-Caltech

Chapter 42
370: Aero Data, Baton Rouge, LA; 371: (top) Bryan Christie Design; (bottom) Maximilian Stock Ltd/PR; 372: Gary Bower, Richard Green (NOAO)/STIS Instrument Definition Team/NASA; 373: Phillip Hayson/PR; 374: NASA/CXC/PSU/S. Park & D. Burrows; 375: Dana Berry/NASA; 376: JPL/NASA; 377: (left) K. Thorne (Caltech), T. Carnahan (NASA/GSFC); (inset) Dana Berry, Sky Works Digital/NASA; 378: MA; 380: (top and bottom) JPL/NASA; 381: Victor Habbick Visions/PR

Chapter 43
383: (top) M. Kulyk/PR; (bottom) ArSciMed/PR; 384: (original image) ArSciMed/PR, (illustrations) MA; (top images) ArSciMed/PR, (illustrations) MA; (bottom left) NASA/ESA/R. Massey (Caltech); (bottom right) NASA/JPL-Caltech/A. Kashlinsky (GFSC) et al.; 386: Sidney Harris; 387: (top) NASA; (bottom inset) AIP/Emilio

Segrè Visual Archives; 388: GSFC/NASA; 389: (top left) COBE Project/DMR/NASA; (top right, bottom left and right) NASA/WMAP Science Team; 390: NASA/WMAP Science Team; 391: Yannick Mellier/IAP/PR

Chapter 44
392: Sidney Harris; 393: LBNL/PR; 394: David Parker/PR; 395: GSFC/NASA; 396: Detlev van Ravensway/PR; 397: David A. Hardy/PR; 398: Published by Dover Publications, Mineola, NY; 399: (top) Lockheed Martin Space Systems; (bottom left and right) Brookhaven National Laboratory; 400: *Inferno, Purgatory and Paradise*, illustration from Dante's "Divine Comedy", 14th century (manuscript), Italian School, British Museum, London, UK/BAL; 401: GSFC/NASA; 402: (top) Mehau Kulyk/PR; (bottom) Fred Tomaselli, *Cyclopticon 2*, 2003, (mixed media, acrylic paint, resin on wood, 24 x 24 x 1½ inches). Image courtesy James Cohan Gallery, New York; 403: Fred Tomaselli, Abductor, 2006, (leaves, photo collage, acrylic and resin on wood panel, 96 x 78 inches). Image courtesy James Cohan Gallery, New York; 404: (1) Josiah McElheny, *The Last Scattering Surface*, 2006, (hand-blown glass, chrome-plated aluminum, rigging, electric lighting, 10h. x 10 x 10 feet). Image courtesy Donald Young Gallery, Chicago; (2) Josiah McElheny, *The Last Scattering Surface*, detail; (3) Matthew Ritchie, *Where I Am Coming From*, 2003, (oil and marker on canvas, 99 x 121 inches). Image courtesy Andrea Rosen Gallery, New York; (4) Matthew Ritchie, Installation view *The Universal Adversary*, Andrea Rosen Gallery, NY September 21–October 28, 2006, (at top) *The Universal Adversary*, 2006 (powder-coated aluminum and stainless steel, approximately 30 feet in diameter), Image courtesy Andrea Rosen Gallery, 2006; 405: (5) Ati Maier, *Dérive*, 2007, (acrylic paint and ink on canvas, 38 x 96 inches), Image courtesy Pierogy Gallery, Brooklyn, New York; (6) Diana Al-Hadid, *A Measure of Ariadne's Love*, 2007, (wood, cast aluminum, fiberglass, cardboard, plaster, resin, Plexiglas, paint, 116 x 85 x 96 inches). Image courtesy Michael Janssen, Berlin; (7) Lee Bontecou, *Untitled*, 1980–1998, (welded steel, porcelain, wire mesh, canvas, and wire, 7 x 8 x 6 feet). Museum of Modern Art, New York, gift of Philip Johnson. Copyright © Lee Bontecou/courtesy Knoedler & Company, New York.

Chapter 45
406: © Deborah Betz Collection/COR; 407: JPL/NASA; 408: MA; 409: MA; 411: Library of Congress, New York World-Telegram and Sun Collection, AIP/Emilio Segrè Visual Archives; 412: Pascal Goetgheluck/PR; 413: (top left) GC; (top right) Ullstein Bild/GC

Chapter 46
415: Courtesy of Harlan Devore, Fayetteville, NC; 416: Chandra: NASA/CXC/University of Utrecht, Germany/J. Vink et al. XMM-Newton: ESA/University of Utrecht, Germany/J. Vink et al.; 417: (top) NASA; (bottom) courtesy of Harlan Devore, Fayetteville, NC; 418: (top) NOAO/AURA/NSF; (bottom left) "Robert Frost, head-and-shoulders portrait, facing front." (Between 1910 and 1920) New York World-Telegram and the Sun Newspaper Photograph Collection, Library of Congress; (bottom right) Omikron/PR; 419: Courtesy of NFAO/AUI/Balz Bietenholz, Michael Bietenholz and Norbert Bartel, York University; 420: NASA/CXC/M. Weiss; 421: NASA/CXC/M. Markevitch et al.; 422: (left) MA; (right) NASA/ESA/R. Massey (Caltech); 423: (top) NASA/A. Riess (STScI); (bottom) NASA/ESA/A. Feild (STScI); 424: AIP/Emilio Segrè Visual Archives; 425: NASA/ESA/Andrew Fruchter (STScI) and the ERO team (STScI + ST-ECF); 426: Lynette Cook/PR; 427: Mark Godfrey/AIP/Emilio Segrè Visual Archives

Chapter 47
428: (left) SSPL; (right) Erich Lessing/AR, NY; 429: JPL/NASA; 430: Mike Agliolo/PR; 431: Computer History Museum, Mountain View, CA; 432: (left) GC; (right) MA; 433: (top) Eric Heller/PR; (bottom) SPEC/CEA, Gif-sur-Yvette Cedex, France; 434: (top) SSPL; (bottom) University of Ulm, Germany; 435: Pair of Pocket Globes, made by Newton, c. 1830, English School, Private Collection/BAL; 436: Research School of Physical Sciences and Engineering, The Australian National University, Canberra; 437: (top) Alfred Pasieka/PR; (bottom) Lawrence Livermore National Laboratory

Chapter 48
438: © Christian Simonpietri/Sygma/COR; 439: (top) Big Wave Productions/PR; (bottom) Smithsonian Institute; 440: NRAO/AUI and Earth image courtesy of the SeaWiFS Project NASA/GSFC/ORBIMAGE; 441: SETI Institute; 442: NASA/JPL-Caltech/R. Hurt (SSC); 443: Bill Ray; 444: (left) Courtesy of Isaac Gary; (right) IAR/Guillermo Lemarchand; 445: SETI@home; (inset) Paul Rapson/PR; 446: (top) NASA; (middle) Jon Lomberg/PR; (bottom) NASA/JPL/Malin Space Science Systems; 447: (left) NASA/ESA/K.Sahu (STScI); (right) NASA/ESA/G. Bacon (STScI); 448: Big Wave Productions/PR449: (top) ESO; (bottom left) NASA/ESA/G. Bacon (STScI); (bottom right) ESA

Chapter 49
450: Michael Donne/PR; 451: (top) NASA/PR; (bottom) IMB (Irvine-Michigan-Brookhaven) Collaboration/PR; 452: NASA/WIYN/NOAO/ESA/Hubble Helix Nebula Team/M. Meixner (STScI)/T.A. Rector (NRAO); 453: Fred Espenak/PR; 454: GC; 455: Image courtesy of NRAO/AUI and Michael Bietenholz, York University, Toronto

Index

WITHDRAWN